Electronic Devices and Circuits

Fourth Edition

Electronic Devices and Circuits

Fourth Edition

Prof. K. Lal Kishore

M.Tech, Ph.D, SMIEEE, FIETE, FIE, FAPAS, LMISTE, LMISHM

Vice-Chancellor

JNTU, Anantapur

Ananthapuramu – 515 002

BS Publications

A unit of **BSP Books Pvt. Ltd.**

4-4-309/316, Giriraj Lane, Sultan Bazar,

Hyderabad - 500 095

Phone : 040 - 23445605, 23445688

Electronic Devices and Circuits, ***Fourth Edition by*** **K. Lal Kishore**

Published by

BS Publications

A unit of **BSP Books Pvt. Ltd.**

4-4-309/316, Giriraj Lane, Sultan Bazar,

Hyderabad - 500 095

Phone : 040 - 23445605, 23445688

e-mail : info@bspbooks.net

ISBN: 978-93-85433-65-8 (HB)

DEDICATED TO

Sri *Saraswati*
the goddess of learning

Preface to Fourth Edition

I am happy that the 4th edition of the book "Electronic Devices and Circuits" is brought out, to meet the requirements of students of JNTUA. Recently the syllabus for this subject has been revised. Accordingly changes are made in this book. The contents and topics are changed so that there is almost one to one match between the syllabus and topics given in this text book. This will enable the students and teachers to feel 'at home' while using the book.

For every chapter, the topics covered are listed and scope is given under the title "In this Chapter". At the end of every chapter, summary of the chapter is given briefly. Objective type questions with answers and essay type questions are also given for every chapter. This will enable students to prepare for examinations very well. This text book will also be very useful for preparing competitive examinations like GATE, Engineering services exams and such other examinations.

This book is also useful for AMIETE, AMIE (Electronics), M.Sc (Electronics), Polytechnic and Diploma programs in Electronics and B.Sc (Electronics) Courses.

Number of textbooks are referred in writing this book. The author thanks, publishers and author of those books. Though number of text books are available in electronics, there is no single text book available, covering all the topics mentioned here and dealing with the topics in a very clear manner, so that, it is useful to teachers as well as students.

Conceptual questions and answers for such questions are given, which is unique only to this text book and no other book gives such information. Teachers are requested to explain those concepts very clearly to students while teaching those topics.

Questions from GATE examinations pertaining to the topics covered in the text book are taken and solutions are given, which will help teachers while teaching and students while preparing for examinations.

The author thanks Prof. M.N. Giriprasad, HOD, ECE Dept., Mr. C. Ravishankar Reddy, Ms. J. Anusha, Mr. K. Srinivasa Rao and Mrs. V. Annapurna, Adhoc Lecturers in ECE Dept., for the help rendered by them in revising the book and preparing answers for objective type questions. The author also thanks the management of M/s B.S. Publications, Sri Nikhil Shah ji, Mr. Naresh Davergave and his team members R. Kavya and B. Kalpana, who have helped in bringing out this book.

- Author

Contents

Chapter 6

Feedback Amplifiers

Symbols

a	:	Acceleration of electrons (m/sec or cm/sec)
B	:	Magnetic field Intensity (Wb/m^2 or Tesla)
C	:	Charge of electrons (Coulombs)
c	:	Velocity of light $= 3 \times 10^8$ m/sec.
d	:	Distance between the plates in a CRT
D	:	Distance between the centre of the deflecting plates and screen.
D	:	Diffusion constant;
D	:	Distortion in output waveform
ε	:	E = Electric field intensity (V/m or V/cm)
f	:	frequency (Hzs/KHzs/MHzs)
F	:	Force experienced by an electron in Newtons
h	:	Plank's constant $= 6.62 \times 10^{-34}$ J-sec.
I	:	D.C. current (mA or μA)
i	:	A.C current (mA or μA)
J	:	Current density (A/m^2 or mA/cm^2)
K	:	Boltzman's constant $= 8.62 \times 10^{-5}$ eV / ^{0}K
	:	Boltzman's constant $= 1.38 \times 10^{-23}$ J / ^{0}K
l	:	Length of deflecting plates of CRT (cms)
L	:	Distance between the centre of the field and screen (cm or m)
L	:	Diffusion length
m	:	Mass of electron (kgs)
M	:	Mutual conductance
n	:	free electron concentration (No./m^3 or No./cm^3)
N_A	:	Acceptor atom concentration (No./m^3 or No./cm^3)
N_D	:	Donor atom concentration (No/m^3 or No/cm^3)
p	:	Hole concentration (No./cm^3 or No./cm^3)
q	:	Q = Charge of an electron in coulombs $= 1.6 \times 10^{-19}$ C
s	:	Spacing between the deflecting plates of CRT (in cms)
S	:	Stability factor

T	:	Period of rotation (secs or μ secs)
V	:	Accelerating potential or voltage (volts)
v	:	Velocity (m/sec or cm/sec)
W	:	Work function or Energy (eV)
y	:	Displacement of electron on the CRT screen (cms or mms)
Y	:	Admittance (in mhos ℧);
Z	:	Impedance (ohms Ω)
K.E	:	Kinetic Energy (eV)
P.E	:	Potential Energy (eV)
L	:	Inductor
C	:	Capacitor
R	:	Resistor
α	:	D.C large signal current gain of BJT $= \frac{I_C}{I_E}$
β'	:	Small signal common emitter forward current gain
β	:	D.C large signal current gain of BJT $= \frac{I_C}{I_B}$
β^*	:	Transportation factor of BJT $= \frac{I_{PC}}{I_{PE}} = \frac{I_{nC}}{I_{nE}}$
γ	:	Emitter efficiency of BJT $= \frac{I_{PE}}{I_E} = \frac{I_{nE}}{I_E}$
γ	:	Ripple factor in filter circuits
	:	Conductivity of p-type semiconductor in (℧/cm or siemens)
	:	Conductivity of n-type semiconductor in (℧/cm or siemens)
ρ	:	Resistivity (Ω - cm)
θ	:	Thermal resistance (in W/cm^2)
θ	:	Angle of deflection
ϕ	:	Volt equivalent of work function (volts)

Δ	:	Incremental value
Ω	:	Resistance (ohms)
℧	:	Conductance (mhos)
η	:	Efficiency (%)
$\in_0 = \varepsilon_0$	:	Permittivity of free space (F/m) $= 8.85 \times 10^{-12}$ F/m
μ	:	Mobility of electrons or holes (m^2/V-sec)
μ_0	:	Permiability of free space (H/m) $= 1.25 \times 10^{-6}$ H/m
σ	:	Wavelength (A^0)
h_i	:	Input resistance or input impedance (Ω)
h_r	:	Reverse voltage gain
h_0	:	Output admittance (℧)
h_f	:	Forward short circuit current gain

Brief History of Electronics

In science we study about the laws of nature and its verification and in technology, we study the applications of these laws to human needs.

Electronics is the science and technology of the passage of charged particles in a gas or vacuum or semiconductor.

Before electronic engineering came into existence, electrical engineering flourished. Electrical engineering mainly deals with motion of electrons in metals only, whereas Electronic engineering deals with motion of charged particles (electrons and holes) in metals, semiconductors and also in vacuum. Another difference is, in electrical engineering the voltages and currents are of very high-kilovolts, and Amperes, whereas in electronic engineering one deals with few volts and mA. Yet another difference is, in electrical engineering, the frequencies of operation are 50 Hertzs/60 Hertzs, whereas in electronics, it is KHzs, MHz, GHzs, (high frequency).

The beginning for Electronics was made in 1895, when H.A. Lorentz postulated the existence of discrete charges called *electrons*. Two years later, J.J.Thomson proved the same experimentally in 1897.

In the same year, Braun built the first tube, based on the motion of electrons, and called it *Cathode ray tube* (CRT).

In 1904, Fleming invented the Vacuum diode called '*valve*'.

In 1906, a semiconductor diode was fabricated but they could not succeed, in making it work. So, semiconductor technology met with premature death and vacuum tubes flourished.

In 1906 itself, De Forest put a third electrode into Fleming's diode and he called it *Triode*. A small change in grid voltage produces large change in plate voltage in this device.

In 1912 Institute of Radio Engineering (IRE) was set up in USA to take care of the technical interests of electronic engineers. Before that, in 1884 Institute of Electrical Engineers was formed and in 1963 both institutes merged into one association called IEEE (Institute of Electrical and Electronic Engineers).

The first radio broadcasting station was built in 1920 in USA.

In 1930, black and white television transmission started in USA.

In 1950, Colour television broadcasting was started.

The electronics Industry can be divided into 4 categories :

Components	:	Transistors, ICs, R, L, C components
Communications	:	Radio, Television, Telephone - wireless, landline communications
Control	:	Industrial electronics, control systems
Computation	:	Computers

Vacuum Tubes ruled the electronic field till the invention of transistors. The difficulty with vacuum tubes was, it generated lot of heat. The filaments get heated to > 2000^0 k, so that electron emission takes place. The filaments get burnt and tubes occupy large space. So in 1945, Solid State Physics group was formed to invent semiconductor devices in Bell Labs, USA.

Major milestones in development of Electronics :

1895 :	H. A. Lorentz - Postulated existance of Electrons	
1897 :	J.J. Thomson - Proved the same	
1904 :	Fleming invented Vacuum Diode	
1906 :	De Forest developed Triode	
1920 :	Radio Broadcasting in USA	
1930 :	Black and White Television Transmission in USA.	
1947 :	Shockley - invented the junction transistor. (BJT)	
1950 :	Colour Television Transmission started in USA.	
1959 :	Integrated circuit concept was announced by Kilby at an IRE convention.	
1969 :	LSI, IC - Large Scale Integration, with more than 1000 but < 10,000 components per chip (integrated or joined together), device was announced.	
1969 :	SSI 10 - 100 components/chip, LOGIC GATES, FFs were developed.	
1970 :	INTEL group announced, chip with 1000 Transistors (4004m)	
1971 :	4 bit Microprocessor was made by INTEL group.	
1975 :	VLSI : Very large scale integration > 10,000 components per chip. ICs were made.	
1975 :	CHMOS - Complimentary High Metal Oxide Semiconductor ICs were announced by INTEL group.	
1975 :	MSI (Multiplenum, Address) 100 - 1000 components/chip was developed.	
1978 :	LSI 8 bit microprocessors (μp), ROM, RAM 1000 - 10,000 components/chip	
1980 :	VLSI	> 1,00,000 components/chip, Ex : 16 bit and 32 bit μPs
1981 :	16 bit μ p	> 1,00,000 components/chip, Ex : 16 bit and 32 bit μPs
1982 :	100,000 Transistors, (80286) was developed	
1984 :	CHMOS	> 2,00,000 components/chip Ex : 16 bit and 32 bit μPs
1985 :	32 bit μ p	> 4,50,000 components/chip Ex : 16 bit and 32 bit μPs
1986 :	64 bit μ p	> 10,00,000 components/chip Ex : 16 bit and 32 bit μPs
1987 :	MMICS	Monolithic Microwave Integrated Circuits
1989 :	i860 Intel's 64 bit CPU developed	
1990s :	ULSI > 500,000 Transistors; Ultra Large Scale Integration GSI > 1,000,000 Transistors; Giant Scale Integration	

1992 : 3 million Transistors, (Pentium series)
1998 : 2 Million Gates/Die
2001 : 5 Million Gates / Die
2002 : 1 Gigabit Memory Chips
2003 : 10 nanometer patterns, line width
2004 : Commercial Super Compter 10T. Flip Flops developed.
2010 : Neuro - Computer Using Logic Structure Based on Human Brain likely

Still Nature is superior. There are 10^7 cells/cm^3 in human brain

Development of VLSI Technology :

3 μ Technology
↓
0.5 μ Technology
↓
0.12 μ Technology
ASICS (Application Specific Integrated Circuits)
HYBRID ICs
BI CMOS
MCMs (Multi Chip Modules)
3-D packages

Table showing predictions made in 1995 on VLSI Technology

	1995	**1998**	**2001**	**2004**	**2007**
Lithography (μ)	0.35	0.25	0.18	0.12	0.1
No. Gates/Die :	800K	2 M	5 M	10 M	20M
No. Bits/Die					
Dram	64 M	256	1 G	4 G	16G
Sram	16 M	64 N	256 M	1G	4G
Wafer Dia (mm)	200	200-400	-400	-400	-400
Power (μW/Die)	15	30	40	40-120	40-200
Power Supply. V.	3.3	2.2	2.2	1.5	1.5
Frequency MHz	100	175	250	350	500

1 PN Junction Diode and its Applications

In this Chapter,

- The basic aspects connected with Semiconductor Physics and the terms like Effective Mass, Semiconductors and their types are introduced, Intrinsic, Extrinsic, and Semiconductors are introduced.
- Atomic Structure, Configurations, Concept of Hole, Conductivity in ***p-type*** and ***n-type*** semiconductors are given. This forms the basis to study Semiconductor Devices in the following chapters.
- After p-type and n-type semiconductors, we shall study semiconductor devices formed using these two types of semiconductors.
- We shall also study the physical phenomena such as conduction, transport mechanism, electrical characteristics, and applications of semiconductor devices such as p-n diode, zener diode, tunnel diode and so on.
- Circuit applications of p-n junction diode namely Half Wave Rectifier (HWR), Full Wave Rectifier (FWR) and Bridge Rectifier circuits, for rectification applications are described.
- Inductor, L-section and π-section filter circuits are explained.

1.1 REVIEW OF SEMICONDUCTOR PHYSICS

1.1.1 Energy Levels and Energy Bands

To explain the phenomenon associated with conduction in metals and semiconductors and the emission of electrons from the surface of a metal, we have to assume that atoms have loosely bound electrons which can be removed from it.

Rutherford found that atom consists of a nucleus with electrons (each electron carrying a unit negative charge) rotating around it. The mass of the atom is concentrated in the nucleus. It consist of protons which are positively charged. In hydrogen atom, there is one positively charged nucleus (a proton) and a single electron. The charge on particle is positive and is equal to that of electron. So hydrogen atom is neutral in charge. The proton in the nucleus carries the charge of the atom, so it is immobile. The electron will be moving around it in a closed orbit. The force of attraction between electron and proton follows Coulombs Law. {directly proportional to product of charges and inversely proportional to (distance)^2}.

Assume that the orbit of the electron around nucleus is a circle. We can calculate the radius of this circle, in terms of total energy 'W' of the electron. The force of attraction between the electron and the nucleus is :

$$F \alpha e^2$$

(Therefore, the nucleus has the proton with electron charge equal to 'e')

$$F \alpha \frac{1}{r^2}$$

(r is the radius of the orbit)

$$\therefore \quad F \alpha \frac{e^2}{r^2}$$

$$F = \frac{e^2}{4\pi \epsilon_o r^2}$$

$$\therefore \quad F \alpha \frac{e^2}{r^2}$$

where ϵ_0 is the permitivity of free space. Its value is $\frac{10^{-9}}{36\pi}$ F/m.

As the electron is moving around the nucleus in a circular orbit with radius r, and with velocity v, then the force of attraction given by the above expression F should be equal to the centripetal force $\frac{mv^2}{r}$ (according to Newton's Second Law of Motion)

(F = m.a; m is the mass of electron and acceleration, $a = \frac{v^2}{r}$)

$\therefore$ The Potential Energy of the electron $= -\dfrac{e^2}{4\pi\varepsilon_0 r^2} \times r = -\dfrac{e^2}{4\pi\varepsilon_0 r}$

(–ve sign is because Potential Energy is by definition work done against the field)

$$\text{Kinetic Energy} = \frac{1}{2}\,mv^2$$

$\therefore$ Total Energy W possessed by the electron $= \dfrac{1}{2}\,mv^2 - \dfrac{e^2}{4\pi\varepsilon_0 r}$

But $mv^2 = \dfrac{e^2}{4\pi\varepsilon_0 r}$ reduces to

$$W = \frac{e^2}{8\pi\varepsilon_0 r} - \frac{e^2}{4\pi\varepsilon_0 r} = -\frac{e^2}{8\pi\varepsilon_0 r}$$

$\therefore$ Energy possessed by the electron $W = -\dfrac{e^2}{8\pi\varepsilon_0 r}$

W is the energy of the electron. Only for Hydrogen atom, W will also be the energy of atom since it has only one electron. The negative sign arises because the Potential Energy of the electron is

$$W = -\frac{e^2}{8\pi\varepsilon_0 r}$$

As radius r increases, Potential Energy decreases. When r is infinity, Potential Energy is zero. Therefore the energy of the electron is negative. If r is $< \infty$, the energy should be less. (Any quantity less than 0 is negative).

The above equation is derived from the classical model of the electron. But according to classical laws of electromagnetism, an accelerated charge must radiate energy. Electron is having charge = e. It is moving with velocity v or acceleration v^2/r around the nucleus. Therefore this electron should also radiate energy. If the charge is performing oscillations with a frequency 'f', then the frequency of the radiated energy should also be the same. Hence the frequency of the radiated energy from the electron should be equal to the frequency with the electron orbiting round the nucleus.

But if the electron is radiating energy, then its total energy 'W' must decrease by the amount equal to the radiation energy. So 'W' should go on decreasing to satisfy the equation,

$$W = -\frac{e^2}{8\pi\varepsilon_0 r}$$

If W decreases, r should also decrease. Since if $-\dfrac{e^2}{8\pi\varepsilon_0 r}$ should decrease, this quantity should become more negative. Therefore, r should decrease. So, the electron should describe smaller and smaller orbit and should finally fall into nucleus. So classical model of atom is not fairly satisfactory.

1.1.2 The Bohr Atom

The above difficulty was resolved by Bohr in 1913. He postulated three fundamental laws.

1. *The atom can possess only discrete energies. While in states corresponding to these discrete energy levels, electron does not emit radiation in stationary state.*

2. *When the energy of the electron is changing from W_2 to W_1 then radiation will be emitted. The frequency of radiation is given by*

$$f = \frac{W_2 - W_1}{h}$$

where 'h' is Plank's Constant.

$$h = 6.626 \times 10^{-34} \text{ J} - \text{sec.}$$

i.e., when the atom is in stationary state, it does not emit any radiation. When its energy changes from W_2 to W_1 then the atom is said to have moved from one stationary state to the other. The atom remains in the new state corresponding to W_1. Only during transition, will some energy be radiated.

3. *A stationary state is determined by the condition that the angular momentum of the electron in this state must be an integral multiple of $h/2\pi$.*

So $$mvr = \frac{nh}{2\pi}$$

where n is an integer, other than zero.

1.1.3 Effective Mass

An electron mass 'm' when placed in a crystal lattice, responds to applied field as if it were of mass m*. The reason for this is the interaction of the electron even within lattice.

E = Kinetic Energy of the Free Electron

p = Momentum

v = Velocity

m = Mass

p = mv

m* = Effective mass of electron

$$E = \frac{p^2}{2m}$$

Electrons in a solid are not free. They move under the combined influence of an external field plus that of a periodic potential of atom cores in the lattice. An electron moving through the lattice can be represented by a wave packet of plane waves grouped around the same value of K which is a wave vector.

Electron velocity falls to zero at each band edge. This is because the electron wave further becomes standing wave at the top and bottom of a band i.e., $v_g = 0$.

Now consider an electronic wave packet moving in a crystal lattice under the influence of an externally applied uniform electric field. If the electron has an instantaneous velocity 'v_g' and moves a distance 'dx' in the direction of an accelerating force 'F', in time 'dt', it acquires energy δE where

$$\delta E = F \times \delta x = F \times v_g \times \delta t$$

$$\delta E = \frac{F}{\hbar} \times \frac{\delta E}{\delta k} \times \delta t$$

$$v_g = \frac{\delta E}{\hbar \delta k}$$

Within the limit of small increments in 'K', we can write,

$$\frac{dK}{dt} = \frac{F}{\hbar} \qquad \text{..........(1.1)}$$

But this is not the case for the electron in a solid because the externally applied force is not the only force acting on the electrons. Forces associated with the periodic lattice are also present.

Acceleration of an electronic wave packet in a solid is equal to the rate of change of its velocity.

$$\text{Acceleration of an electron in a solid} = \frac{dv_g}{dt} = \frac{d}{dt}\left(\frac{dE}{dp}\right) = \frac{d^2E}{(dp)(dt)}$$

$$\because \qquad v_g = \frac{dE}{dp}$$

$$\frac{dk}{dt} = \frac{F}{\hbar}$$

$$F = \hbar \times \frac{dk}{dt}$$

But $$\frac{dk}{dt} = \frac{F}{\hbar} \qquad \text{from Eq. (1.1)}$$

$$\therefore \qquad \frac{dv_g}{dt} = \left(\frac{dp}{dt}\right)\frac{d^2E}{dp^2} = \frac{F}{\hbar^2} \cdot \frac{d^2E}{dp^2}$$

or $$F = \hbar^2 \times \left(\frac{d^2E}{dp^2}\right)^{-1} \times \frac{dv_g}{dt}$$

This is of the form, F = ma, from Newton's Laws of Motion,

where $$m^* = \hbar^2 \times \left(\frac{d^2E}{dp^2}\right)^{-1} \text{ and } a = \frac{dv_g}{dt}$$

$$\therefore \qquad F = m^* \times \frac{dv_g}{dV}$$

where m* is the effective mass.

If an electric field 'ε' is impressed, the electron will accelerate and its velocity and energy will increase. Hence the electron is said to have positive mass. On the other hand, if an electron is at the upper end of a band, when the field is applied, its energy will increase and its velocity decreases. So the electron is said to have negative mass.

In an atom,

$$\text{Coulombs force of attraction} = \frac{e^2}{4\pi \epsilon_0 r^2}$$

$$\text{Centripetal Force} = \frac{mv^2}{r}$$

Equating these two forces in an atom,

$$\frac{e^2}{4\pi \epsilon_0 r^2} = \frac{mv^2}{r} \quad v = \text{velocity; } r = \text{radius of Orbit}$$

$$\therefore \quad \frac{e^2}{4\pi \epsilon_0} = \frac{mv^2}{r} \times r^2 = mv^2 r$$

$$\therefore \quad r = \frac{e^2}{4\pi \epsilon_0 mv^2} \qquad \text{..........(1.2)}$$

But $\quad mvr = \dfrac{nh}{2\pi} \quad$ h = Plank's Constant; n = Principle Quantum Number.

$$\therefore \quad v = \frac{nh}{2\pi mr} \qquad \text{..........(1.3)}$$

Substituting the value of v in eq. (1.2),

$$r = \frac{e^2 \times 4\pi^2 m^2 r^2}{4\pi \epsilon_0 .m \times n^2 h^2}$$

$$\therefore \quad \boxed{r = \frac{n^2 h^2 \epsilon_0}{\pi m e^2}} \qquad \text{..........(1.4)}$$

This is the expression for radii of stable states.

Energy possessed by the atom in the stable state is

$$W = \frac{e^2}{8\pi \epsilon_0 r} \qquad \text{..........(1.5)}$$

Substituting the value of r in Eq. (1.5), then

$$W = -\frac{e^2 \times \pi m e^2}{8\pi \epsilon_0 n^2 h^2 \epsilon_0}$$

$$W = -\frac{me^4}{8h^2 \epsilon_0^2} \cdot \frac{1}{n^2}$$

Thus energy 'W' corresponds to only the coulombs force due to attraction between ground electron (negative charge) and proton (positive charge).

Problem 1.1

Determine the radius of the lowest state of Ground State.

Solution

$$n = 1$$

$$\because \quad r = \frac{n^2 h^2 \epsilon_0}{\pi m e^2}$$

Plank's Constant, $h = 6.626 \times 10^{-34}$ J-sec

Permitivity, $\epsilon_o = 10^{-9}/36\pi$

Substituting the values and simplifying,

$$\therefore \quad r = 0.58 \text{ A}^o$$

1.1.4 Atomic Energy Levels

For different elements, the value of the free electron concentration will be different. By spectroscopic analysis we can determine the energy level of an element at different wavelengths. This is the characteristic of the given element.

The lowest energy state is called the normal level or ground level. Other stationary states are called ***excited***, ***radiating***, ***critical*** or ***resonance*** levels.

Generally, the energy of different states is expressed in (eV) electrovolts rather than in Joules (J), and the emitted radiation is expressed by its wavelength λ rather than by its frequency. This is only for convenience since Joule is a larger unit and the energy is small and is in electron volts.

$$F = \frac{W_2 - W_1}{h}$$

$$f = \frac{C}{\lambda}$$

C = Velocity of Light = 3×10^{10} cm / sec.

h = Plank's Constant = 6.626×10^{-34} J - sec.

f = Frequency of Radiation

λ = Wavelength of emitted radiation

W_1 and W_2 are the energy levels in Joules.

$$\boxed{1 \text{ eV} = 1.6 \times 10^{-19} \text{ Joules}}$$

$$\therefore \quad \frac{C}{\lambda} = \frac{(E_2 - E_1) \times 1.6 \times 10^{-19}}{h}$$

$$\frac{3 \times 10^{10}}{\lambda} = \frac{(E_2 - E_1) \times 1.6 \times 10^{-19}}{6.626 \times 10^{-34}}$$

E_1 and E_2 are energy levels in eV.

or
$$\lambda = \frac{3\times 10^{10}\times 6.626\times 10^{-34}}{(E_2 - E_1)\times 1.6\times 10^{-19}}$$

$$\boxed{\lambda = \frac{12,400}{(E_2 - E_1)} A^o}$$

where 'λ' is in Armstrong, $1 A^o = 10^{-8}$ cm $= 10^{-10}$ m, and E_2 and E_1 in eV.

1.1.5 Photon Nature of Light

An electron can be in the excited state for a small period of 10^{-7} to 10^{-10} sec. Afterwards it returns back to the original state. When such a transition occurs, the electrons will loose energy equal to the difference of energy levels ($E_2 - E_1$). This loss of energy of the atom results in radiation of light. The frequency of the emitted radiation is given by

$$f = \frac{(E_2 - E_1)}{h}.$$

According to classical theory it was believed that atoms continuously radiate energy. But this is not true. Radiation of energy in the form of photons takes place only when the transition of electrons will take place from higher energy state to lower energy state, so that ($E_2 - E_1$), is positive. This will not occur if the transition is from lower energy state to higher energy state. When such photon radiation takes place, the number of photons liberated is very large. This is explained with a numercial example given below.

Problem 1.2

For a given 50 W energy vapour lamp. 0.1% of the electric energy supplied to the lamp, appears in the ultraviolet line 2,537 A^0. Calculate the number of photons per second, of this wavelength emitted by the lamp.

Solution

$$\lambda = \frac{12,400}{(E_2 - E_1)}$$

λ = Wavelength of the emitted radiation.

E_1 and E_2 are the energy levels in eV.

($E_2 - E_1$) is the energy passed by each photon in eV of wavelength λ. In the given problem, $\lambda = 2,537\ A^o$. $(E_2 - E_1) = ?$

$$\therefore \quad (E_2 - E_1) = \frac{12,400}{2,537} = 4.88 \text{ eV/photon}$$

0.1% of 50W energy supplied to the lamp is,

i.e.,
$$\frac{0.1}{100}\times 50 = 0.05 \text{ W} = 0.05 \text{ J/Sec}$$

$$\because \quad 1\text{W} = 1 \text{ J/sec}$$

Converting this into the electron Volts,

$$\frac{0.05\ \text{J/sec}}{1.6\times10^{-19}\ \text{J/eV}} = 3.12 \times 10^{17}\ \text{eV/sec}$$

This is the total energy of all the photons liberated in the λ = Wavelength of 2,537 A^o.

$$\therefore \quad \text{Number of photons emitted per sec} = \frac{\text{total energy}}{\text{energy / photon}}$$

$$= \frac{3.12\times10^{17}\ \text{ev/sec}}{4.88\ \text{ev/photon}}$$

$$= 6.4 \times 10^{16}\ \text{photon/sec}$$

The lamp emits 6.4×10^{16} photons / sec of wavelength $\lambda = 2,537 A^o$.

1.1.6 Ionization Potential

If the most loosely bound electron (free electron) of an atom is given more and more energy, it moves to stable state (since it is loosely bound, its tendency is to acquire a stable state. Electrons orbiting closer to the nucleus have stable state, and electron orbiting in the outermost shells are loosely bound to the nucleus). But the stable state acquired by the electron is away from the nucleus of the atom. If the energy supplied to the loosely bound electron is enough large, to move it away completely from the influence of the parent nucleus, it becomes detached from it. ***The energy required to detach an electron is called Ionization Potential.***

1.1.7 Collisions of Electrons with Atoms

If a loosely bound electron has to be liberated, energy has to be supplied to it. Consider the case when an electron is accelerated and collides with an atom. If this electron is moving slowly with less energy, and collides with an atom, it gets deflected, i.e., its direction changes. But no considerable change occurs in energy. This is called ***Elastic Collision***.

If the electron is having much energy, then this electron transfers its energy to the loosely bound electron of the atom and may remove the electron from the atom itself. So another free electron results. If the bombarding electron is having energy greater than that required to liberate a loosely bound electron from atom, the excess energy will be shared by the bombarding and liberated electrons.

Problem 1.3

Argon resonance radiation, corresponding to an energy of 11.6 eV falls upon sodium vapor. If a photon ionizes an unexcited sodium atom, with what speed is the electron ejected ? The ionization potential of sodium is 5.12 eV.

Solution

Ionization potential is the minimum potential required to liberate an electron from its parent atom. Argon energy is 11.6 eV. Ionization Potential of Na is 5.12 eV.

$\therefore$ The energy possessed by the electron which is ejected is

$$11.6 - 5.12 = 6.48\ \text{eV}$$

or Potential, $V = 6.48$ volts $\quad$ ($\because$ 1eV energy, potential is 1V)

Its velocity $\quad v = \sqrt{\frac{2eV}{m}} = 5.93 \times 10^5 \sqrt{6.48} = 1.51 \times 10^6$ m/sec.

Problem 1.4

With what speed must an electron be travelling in a sodium vapor lamp in order to excite the yellow line whose wavelength is 5,893 A°.

Solution

$$Ee \geq \frac{12{,}400}{5{,}893} \geq 2.11 \text{ eV}$$ (Corresponding to evergy of 2.11eV, the potential is 2.11 Volts)

$$V = 2.11 \text{ Volts}$$

$$\therefore \text{ Velocity, } v = \sqrt{\frac{2eV}{m}} = 5.93 \times 10^5 \sqrt{2.11} = 8.61 \times 10^5 \text{ m/sec.}$$

Problem 1.5

A radio transmitter radiates 1000 W at a frequency of 10 MHz.

(a) *What is the energy of each radiated quantum in ev ?*

(b) *How many quanta are emitted per second ?*

(c) *How many quanta are emitted in each period of oscillation of the electromagnetic field ?*

Solution

(a) Energy of each radiated quantum = E = hf

$$f = 10 \text{ MHz} = 10^7 \text{ Hz}, h = 6.626 \times 10^{-34} \text{ Joules / sec}$$

$$\therefore \quad E = 6.626 \times 10^{-34}\ 10^7 = 6.626 \times 10^{-27} \text{ Joules / Quantum}$$

$$= \frac{6.626 \times 10^{-27}}{1.6 \times 10^{-19}} = 4.14 \times 10^{-8} \text{ eV / Quantum}$$

(b) 1W = 1 Joule/sec

1000 W = 1000 Joules/sec = Total Energy

Energy possessed by each quantum = 6.626×10^{-27} Joules/Quantum.

$$\therefore \quad \text{Total number of quanta per sec, } N = \frac{1000}{6.626 \times 10^{-27}} = 1.5 \times 10^{29}\text{/sec}$$

(c) One cycle = 10^{-7} sec

$$\therefore \quad \text{Number of quanta emitted per cycle} = 10^{-7} \times 1.51 \times 10^{29}$$

$$= 1.51 \times 10^{22} \text{ per cycle}$$

Problem 1.6

(a) *What is the minimum speed with which an electron must be travelling in order that a collision between it and an unexcited Neon atom may result in ionization of this atom? The Ionization Potential of Neon is 21.5 V.*

(b) *What is the minimum frequency that a photon can have and still be able to cause Photo-Ionization of a Neon atom?*

Solution

(a) Ionization Potential is 21.5 V

$\therefore$ $$\text{Velocity} = \sqrt{\frac{2eV}{m}} = 5.93 \times 10^5 \sqrt{21.5} = 2.75 \times 10^6 \text{ m/sec}$$

Wavelength of radiation,

(b) $$\lambda = \frac{12{,}400}{E_2 - E_1} = \frac{12{,}400}{21.5} = 577 \text{ A}^\circ$$

Frequency of radiation,

$$f = \frac{C}{\lambda} = \frac{3 \times 10^8}{577 \times 10^{-10}} = 5.2 \times 10^{15} \text{ Hz}$$

Problem 1.7

Show that the time for one revolution of the electron in the hydrogen atom in a circular path around the nucleus is

$$T = \frac{m^{1/2} e^2}{4\sqrt{2}\, \epsilon_0 (-W)^{\frac{3}{2}}}$$

Solution

$$v = \sqrt{\frac{e^2}{4\pi m \epsilon_0 r}}$$

$$T = \frac{2\pi r}{v} = \frac{2\pi r (4\pi m \epsilon_0 r)^{1/2}}{e} = \frac{2\pi r^{\frac{3}{2}} (4\pi m \epsilon_0)^{1/2}}{e}$$

But radius, $$r = \frac{e^2}{4\pi \epsilon_0 W}$$

$\therefore$ $$T = \frac{m^{1/2} e^2}{4\sqrt{2}\, \epsilon_0 (-W)^{3/2}}$$

Problem 1.8

A photon of wavelength 1,400 A° is absorbed by cold mercury vapor and two other photons are emitted. If one of these is the 1,850 A° line, what is the wavelength λ of the second photon ?

Solution

$$(E_2 - E_1) = \frac{12{,}400}{\lambda} = \frac{12{,}400}{1400} = 8.86 \text{ eV}$$

1850^0 A° line is from 6.71 eV to 0 eV.

$\therefore$ The second photon must be from 8.86 to 6.71 eV.

So, $$\Delta E = 2.15 \text{ eV.}$$

$$\lambda = \frac{12{,}400}{(E_2 - E_1)}$$

$$\lambda = \frac{12{,}400}{2.15} = 5767 \text{ A}^\circ$$

1.1.8 Metastable States

An atom may be elevated to an excited energy state by absorbing a photon of frequency '*f*' and thereby move from the level of energy W_1 to the higher energy level W_2 where $W_2 = W_1 + hf$. But certain states may exist which can be excited by electron bombardment but not by photo excitation (absorbing photons and raising to the excited state). ***Such levels are called metastable states***. A transition from a metastable level to a normal state with the emission of radiation has a very low probability of occurence. Transition from higher level to a metastable state are permitted, and several of these will occur.

An electron can be in the metastable state for about 10^{-2} to 10^{-4} sec. This is the mean life of a metastable state. Metastable state has a long lifetime because they cannot come to the normal state by emitting a photon. Then if an atom is in metastable state how will it come to the normal state? It cannot release a photon to come to normal state since this is forbidden. Therefore an atom in the metastable state can come to normal state only by colliding with another molecule and giving up its energy to the other molecule. Another possibility is the atom in the metastable state may receive additional energy by some means and hence may be elevated to a higher energy state from where a transition to normal state can occur.

1.1.9 Wave Properties of Matter

An atom may absorb a photon of frequency '*f*' and move from the energy level 'W_1' to the higher energy level 'W_2' where $W_2 = W_1 + hf$.

Since a photon is absorbed by only one atom, the photon acts as if it were concentrated in one point in space. So wave properties can not be attributed to such atoms and they behave like particles.

Therefore according to 'DeBroglie' hypothesis, dual character of wave and particle is not limited to radiation alone, but is also exhibited by particles such as electrons, atoms, and molecules. He calculated that a particle of mass 'm' travelling with a velocity 'v' has a wavelength 'λ' given by $\lambda = \frac{h}{mv} = \frac{h}{p}$ (λ is the wavelength of waves consisting of these particles).

Where 'p' is the momentum of the particle. Wave properties of moving electrons can be made use to explain Bohr's postulates. A stable orbit is one whose circumference is exactly equal to the wavelength 'λ' or 'nλ' where 'n' is an integer other than zero.

Thus, $2\pi r = n\lambda$ r = radius of orbit, n = Principle Quantum Number.

But according to DeBroglie,

$$\lambda = \frac{h}{mv}$$

$$\therefore \quad 2\pi r = \frac{nh}{mv}$$

This equation is identical with Bohr's condition, $mvr = \frac{nh}{2\pi}$

1.1.10 Schrodinger Equation

Schrodinger carried the implications of the wave nature of electrons. A branch of physics called ***Wave Mechanics*** or ***Quantum Mechanics*** was developed by him. If deBroglie's concept of wave nature of electrons is correct, then it should be possible to deduce the properties of an electron system from a mathematical relationship called the ***Wave Equation*** or ***Schrodinger Equation***. It is

$$\nabla^2\phi - \frac{1}{v^2}\cdot\frac{\partial^2\phi}{\partial t^2} = 0, \qquad \text{..........(1.6)}$$

where $$\nabla^2 = \frac{\partial^2}{\partial x^2} + \frac{\partial^2}{\partial y^2} + \frac{\partial^2}{\partial z^2}$$

'ϕ' can be a component of electric field or displacement or pressure. 'v' is the velocity of the wave, and 't' is the time.

The variable can be eliminated in the equation by assuming a solution.

$$\phi\,(x, y, z, t) = \psi\,(x, y, z)e^{jwt}$$

This represents the position of a particle at 't' in 3-D Motion.

ω = Angular Frequency = 2 πf

ω is regarded as constant, while differentiating.

ϕ is a function of x, y, z and t.

But ψ is a function of x, y and z only.

To get the independent ***Schrodinger Equation***

$$\frac{\partial\phi}{\partial t} = j\omega\psi(x, y, z)e^{j\omega t}$$

$$\frac{\partial^2\phi}{\partial t^2} = -\omega^2\psi(x, y, z)e^{j\omega t}$$

But $\omega = 2\pi f$

So, $\omega^2 = 4\pi^2 f^2$

Substituting this in the original ***Schrodinger's Equation***,

$$e^{jwt}\left\{\nabla^2\psi + \frac{1}{v^2}\times 4\pi^2 f^2\psi\right\} = 0$$

But $\lambda = \frac{v}{f}$; Wavelength = $\frac{\text{Velocity}}{\text{Frequency}}$

$$\therefore \quad \nabla^2\psi + \frac{4\pi^2}{\lambda^2}\psi = 0 \qquad \text{..........(1.7)}$$

But $\lambda = \frac{h}{mv} = \frac{h}{p}$ v = Velocity.

$\because$ p = mv

This is deBrogile's Relationship.

$$\frac{1}{\lambda^2} = \frac{p^2}{h^2}$$

But $p^2 = m^2 v^2 = 2\,(\text{Kinetic Energy})\,m \qquad \because \text{K.E.} = \frac{1}{2}mv^2$

Kinetic Energy = Total Energy (W) – Potential Energy (U)

$\therefore \quad p^2 = 2\,(W - U)\,.\,m$

$$\therefore \quad \frac{p^2}{h^2} = \frac{2m}{h^2}\,(W - U) \qquad \text{..........(1.8)}$$

Substituting Eq. (1.8) in Eq. (1.7) we get,

$$\nabla^2\,\Psi + \frac{8\pi^2 m}{h^2}\,(W - U)\,\Psi = 0.$$

This is the Time Independent Schrodinger Equation Ψ is a function of 't', But this equation as such is not containing the term 't'.

1.1.11 Wave Function

'Ψ' is called as the wave function, which describes the behavior of the particle. 'Ψ' is a quantity whose square gives the probability of finding an electron. $|\Psi|^2(dx \times dy \times dz)$ is proportional to the probability of finding an electron in volume dx, dy and dz at point P(x, y, z).

Four quantum numbers are required to define the wave function. They are :

1. ***The Principal Quantum Number 'n' :***

   It is an integer 1, 2, 3, ... This number determines the total energy associated with a state. It is same as the quantum number 'n' of Bohr atom.

2. ***The Orbital Angular Momentum Quantum Number l :***

   It takes values 0, 1, 2...l... (n – l)

   The magnitude of this angular momentum is $\sqrt{(l)(l+1)} \times \frac{h}{2\pi}$

   It indicates the shape of the classical orbit.

3. ***The orbital magnetic number m_l :***

   This will have values 0, ± 1, ±2 .... ± l. This number gives the orientation of the classical orbit with respect to an applied magnetic field.

   The magnitude of the Angular Momentum along the direction of magnetic field $= m_\ell\left(\frac{h}{2\pi}\right)$.

4. ***Electron Spin :***

   It was found is 1925 that in addition to assuming that electron orbits round the nucleus, it is also necessary to assume that electron also spins around itself, in addition to orbiting round the nucleus. This intrinsic electronic angular momentum is called ***Electron Spin***.

   When an electron system is subjected to a magnetic field, the spin axis will orbit itself either parallel or anti-parallel to the direction of the field. The electron angular momentum is given by $m_s\left(\frac{h}{2\pi}\right)$ where the spin quantum m_s number may have values $+\frac{1}{2}$ or $-\frac{1}{2}$.

1.1.12 ELECTRONIC CONFIGURATION

PAULI'S EXCLUSION PRINCIPLE

No two electrons in an electron system can have the same set of four quantum numbers n, l, m_l and m_s.

Electrons will occupy the lower most quantum state.

ELECTRONIC SHELLS (PRINCIPLE QUANTUM NUMBER)

All the electrons which have the same value of 'n' in an atom are said to belong to the same electron shell. These shells are identified by letters K, L, M, N corresponding to n = 1, 2, 3, 4..... A shell is subdivided into sub shells corresponding to values of l and identified as s, p, d, f, g, h,. for l = 0, 1, 2, 3... respectively. This is shown in Table 1.1.

Table 1.1

Shell	K	L		M			N			
n	1	2		3			4			
l	0	0	1	0	1	2	0	1	2	3
subshell	s	s	p	s	p	d	s	p	d	f
No. of	2	2	6	2	6	10	2	6	10	14
electron	2	8		18			32			

ELECTRON SHELLS AND SUBSHELLS

Number of Electrons in a sub shell = $2(2l + 1)$

n = 1 corresponds to K shell

$l = 0$ corresponds to s sub shell

$\because$ $l = 0, \ldots, (n - 1)$ if $n = 1$,

$l = 0$ is the only possibility

Number of electrons in K shell = $2(2l + 1) = 2(0 + 1) = 2$ electrons.

$\therefore$ K shell will have 2 electrons.

This is written as $1s^2$ pronounced as "***one s two***" (1 corresponds to K shell, n = 1; s is the sub shell corresponds to $l = 0$ number of electron is 2. Therefore, $1s^2$)

If n = 2, it is 'l' shell

If $l = 0$, it is 's' sub shell

If $l = 1$, it is 'p' sub shell

Number of electron in 's' sub shell (i.e., $l = 0$) = $2 \times (2l + 1) = 2(0 + 1) = 2$

Number of electron is 'p' sub shell (i.e., $l = 1$) = $2[(1 \times 2) + 1)] = 6$

In l shell there are two sub shells, s and p.

$\therefore$ Total number of electron in l shell = 2 + 6 = 8

This can be represented as $2s^2\ 2p^6$

If n = 3, it is M shell

It has 3 sub shell s, p, d, corresponding to $l = 0, 1, 2$

In 's' sub shell number of electrons ($l = 0$) = 2($\therefore$ $l = 0$) $2(2l + 1)$

In 'p' sub shell number of electrons ($l = 1$) = $2(2 + 1) = 6$

In 'd' sub shell number of electrons ($l = 2$) = $2(2 \times 2 + 1) = 10$

$\therefore$ Total number of electrons in M shell = 10 + 6 + 2 = 18

This can be represented as $3s^2\ 3p^6\ 3d^{10}$

$\therefore\ 1s^2\ 2s^2\ 2p^6\ 3s^2\ 3p^6\ 3d^{10}\ 4s^2\ 4p^6\ 4d^{10}$

ELECTRONIC CONFIGURATION

Atomic number 'z' gives the number of electrons orbiting round the nucleus. So from the above analysis, electron configuration can be given as $1s^2\ 2s^2\ 2p^6\ 3s^1$. First k shell (n = 1) is to be filled. Then *l* shell (n = 2) and so on.

In 'K' shell there is one sub shell (s) ($l = 0$). This has to be filled.

In 'L' shell there are two sub shells and p. First – s and then p are to be filled and so on.

The sum of subscripts 2 + 2 + 6 + 1 = 11. It is the atomic number represented as Z.

For Carbon, the Atomic Number is 6., i.e., Z = 6.

$\therefore$ The electron configuration is $1s^2\ 2s^2\ 2p^2$

For the Ge Z = 32. So the electronic configuration of Germanium is,

$\therefore$ $1s^2\ 2s^2\ 2p^6\ 3s^2\ 3p^6\ 3d^{10}\ 4s^2\ 4p^2$

For the Si Z = 14. So the electronic configuration of Silicon is,

$\therefore$ $1s^2\ 2s^2\ 2p^6\ 3s^2\ 3p^2$

1.1.13 TYPES OF ELECTRON EMISSION

Electrons at absolute zero possess energy ranging from 0 to E_F the fermi level. It is the characteristic of the substance. But this energy is not sufficient for electrons to escape from the surface. They must posses energy $E_B = E_F + E_W$ where E_W is the work function in eV.

E_B = Barrier's Energy

E_F = Fermi Level

E_W = Work Function.

Different types of Emission by which electrons can emit are

(1) *Thermionic Emission*

(2) *Secondary Emission*

(3) *Photoelectric Emission*

(4) *High field Emission.*

1. THERMIONIC EMISSION

Suppose, the metal is in the form of a filament and is heated by passing a current through it. As the temperature is increased, the electron energy distribution starts in the metal changes. Some electrons may acquire energy greater than E_B sufficient to escape from the metal.

E_W : ***Work function of a metal. It represents the amount of energy that must be given for the electron to be able to escape from the metal.***

It is possible to calculate the number of electrons striking the surface of the metal per second with sufficient energy to be able to surmount the surface barriers and hence escape. Based upon that, the thermionic current is,

$$I_{th} = S \times Ao\, T^2\, e^{-E_w/KT} \qquad \text{.......... (1.9)}$$

It is also written as

$$\frac{I_{th}}{S} = J = AoT^2\, e^{\frac{-E_w}{kT}} = AT^2\, e^{\frac{-B}{T}}$$

where $A = A_o$ and $B = \dfrac{E_W}{k}$

where
- S = Area of the filament in m^2 (Surface Area)
- Ao = Constant whose dimensions are $A/m^2\,{}^oK$
- T = Temperature in ^oK
- K = Boltzman's constant $eV/{}^oK$
- E_w = Work function in eV

This equation is called ***Thermionic Emission Current*** or ***Richardson - Dushman Equation***. E_w is also called as latent heat of evaporation of electrons similar to evaporation of molecules from a liquid.

Taking logarithms

$$\log I_{th} = \log (S\, A_o) - \frac{E_w}{kT} \log e + \log T^2$$

$$\log I_{th} - 2 \log T = \log S\, Ao - 0.434 \left(\frac{E_W}{kT}\right)$$

$\because$ $\log e = 0.434$

So if a graph is plotted between ($\log I_{th} - 2 \log T$) Vs $\dfrac{1}{T}$, the result will be a straight line having a slope $= -0.434 \left(\dfrac{E_W}{kT}\right)$ from which E_W can be determined.

$\because$ I_{th} and T can be determined experimentally.

I_{th} is a very strong function of T. For Tungsten, $E_W = 4.52$ eV.

CONTACT POTENTIAL

Consider two metals in contact with each other forming junction at 'C' as in Fig 1.1 . The contact difference of potential is defined as the Potential Difference V_{AB} between a point A, just out side metal1 and a point B just outside metal2. The reason for the difference of potential is, when two metals are joined, electrons will flow from the metal of lower Work Function, say 1 to the metal of higher Work Function say 2. ($\because E_W = E_B - E_F$ electrons of lower work function means E_W is small or E_F is large). Flow of electrons from metal 1 to 2 will continue till metal 2 has acquired sufficient negative charge to repel extra new electrons. 'E_B' value will be almost same for all metals. But E_F differs significantly.

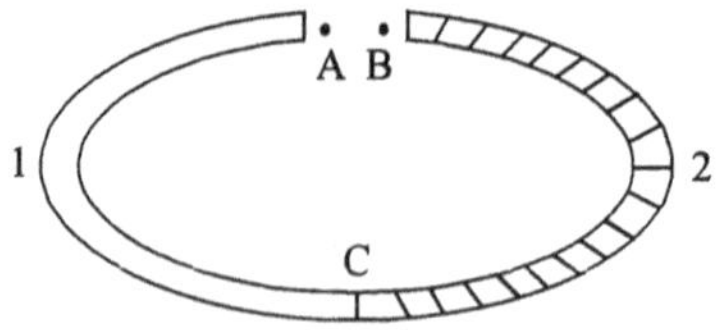

Fig 1.1 Contact Potential

In order that the fermi levels of both the metals are at the same level, the potential energy difference $E_{AB} = E_{W2} - E_{W1}$, that is, ***the contact difference of potential energy between two metals is equal to the difference between their work functions.***

If metals 1 and 2 are similar then contact potential is zero. If they are dissimilar, the metal with lower work function becomes positive, since it looses electrons.

ENERGIES OF EMITTED ELECTRONS

At absolute zero, the electrons will have energies ranging from zero to E_F. So the electrons liberated from the metal surface will also have different energies. The minimum energy required is the barrier potential to escape from the metal surface, But the electrons can acquire energy greater than E_B to escape from the surface. This depends upon the initial energy, the electrons are possessing at room temperature.

Consider a case where anode and cathode are plane parallel. Suppose the voltage applied to the cathode is lower, and the anode is indirectly heated. Suppose the minimum energy required to escape from the metal surface is 2 eV. As collector is at a potential less than 2V, electrons will be collected by the collector. If the voltage of cathode is lower, below 2V, then the current also decreases exponentially, but not abruptly. If electrons are being emitted from cathode with 2eV energy, and the voltage is reduced below 2V, then there must be abrupt drop of current to zero. But it doesn't happen This shows that electrons are emitted from surface of the emitter with different velocities. The decrease of current is given by $I = I_{th}.e^{-V}{}_T$ where V_T is the retarding potential applied to the collector and V_T is Volt equivalent of temperature.

$$V_T = \frac{kT}{e} = \frac{T}{11{,}600} \qquad \text{.......... (1.10)}$$

T = Temperature in ^{o}K

K = Boltzman's Constant in J/^{o}K

SCHOTTKY EFFECT

If a cathode is heated, and anode is given a positive potential, then there will be electron emission due to thermionic emission. There is accelerating field, since anode is at a positive potential. This accelerating field tends to lower the Work Function of the cathode material. It can be shown that under the condition of accelerating field E is

$$I = I_{th}\, e^{+0.44\ \varepsilon 1/2/T}$$

where I_{th} is the zero field thermionic current and T is cathode temperature in 0 ^{o}K.

The effect that thermionic current continues to increase as E is increased (even though T is kept constant) is known as ***Schottky Effect.***

2. Secondary Emission

This emission results from a material (metal or dielectric) when subjected to electron bombardment.

It depends upon,

1. *The energy of the primary electrons.*
2. *The angle of incidence.*
3. *The type of material.*
4. *The physical condition of surface ; whether surface is smooth or rough.*

Yield or secondary emission ratio S is defined as the ratio of the number of secondary electrons to primary electrons. It is small for pure metals, the value being 1.5 to 2. By contamination or giving a coating of alkali metal on the surface, it can be improved to 10 or 15.

3. Photo Electric Emission

Photo-Electric Emission consists of liberation of electrons by the incidence of light, on certain surfaces. The energy possessed by photons is 'hf' where 'h' is Plank's Constant and 'f' is frequency of incident light. When such a photon impinges upon the metal surface, this energy **hf** gets transferred to the electrons close to the metal surface whose energy is, very near to the barrier potential. Such electrons gain energy, to overcome the barrier potential and escape from the surface of the metal resulting in photo electric emission.

For photoelectric emission to take places, the energy of the photon must at least be equal to the work function of the metal. That is $hf \geq e\phi$ where ϕ is the voltage equivalent Work Function, (i.e. Work Function expressed in Volts). ***The minimum frequency that can cause photo-electric emission is called threshold frequency*** and is given by

$$f_T = \frac{e\phi}{h}$$

The wavelength corresponding to threshold frequency is called the ***Threshold Wavelength***.

$$\lambda_t = \frac{C}{f_t} = \frac{h_C}{\phi e}$$

If the frequency of radiation is less than f_T, then additional energy appears as kinetic energy of the emitted electron

$$\therefore \quad hf = \phi e + \tfrac{1}{2} m v^2$$

ϕ is the Volt equivalent of Work Function.

$$v = \sqrt{\frac{2eV}{m}}$$

v is the Velocity of electrons in m/sec.

$$\therefore \quad \boxed{hf = = \phi e + eV} \qquad \text{.......... (1.11)}$$

Laws of Photo electric Emission

1. *For each photo sensitive material, there is a threshold frequency below which emission does not take place.*
2. *The amount of photo electric emission (current) is proportional to intensity.*

3. *Photo electric emission is instantaneous. (But the time lag is in nano-sec).*

4. *Photo electric current in amps/watt of incident light depends upon 'f'.*

4. HIGH FIELD EMISSION

Suppose a cathode is placed inside a very intense electric field, then the Potential Energy is reduced. For fields of the order of 10^9 V/m, the barrier may be as thin as 100 A°. So the electron will travel through the barrier. This emission is called as ***High Field Emission*** or ***Auto Electronic Emission***.

Problem 1.9

Estimate the percentage increase in emission from a tungsten filament when its temperature is raised from 2400 to 2410 °K.

$$A_o = 60.2 \times 10^4 \text{ A/m}^2/\text{°K}^2$$
$$B = 52{,}400 \text{ °K}$$

A and B are constants in the equation for current density J

Solution

$$JS_1 = A\,T_1^{\,2} = 1142 \text{ A / m}^2$$
$$JS_2 = AT_2^{\,2}\, e^{-B/T} = 1261 \text{ A / m}^2.$$

$$\text{Percentage Increase} = \frac{1261-1142}{1142} \times 100 = 10.35\%$$

Problem 1.10

A photoelectric cell has a cesium cathode. When the cathode is illuminated with light of $\lambda = 5500 \times 10^{-10}$ m, the minimum anode voltage required to inhibit built anode current is 0.55 V. Calculate

(a) *The work function of cesium*

(b) *The longest λ for which photo cell can function.*

by applying –0.55V to anode the emitted electrons are repelled. So the current can be inhibited

Solution

(a) $hf = e\phi + eV$ V = 0.55 volts ϕ = Work Function (WF) = ?

$$f = \frac{C}{\lambda}$$

$$\therefore \quad \frac{h.C}{\lambda} = e(\phi + V)$$

Plank's Constant, $h = 6.63 \times 10^{-34}$ J sec.

Charge of Electron, $e = 1.6 \times 10^{-19}$ C

Velocity of Light, $C = 3 \times 10^8$ m/sec

$$= \frac{6.63 \times 10^{-34} \times 3 \times 10^8}{5500 \times 10^{-10}} = 1.6 \times 10^{-19}(\phi + 0.55)$$

$$\phi = 1.71 \text{ Volts}$$

(b) ***Threshold Wavelength :***

$$\lambda_o = \frac{Ch}{e\phi}$$

$$\therefore \qquad \lambda_o = \frac{12{,}400}{\phi} = \frac{12{,}400}{1.71} = 7{,}250\ A^o$$

Problem 1.11

If the temperature of a tungsten filament is raised from 2300 to 2320 ^oK, by what percentage will the emission change ? To what temperature must the filament be raised in order to double its energy at 2300 ^oK. E_W for Tungsten = 4.52 eV. Boltzman's Constant K = 8.62×10^5 eV/^oK.

Solution

(a) $$I_{th} = S.A_o\ T^2\ e^{-Ew/kT}$$

Taking Logarithms,

$$ln\ I_{th} - 2 \log T = ln\ S\ A_o - \frac{E_w}{kT}$$

Differentiating, $$\left(\frac{2}{T} + \frac{E_W}{KT^2}\right)\frac{dT}{T^2}$$

$$\frac{dI_{th}}{I_{th}} = \left(2 + \frac{E_W}{kT}\right)\frac{dT}{T} = \left(2 + \frac{4.52}{(8.62 \times 10^{-5})2310}\right)\frac{20}{2310} = 21.4\%$$

(b) $$I_{th} = S.Ao\ (2300)^2\ e^{\frac{-4.52}{8.62\times10^{-5}\times2300}}$$

$$2I_{th} = S\ Ao\ (T)^2\ e^{\frac{-4.52}{8.62\times10^{-5}}}$$

Ratio of these two equation is

$$2 = \left(\frac{T}{2300}\right)^2\ e^{-\frac{52{,}400}{T}+22.8}$$

Taking log to the base 10,

$$\log 2 = 2 \log\left(\frac{T}{2300}\right) - \left(\frac{52{,}400}{T} + 22.8\right)\log e$$

$$\log 2 = 2 \log\left(\frac{T}{2300}\right) - \frac{52400}{T} \times 0.434 - 22.8 \times 0.434$$

$$9.6 + 2 \log\left(\frac{T}{2300}\right) = \frac{22{,}800}{T}$$

This is solved by Trail and Error Method to get $T = 2370^0K$

Problem 1.12

In a cyclotron, the magnetic field applied is 1 Tesla. If the ions (electrons) cross the gap between the D shaped discs dees twice in each cycle, determine the frequency of the R.F. voltage. If in each passage through the gap, the potential is increased by 40,000 volts how many passages are required to produce a 2 million volts particle? What is the diameter of the last semicircle?

Solution

$$B = 1 \text{ Tesla} = 1 \text{Wb/m}^2$$

Time taken by the particle to describe one circle is $T = \dfrac{35.5 \mu \sec}{B}$

$$T = \frac{35.5 \times 10^{-6}}{1} \text{ sec,}$$

when it describes one circle, the particle passes through the gap twice.

∴ The frequency of the R.F voltage should be the same.

$$\therefore \quad f = \frac{1}{T} = \frac{1}{35.5 \times 10^{-6}} = 2.82 \times 10^{10} \text{ Hz}$$

Number of passages required :

In each passage it gains 40,000 volts.

To gain 2 million electron volts $= \dfrac{2 \times 10^6 \text{ V}}{40 \times 10^3 \text{ V}} = 50$

Diameter of last semicircle :

The initial velocity for the 50th revolution is the velocity gained after 49th revolution.

∴ Accelerating potential $V_o = 49 \times 40{,}000 = 1{,}960$ KV 49th revolution

Radius of the last semicircle $= R = \dfrac{3.37 \times 10^{-6}}{B} \sqrt{V_o}$

$$R = \frac{3.37 \times 10^{-6}}{1} \sqrt{1960 \times 10^3} = 47.18 \times 10^{-4} \text{ m}$$

$$\text{Diameter} = 2R = 94.36 \times 10^{-4} \text{ m}$$

Problem 1.13

The radiated power density necessary to maintain an oxide coated filament at 110 °K is found to be 5.8×10^4 W/m^2. Assume that the heat loss due to conductor is 10% of the radiation loss. Calculate the total emission current and the cathode efficiency η in ma/w.

Take

$E_W = 1.0$ eV

$A_o = 100$ A/m^2 / °K^2

$S = 1.8$ cm^2

$\bar{K}$ = Boltzman's Constant in eV / °K = 8.62×10^{-5}eV / °K

Solution

Power Density = 5.8×10^4 W/m^2

If 10% is lost by conductor, then the total input power density is $1.1 \times 5.8 \times 10^4 = 6.38$ W/m^2 and the total input power is $6.38 \times 10^4 \times$ Area $(1.8 \times 10^{-4}) = 11.5$ W.

$$I_{th} = S\,Ao\,T^2\,e^{-Ew/kT}$$

$$= 1.8 \times 10^{-4} \times 100 \times (1100)^2\,e^{-\frac{1}{8.62\times10^{-5}\times1100}}$$

$$I_{th} = 2.18 \times 10^4\,e^{-10.55}$$

$$I_{th} = 0.575\text{ A}$$

$$\text{Cathode Efficiency, } \eta = \frac{0.575}{11.5} = 50\text{ mA}/{}^{o}\text{K}$$

1.2 n-TYPE AND p-TYPE SEMICONDUCTORS

1.2.1 n-Type or Donor Type Semiconductors

Intrinsic or pure semiconductor is of no use since its conductivity is less and it can not be charged much. If a pure semiconductor is doped with impurity it becomes extrinsic. Depending upon impurity doped, the semiconductor may become ***n-type, where electrons are the majority carriers or donor type***, since it donates an electron. On the other hand ***if the majority carriers are holes, it is p-type or acceptor type semiconductor***, because it accepts an electron to complete the broken covalent bond.

Germanium atom with its electrons arranged in shells will have configuration as

$$1s^2\ 2s^2\ 2p^2\ 3s^2\ 3p^6\ 3d^{10}\ 4s^2\ 4p^2$$

Ge is tetravalent (4). 'Ge' becomes *n-type* if a pentavalent (5), impurity atoms such as Phosphorus (P), or Arsenic are added to it.

The impurity atoms have size of the same order as that of Ge atoms. Because of the energy supplied while doping, the impurity atom dislodges one from its normal position in the crystal lattices takes up that position. But since the concentration of impurity atoms is very small (about 1 atom per million of Ge atoms), the impurity atom is surrounded by Ge atoms. The impurity atom is pentavalent. That is, it has 5 electrons in the outermost orbit (5 valence electrons). Now 4 of these are shared by Ge atoms, surrounding the impurity atom and they form covalent bonds. So one electron of the impurity atom is left free. The energy required to dislodge this fifth electron from its parent impurity atom is very little of the order of 0.01 eV to 0.05 eV. This free electron is in excess to the free electrons that will be generated because of breaking of covalent bonds due to thermal agitation. Since an excess electron is available for each impurity atom, or it can ***denote an electron it is called n-type, or donor type semiconductor.***

1.2.2 p-Type or Acceptor Type Semiconductors

An intrinsic semiconductor when doped with trivalent (3) impurity atoms like Boron, Gallium Indium, Aluminium etc., becomes p-type or acceptor type.

Because of the energy supplied while doping, the impurity atom dislodges one Ge atom from the crystal lattice. The doping level is low, i.e., there is one impurity atom for one million Ge atoms, the impurity atom is surrounded by Ge atom. Now the three valence electrons of impurity atom are shared by 3 atoms. The fourth Ge atom has no electron to share with the impurity atom. So the covalent bond is not filled or a hole exists. The impurity atom tries to steal one electron from the neighboring Ge atoms and it does so when sufficient energy is supplied to it. So hole moves.

There will be a natural tendency in the crystal to form 4 covalent bonds. The impurity atom (and not just 3) since all the other Ge atoms have 4 covalent bonds and the structure of Ge semiconductor is crystalline and symmetrical. The energy required for the impurity atom to steal one Ge electron is 0.01 eV to 0.08 eV. This hole is in excess to the hole created by thermal agitation.

1.3 MASS ACTION LAW

In an intrinsic Semiconductor number of free electrons $n = n_i$ = No. of holes $p = p_i$

Since the crystal is electrically neutral, $n_i p_i = n_i^2$.

Regardless of individual magnitudes of n and p, the product is always constant,

$$\therefore \qquad np = n_i^2$$

$$n_i = AT^{\frac{3}{2}} e^{\frac{-E_{GO}}{2KT}} \qquad \text{.......... (1.12)}$$

This is called ***Mass Action Law***.

1.4 CONTINUITY EQUATION

Thus Continuity Equation describes how the carrier density in a given elemental volume of crystal varies with time.

If an intrinsic semiconductor is doped with n-type material, electrons are the majority carriers. Electron - hole recombination will be taking place continuously due to thermal agitation. So the concentration of holes and electrons will be changing continuously and this varies with time as well as distance along the semiconductor. We now derive the differential. equation which is based on the fact ***that charge is neither created nor destroyed***. This is called ***Continuity Equation***.

Consider a semiconductor of area A, length dx ($x + dx - x = dx$) as shown in Fig. 1.2. Let the average hole concentration be 'p'. Let E_p is a factor of x. that is hole current due to concentration is varying with distance along the semiconductor. Let 'I_p' is the current entering the volume at 'x' at time 't', and ($I_p + dI_p$) is the current leaving the volume at (x + dx) at the same instant of time 't'. So when only I_p colombs is entering, ($I_p + dI_p$) colombs are leaving. Therefore effectively there is a decrease of ($I_p + dI_p - I_p$) = dI_p colombs per second within the volume. Or in other words, since more hole current is leaving than what is entering, we can say that more holes are leaving than the no. of holes entering the semiconductor at 'x'.

If dI_p is rate of change of total charge that is

$$dI_p = d\left(\frac{n \times q}{t}\right)$$

$\frac{dI_p}{q}$ gives the decrease in the number of holes per second with in the volume A × dx. Decrease in holes per unit volume (hole concentration) per second due to I_p is

$$\frac{dI_p}{A \times dx} \times \frac{1}{q}$$

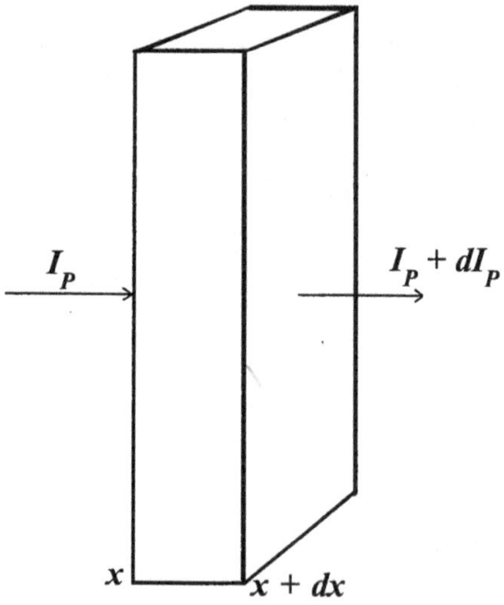

Fig 1.2 Charge flow in semiconductor

But $\frac{dI_p}{A}$ = Current Density

$$= \frac{1}{q} \times \frac{dJ_p}{dx}.$$

But because of thermal agitation, more number of holes will be created. If 'p_0' is the thermal equilibrium concentration of holes, (the steady state value reached after recombination), then, the increase per second, per unit volume due to thermal generation is,

$$g = \frac{p}{\tau_p}$$

Therefore, increase per second per unit volume due to thermal generation,

$$g = \frac{p_o}{\tau_p}$$

But because of recombination of holes and electrons there will be decrease in hole concentration.

The decrease $= \frac{p}{\tau_p}$

Charge can be neither created nor destroyed. Because of thermal generation, there is increase in the number of holes. Because of recombination, there is decrease in the number of holes. Because of concentration gradient there is decrease in the number of holes.

So the net increase in hole concentration is the algebraic sum of all the above.

$$\frac{\partial p}{\partial t} = \frac{p_0 - p}{\tau_p} - \frac{1}{q} \times \frac{\partial J_p}{\partial x}$$

Partial derivatives are used since 'p' and 'Jp' are functions of both time t and distance x.

$\frac{dp}{dt}$ gives the variation of concentration of carriers with respect to time 't'.

If we consider unit volume of a semiconductor (*n-type*) having a hole density p_n, some holes are lost due to recombination. If p_{no} is equilibrium density, (i.e., density in the equilibrium condition when number of electron = holes).

The recombination rate is given as $\frac{p_n - p_{no}}{\tau}$. ***The expression for the time rate of change in carriers density is called the Continuity Equation.***

$$\text{Recombination rate } R = \frac{dp}{dt}$$

Life time of holes in n-type semiconductors

$$\tau_p = \frac{\Delta P}{R} = \frac{p_n - p_{no}}{dp/dt}$$

or

$$\frac{dp}{dt} = \left(\frac{p_n - p_{no}}{\tau_p} \right)$$

where p_n is the original concentration of holes in n-type semiconductors and p_{no} is the concentration after holes and electron recombination takes place at the given temperature. In other word p_{no} is the thermal equilibrium minority density. Similarly for a p-type semiconductors, the life time of electrons

$$\tau_n = \frac{n_p - n_{po}}{dx/dt} \; ; \; \frac{dx}{dt} = \frac{n_p - n_{no}}{\tau_n}.$$

1.5 THE HALL EFFECT

If a metal or semiconductor carrying a current I is placed in a perpendicular magnetic field B, an electric field E is induced in the direction perpendicular to both I and B. This *phenomenon is known as the* ***Hall Effect****. It is used to determine whether a semiconductor is p-type or n-type. By measuring conductivity σ, the mobility μ can be calculated using* ***Hall Effect.***

In the Fig. 1.3 current 'I' is in the positive 'X'-direction and 'B' is in the positive 'Z'-direction. So a force will be exerted in the negative 'Y'-direction. If the semiconductor is *n-type*, so that current is carried by electrons, these electrons will be forced downward toward side 1. So side 1 becomes negatively charged with respect to side 2. Hence a potential V_H called the ***Hall Voltage*** appears between the surface 1 and 2.

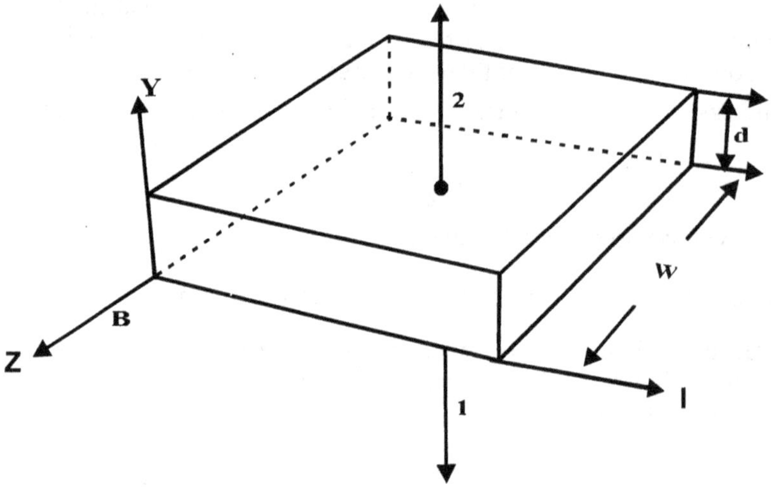

Fig 1.3 Hall effect.

In the equilibrium condition, the force due to electric field intensity 'E', because of Hall effect should be just balanced by the magnetic force or

$$e\varepsilon = B\,ev$$

v = Drift Velocity of carriers in m / sec

B = Magnetic Field Intensity in Tesla (wb/m^2)

or

$$\varepsilon = Bv \qquad \text{.......... (a)}$$

But $\varepsilon = V_H/d$

where V_H = Hall Voltage

d = Thickness of semiconductors.

J = nev or $J = \rho v$

ρ = charge density.

J = Current Density (Amp / m^2)

or $$J = \frac{I}{\omega . d}$$

ω = width of the semiconductor; ωd = cross sectional area

I = current

$\therefore$ $$J = \text{Current Density} = \frac{I}{\omega d}$$

$\varepsilon = V_H/d$

or $V_H = \varepsilon d$

But $\varepsilon = Bv$ From Equation (a)

$\therefore$ $V_H = B \times v \times d$ But $v = J/\rho$

$$= \frac{B.J.d}{\rho} \quad \text{But } J = \frac{I}{\omega d}$$

$$V_H = \frac{B.I.d}{\rho.\omega.d} = \frac{B.I}{\rho\omega}$$

$$V_H = \frac{B.I}{\rho\omega} \qquad \text{.......... (1.13)}$$

If the semiconductor is n-type, electrons the majority carriers under the influence of electric field will move towards side 1, side 2 becomes positive and side 1 negative. If on the other hand terminal 1 becomes charged positive then the semiconductor is p-type.

$\rho = n \times e$ (For n - type semiconductor)

or $\rho = p \times e$ (for p-type semiconductor)

and ρ = Charge density.

$\because$ $$V_H = \frac{B.I}{\rho\omega}$$

$$\rho = \frac{B.I}{V_H.\omega}$$

$\therefore$ $$R_H = \frac{V_H.\omega}{BI} \qquad \text{.......... (1.14)}$$

The Hall Coefficient, R_H is defined as $R_H = \frac{1}{\rho}$. Units of R_H are m^3 / coulombs

If the conductivity is due primarily to the majority carriers conductivity, $\sigma = ne\mu$ in n-type semiconductors.

$$n.e = \rho = \text{charge density.}$$

$$\therefore \quad \sigma = \rho \times \mu$$

But $$\frac{1}{\rho} = R_H$$

$$\therefore \quad \sigma = \frac{1}{R_H} \times \mu$$

or $$\mu = R_H \times \sigma = \frac{V_H.\omega}{B.I} \times \sigma$$

We have assumed that the drift velocity 'v' of all the carriers is same. But actually it will not be so. Due to the thermal agitation they gain energy, their velocity increases and also collision with other atoms increases. So for all particles v will not be the same. Hence a correction has to be made and it has been found that satisfactory results will be obtained if $\frac{1}{R_H}$ is taken as $\frac{3\pi}{8\rho}$.

$$\therefore \quad \mu = \left(\frac{8\sigma}{3\pi}\right) R_H \qquad \text{.......... (1.15)}$$

Multiply R_H by $\frac{8}{3\pi}$. Then it becomes ***Modified Hall Coefficient***. Thus mobility of carriers (electrons or holes) can be determined experimentally using Hall Effect.

The product (Bev) is the ***Lorentz Force***, because of the applied magnetic field B and the drift velocity 'v'. So the majority carriers in the semiconductors, will tend to move in a direction perpendicular to B. But since there is no electric field applied in that particulars direction, there will develop a Hall voltage or field which just opposes the Lorentz field.

So with the help of Hall Effect, we can experimentally determine

1. *The mobility of Electrons or Holes.*
2. *Whether a given semiconductor is* p-type *or* n-type *(from the polarity of Hall voltage V_H)*

Problem 1.14

The Hall Effect is used to determine the mobility of holes in a p-type Silicon bar. Assume the bar resistivity is 200,000 Ω-an, the magnetic field $B_z = 0.1$ Wb/m^2 and d = w = 3mm. The measured values of the current and Hall voltage are 10mA and 50 mv respectively. Find μ_p mobility of holes.

Solution

$B = 0.1\ Wb / m^2$ (or Tesla)

$V_H = 50$ mv.

$I = 10$ mA;

$\rho = 2 \times 10^5\ \Omega$ - cm ;

$d = w = 3mm = 3 \times 10^{-3}$ meters

$$\frac{1}{R_H} = \frac{B.I}{V_H.w} = \frac{0.1\times10\times10^{-3}}{50\times10^{-2}\times3\times10^{-3}} = \frac{1}{150} = 0.667.$$

$$\text{Conductivity} = \frac{1}{\rho} = \frac{1}{2\times10^5\times10^{-2}} = \frac{1}{2000}\ \text{mhos / meter.}$$

$$\mu = \sigma \times R_H$$

$$\mu_p = \frac{1}{0.667}\times\frac{1}{2000} = 750\ cm^2/\ V\text{ - sec}$$

1.6 FERMI LEVEL IN INTRINSIC AND EXTRINSIC SEMICONDUCTORS

1.6.1 Fermi Level

Named after ***Fermi***, ***it is the Energy State, with 50% probability of being filled if no forbidden band exists***. In other words, it is the mass energy level of the electrons, at 0^oK.

If $E = E_f$,

$$f(E) = ½$$

If a graph is plotted between ($E - E_F$) and $f(E)$, it is shown in Fig. 1.4

At $T = 0^oK$, if $E > E_F$ then, $f(E) = 0$.

That is, there is no probability of finding an electron having energy $> E_F$ at $T = 0^0$ K. Since fermi level is the max. energy possessed by the electrons at 0^ok. $f(E)$ varies with temperature as shown in Fig. 1.4.

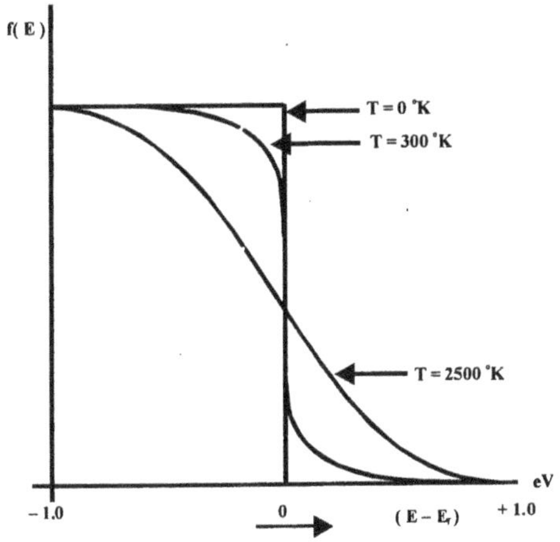

Fig. 1.4 Fermi level variation with temperature.

1.6.2 The Intrinsic Concentration

$$f(E) = \frac{1}{1+e^{(E-E_F)/KT}} \qquad \text{.......... (1.16)}$$

$$1 - f(E) = \frac{e^{(E-E_F)/KT}}{1+e^{(e-E_F)/KT}} \simeq e^{-(E_F-E)/KT}$$

$$e^{(E-E_F)/KT}\left[1+e^{(E-E_f)/KT}\right]^{-1} \simeq e^{-(E_f-E)/KT}$$

Fermi function for a hole = $1 - f(E)$.

$(E_F - E) >> KT$ for $E \leq Ev$

the number of holes per m^3 in the Valence Band is,

$$p = \int_{-\infty}^{E_V} \gamma(E_V - E)^{1/2} e^{-(E_F-E)/KT} \times dE$$

$$= N_V \times e^{-(E_F-E_V)/KT}$$

where $$N_V = 2\left(\frac{2\pi m_p \overline{K}T}{h^2}\right)^{\frac{3}{2}}$$

Similarly $$n = N_e\ e^{-(E_C-E_F)/KT}$$

$$n \times p = N_V \times N_e\ e^{-E_G/KT}$$

Substituting the values of N_C and N_V,

$$n \times p = n_i^2 = AT^3\ e^{-E_G/KT}$$

1.6.3 Carrier Concentrations in a Semiconductor

$d\eta = N(E) \times f(E) \times dE$

dn = number of conduction electrons per cubic meter whose energy lies between E and E + dE

$f(E)$ = The probability that a quantum state with energy E is occupied by the electron.

N(E) = Density of States.

$$f(E) = \frac{1}{1+e^{(E-E_F)/KT}}$$

The concentration of electrons in the conduction band is

$$n = \int_{E_C}^{\infty} N(E) \times f(E) \times dE$$

for $E \geq E_C$

$(E - E_s)$ is >> KT.

$\therefore$ $f(E) = e^{-(E-E_f)/KT}$

$\because \quad e^{(E-E_F)/KT} >> 1$

$$\therefore \quad n = \int_{E_C}^{\infty} \gamma (E-E_C)^{1/2} \times e^{-(E-E_F)KT} . dE$$

Simplifying this integral, we get,

$$n = N_C \times e^{-(E_C-E_F)/KT}$$

where $N_C = 2\left(\frac{2\pi m_n \overline{K}T}{h^2}\right)^{\frac{3}{2}}$

$\overline{K}$ = Boltzman's Constant in J/^oK

'N_i' is constant.

where m_n = effective mass of the electron.

Similarly the number of holes / m^3 in the Valence Band

$$p = N_V \; e^{-(E_F-E_V)/KT}$$

where $N_v = 2\left(\frac{2\pi m_p \overline{K} \times T}{h^2}\right)^{\frac{3}{2}}$

Fermi Level is the maximum energy level that can be occupied by the electrons at 0 ^oK. ***Fermi Level*** or characteristic energy represents the energy state with 50% probability of being filled if no forbidden bond exists. If $E = E_F$, then $f(E) = ½$ for any value of temperature. $f(E)$ is the probability that a quantum state with energy E is occupied by the electron.

1.6.4 Fermi Level in Intrinsic Semiconductor

$$n = p = n_i$$

$$n = N_C \; e^{-(E_C-E_F)/KT}$$

$$p = N_V \; e^{-(E_F-E_V)/KT}$$

$$n = p$$

or $$N_C \; e^{-(E_C-E_F)/KT} = N_V \; e^{-(E_F-E_V)/KT}$$

Electrons in the valence bond occupy energy levels up to 'E_F'. 'E_F' is defined that way. Then the additional energy that has to bo supplied so that free electron will move from valence band to the conduction band is E_C

$$\frac{N_C}{N_V} = e^{\frac{-(E_F-E_V)}{KT} + \frac{(E_C+E_F)}{KT}}$$

$$= e^{\frac{-2E_F+E_C+E_V}{KT}}$$

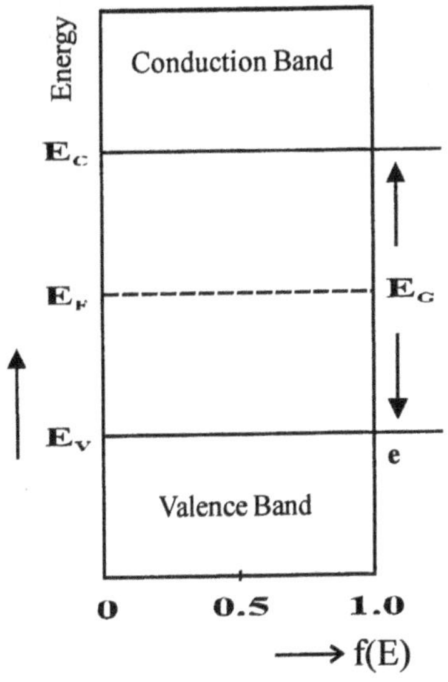

Fig. 1.5 Energy band diagram.

Taking logarithms on both sides,

$$\ln \frac{N_C}{N_V} = \frac{E_C + E_V - 2E_F}{KT}$$

$$\therefore \quad E_F = \frac{E_C + E_V}{2} - \frac{KT}{2} \ln \frac{N_C}{N_V}$$

$$N_C = 2\left(\frac{2\pi m_n \overline{KT}}{h^2}\right)^{\frac{3}{2}} \qquad \text{.......... (1.17)}$$

$$N_V = 2\left(\frac{2\pi m_p .\overline{KT}}{h^2}\right)^{\frac{3}{2}} \qquad \text{.......... (1.18)}$$

where $m_n$and m_p are effective masses of holes and electrons. If we assume that $m_n = m_p$, (though not valid),

$$N_C = N_V$$

$$\therefore \quad \ln \frac{N_C}{N_V} = 0$$

$$\therefore \quad E_F = \frac{E_C + E_V}{2} \qquad \text{.......... (1.19)}$$

The graphical representation is as shown in Fig. 1.5. ***Fermi Level in Intrinsic Semiconductor lies in the middle of Energy gap*** E_G.

Problem 1.15

In p-type Ge at room temperature of 300 °K, for what doping concentration will the fermi level coincide with the edge of the valence bond ? Assume $\mu_p = 0.4$ m.

Solution

$$E_F = E_V$$

when $\quad N_A = N_V$

$$\therefore \quad E_F = E_V + kT \ln . \frac{N_V}{N_A}$$

$$\therefore \quad N_V = 4.82 \times 10^{15}\left(\frac{mp}{m}\right)^{\frac{3}{2}} \times T^{3/2} = 4.82 \times 10^{15}(0.4)^{3/2}(300)^{3/2}$$

$$= 6.33 \times 10^{18}.$$

$\therefore$ Doping concentration $N_A = 6.33 \times 10^{18}$ atoms/cm^3.

Problem 1.16

If the effective mass of an electron is equal to twice the effective mass of a hole, find the distance in electron volts (ev) of fermi level in as intrunsic semiconductor from the centre of the forbidden bond at room temperature.

Solution

For Intrunsic Semiconductor,

$$E_F = \left[\left(\frac{E_C + E_V}{2}\right) - \frac{KT}{2}\ln\left(\frac{N_C}{N_V}\right)\right]$$

If $m_p = m_n$

then $N_C = N_V$.

Hence E_F will be at the centre of the forbidden band. But if $m_p \neq m_n$. E_F will be away from the centre of the forbidden band by

$$\frac{KT}{2}\ln . \frac{N_C}{N_V} = 3/4\,\frac{kT}{2}.\ln\frac{m_n}{m_p}$$

$$\therefore \quad N_C = 2\left(\frac{2\pi m_n \bar{K}T}{n^2}\right)^{3/2}$$

$$N_V = 2\left(\frac{2\pi m_p \bar{k}T}{n^2}\right)^{3/2}$$

$$= \frac{3}{4} \times 0.026\, ln\,(2)$$

$$= 13.5 \text{ m. eV}$$

1.7 p-n DIODE EQUATION

The hole current in the *n-side* ***Ipn(x)*** is given as

$$\boldsymbol{Ipn(x)} = \frac{Ae \times Dp}{L_p} p_n(0)\, e^{-x/L_p}$$

But $\quad p_n(0) = p_{no}\left(e^{V/V_T} - 1\right)$

$$\boldsymbol{Ipn(0)} = \frac{Ae \times Dp}{L_p} \times p_{no}\left(e^{V/V_T} - 1\right)$$

D_p = Diffusion coefficient of holes

D_n = Diffusion coefficient of electrons.

$\therefore \quad e^{-x/L_p}$ at $x = 0$ is 1.

Similarly the electron current due to the diffusion of electrons from *n-side* to *p-side* is obtained from the above equation itself, by interchanging n and p.

$$\therefore \quad \boldsymbol{Inp(0)} = \frac{Ae \times Dn}{L_n} \times n_{po}\left(e^{V/V_T} - 1\right)$$

The total diode current is the sum of ***Ipn (0)*** and ***Inp(0)***

or $$I = I_o\left(e^{V/V_T} - 1\right)$$ .... (1.20)

where $$I_0 = \frac{AeDp}{Lp} \times p_{no} + \frac{AeDn}{L_n} \times n_{po}$$

In this analysis we have neglected charge generation and recombination. Only the current that results as a result of the diffusion of the carriers owing to the applied voltage is considered.

Reverse Saturation Current

$$I = I_0 \times \left(e^{V/V_T} - 1\right)$$

This is the expression for current I when the diode is forward biased. If the diode is reverse biased, V is replaced by –V. V_T value at room temperature is ~ 26 mV. If the reverse bias voltage is very large, $e^{\frac{-V}{V_T}}$ is very small. So it can be neglected.

$$\therefore \quad I = -I_0$$

I_0 will have a small value and I_0 is called the ***Reverse Saturation Current.***

$$I_0 = \frac{AeD_p p_{no}}{L_p} + \frac{AeD_n n_{po}}{L_n}.$$

In *n-type* semiconductor,

$$n_n = N_D$$

But $$n_n \times p_n = n_i^2 \qquad \therefore p_n = \frac{n_i^2}{N_D}$$

In *p-type* semiconductor,

$$p_p = N_A \qquad \therefore n_p = \frac{n_i^2}{N_A}$$

Substituting these values in the expression for I_0,

$$I_0 = Ae\left(\frac{D_p}{L_p N_D} + \frac{D_n}{L_n . N_A}\right) \times n_i^2$$

where $$n_i^2 = A_0\, T^3\, e^{-E_G/KT}$$

E_G is in electron volts = e.V_G, where V_G is in Volts.

$$\therefore \quad n_i^2 = A_0\, T^3\, e^{-\frac{E_{GO}}{KT}}$$

$$E_{GO} = V_G \cdot e$$

But $$\frac{KT}{e} = \text{Volt equivalent of Temperature } V_T.$$

For Germanium, D_p and D_n decrease with temperature and n_i^2 increases with T. Therefore, temperature dependance of I_0 can be written as,

$$I_0 = K_1 T^2 \, e^{-V_G/V_T}$$

For Germanium, the current due to thermal generation of carriers and recombination can be neglected. But for Silicon it cannot be neglected. So the expression for current is modified as

$$I = I_0 \left(e^{\frac{V}{\eta V_T}} - 1 \right)$$

where n = 2 for small currents and n = 1 for M large currents.

1.8 VOLT-AMPERE (V-I) CHARACTERISTICS

The general expression for current in the *p-n junction* diode is given by

$$I = I_0 \left(e^{\frac{V}{\eta V_T}} - 1 \right)$$

$\eta = 1$ for Germanium and 2 for Silicon. For Silicon, 'I' will be less than that for Germanium.

$$V_T = 26 \text{ mV}.$$

If 'V' is much larger than V_T, 1 can be neglected. So 'I' increases exponentially with forward bias voltage 'V'. In the case of reverse bias, if the reverse voltage $-V >> V_T$, then e^{-V/V_T} can be neglected and so reverse current is $-I_0$ and remains constant independent of 'V'. So the characteristics are as shown in Fig. 1.6 and not like theoretical characteristics. The difference is that the practical characteristics are plotted at different scales. If plotted to the same scale, (reverse and forward) they may be similar to the theoretical curves. Another point is, in deriving the equations the breakdown mechanism is not considered. As 'V' increases **Avalanche multiplication** sets in. So the actual current is more than the theoretical current.

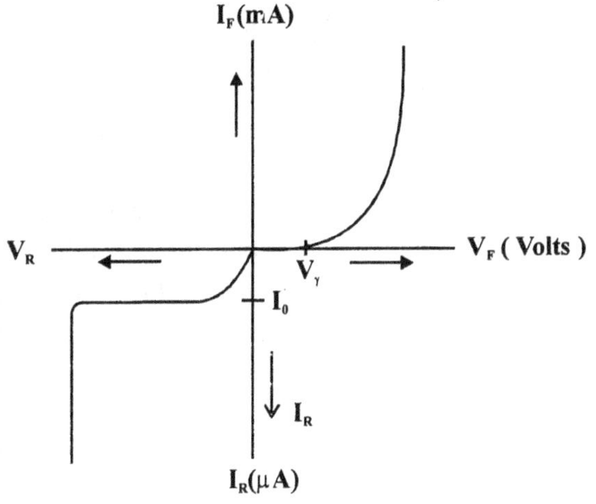

Fig 1.6 V-I Characteristics of p-n junction diode.

CUT IN VOLTAGE V_γ

In the case of Silicon and Germanium, diodes there is a ***Cut In*** or ***Threshold*** or ***Off Set*** or ***Break Point Voltage***, below which the current is negligible. It's magnitude is 0.2V for Germanium and 0.6V for Silicon (Fig. 1.7).

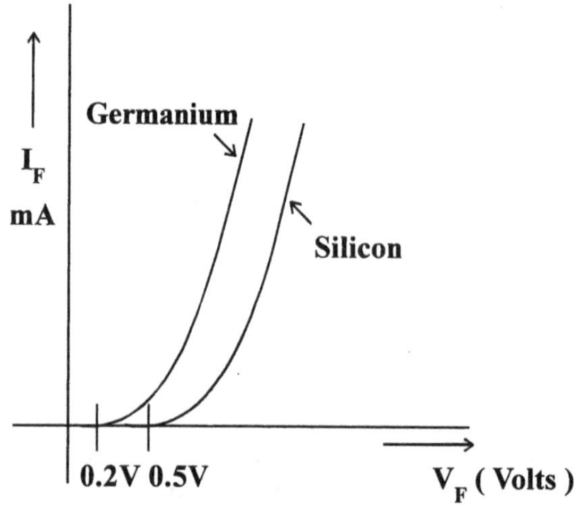

Fig 1.7 Forward characteristics of a diode.

1.9 TEMPERATURE DEPENDENCE OF V-I CHARACTERISTICS

$$\sigma = (\mu_n n + \mu_p p)e,$$

So the electrical characteristics of a semiconductor depends upon '***n***' and '***p***', the concentration of holes and electrons.

The expression for $n = N_C \; e^{-(E_C - E_F)/KT}$ and the expression for $p = N_V \; e^{-(E_F - E_V)/KT}$

These are valid for both intrinsic and extrinsic materials.

The electrons and holes, respond to an external field as if their mass is m*(m* = 0.6m) and not 'm'. So this m^* is known as ***Effective Mass***.

With impurity concentration, only E_F will change. In the case of intrinsic semiconductors, E_F is in the middle of the energy gap, indicating equal concentration of holes and electrons.

If donor type impurity is added to the intrinsic semiconductor it becomes n-type. So assuming that all the atoms are ionized, each impurity atom contributes at least one free electron. So the first N_D states in the conduction band will be filled. Then it will be more difficult for the electrons to reach Conduction Band, bridging the gap between Covalent Bond and Valence Bond. So the number of electron hole pairs, thermally generated at that temperature will be decreased. ***Fermi level is an indiction of the probability of occupancy of the energy states.*** Since Because of doping, more energy states in the ConductionBand are filled, the fermi level will move towards the Conduction Band.

EXPRESSION FOR E_G

$$n = N_C \times e^{\frac{-(E_C - E_F)}{KT}}$$

$$p = N_V \cdot e^{\frac{-(E_F - E_V)}{KT}}$$

$$n \times p = N_C \times N_V \times e^{\frac{-(E_C - E_V)}{KT}}$$

But $(E_C - E_V) = E_G$

and $n \times p = n_i^2$

$$\therefore \quad n_i^2 = N_C \times N_V \times e^{-\frac{E_G}{KT}}$$

$$\frac{n_i^2}{N_C \times N_V} = e^{-\frac{E_G}{KT}}$$

Taking logarithms,

$$\ln\left(\frac{n_i^2}{N_C N_V}\right) = -\frac{E_G}{KT}$$

or

$$-\ln\left(\frac{N_C N_V}{n_i^2}\right) = -\frac{E_G}{KT}$$

or

$$E_G = KT \ln\left(\frac{N_C N_V}{n_i^2}\right) \qquad \text{.......... (1.21)}$$

The position of Fermi Level is as shown in Fig. 1.8.

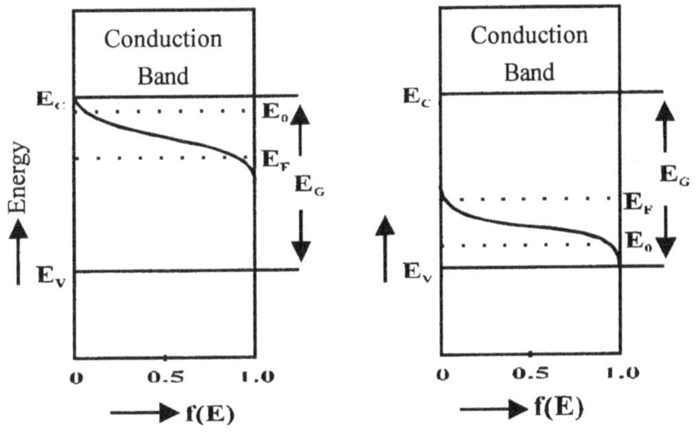

For n-type semiconductor ***For p-type semiconductor***

Fig. 1.8

Similarly, in the case of *p-type* materials, the Fermi level moves towards the valence band since the number of holes has increased. In the case of *n-type* semiconductor, the number of free electrons has increased. So energy in Covalent Bond has increased. Fermi Level moves towards conduction band. ***Similarly in p-type semiconductors, fermi Level moves towards Valence Band.*** So it is as shown in Fig. 1.8.

To calculate the exact position of the Fermi Level in n-type Semiconductor:

In *n-type* semiconductor,

$$n \simeq N_D$$

But
$$n = N_C \times e^{-(E_C - E_F)/KT}$$

$\therefore$
$$N_D = N_C \times e^{-(E_C - E_F)/KT}$$

or
$$\frac{N_D}{N_C} = e^{-(E_C - E_F)/KT}$$

Taking logarithms,

$$\ln \frac{N_D}{N_C} = \frac{-(E_C - E_F)}{KT}$$

or
$$KT \times \left\{ \ln \frac{N_D}{N_C} \right\} = -(E_C - E_F)$$

or
$$E_F = E_C - KT \times \left\{ \ln \frac{N_C}{N_D} \right\} \qquad \text{.......... (1.22)}$$

So Fermi Level E_F is close to Conduction Band E_C in n-type semiconductor.

Similarly for p-type material,

$$p = N_A$$

But
$$p = N_V \times e^{-(E_F - E_V)/KT}$$

$\therefore$
$$\frac{N_A}{N_V} = e^{-(E_F - E_V)/KT}$$

Taking Logarithms,

$$\ln \frac{N_A}{N_V} = -\frac{(E_F - E_V)}{KT}$$

$$KT \times \ln \frac{N_A}{N_V} = E_V - E_F$$

or
$$E_F = E_V + KT \times \ln \frac{N_V}{N_A} \qquad \text{.......... (1.23)}$$

$\therefore$
$$N_A = N_V$$

Fermi Level is close to Valance Band E_V in p-type semiconductor.

Problem 1.17

In n type silicon, the donor concentration is 1 atom per 2×10^8 silicon atoms. Assuming that the effective mass of the electron equals true mass, find the value of temperature at which, the fermi level will coincide with the edge of the conduction band. Concentration of Silicon = 5×10^{22} atom/cm3.

Solution

Donor atom concentration = 1 atom per 2×10^8 Si atom.

Silicon atom concentration = 5×10^{22} atoms/cm^3

$$\therefore \quad N_D = \frac{5\times10^{22}}{2\times10^8} = 2.5 \times 10^{14}/\text{cm}^3.$$

For n-type, Semiconductor,

$$E_F = E_C - KT \ln\left(\frac{N_C}{N_D}\right)$$

If E_F were to coincide with E_C, then

$$N_C = N_D$$

$$N_D = 2.5 \times 10^{14}\ /\text{cm}^3.$$

$$N_C = 2\left\{\frac{2\pi m_n \overline{K} T}{h^2}\right\}^{\frac{3}{2}}$$

h = Plank's Constant; $\overline{K}$ = Boltzman Constant

m_n the effective mass of electrons to be taken as = m_E

$$\therefore \quad N_C = 2\left\{\frac{2\times3.14\times9.1\times10^{-31}\times\overline{K}\times T}{h^2}\right\}^{\frac{3}{2}}$$

$$= 4.28 \times 10^{15}\ T^{\frac{3}{2}}$$

$$\because \quad N_C = N_D,$$

$$4.28 \times 10^{15}\ T^{\frac{3}{2}} = 2.5 \times 10^{14}$$

$$\therefore \quad T = 0.14\ ^oK$$

that *p-side* of the p-n junction is heavily doped. If *n-side* is heavily doped, it would be N_D.

1.10 IDEAL VS PRACTICAL

Ideal: The p-n junction diode allows flow of heavy current during forward bias condition and allows a small level of current when it is reverse biased. The direction of the arrow in the following Fig.1.9(a) depicts the direction of flow of current when the diode is forward biased. Similarly the direction of the arrow in Fig.1.9 (b) depicts the flow of current direction when the diode is reversed biased.

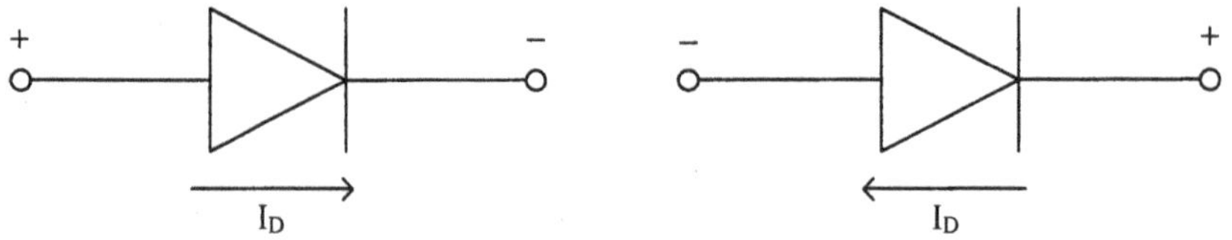

Fig.1.9(a) Illustration of flow of current when the diode is forward biased

Fig. 1.9(b) Illustration of flow of current when the diode is Reverse biased

ANALOGY

A common analogy that can be employed to describe the behaviour of a p-n junction diode is Mechanical switch.

During the forward bias the p-n junction diode allows flow of heavy currents through the junction. And hence the amount of resistance offered by the p-n junction diode during forward bias condition is very small and assumed to be zero. Thus the pn junction diode acts as a closed switch when the diode is forward biased. On the other hand when the diode is reverse biased it offers a flow very small level of current across the junction which is assumed to be zero. In otherwords it offers very high amount of resistance during its reverse bias (prevents flow of current across the junction i.e., offer ∞ resistance). Thus the p-n junction diode acts on a open switch during the reverse bias.

The following Fig.1.10(a) shows the analogy of closed switch of p-n junction diode and Fig. 1.10(b) shows the analogy of open switch of p-n junction diode.

Fig. 1.10(a) Analogy of p-n junction diode as closed switch

Fig.1.10(b) Analogy of p-n junction diode as open switch

PRACTICAL

Even though the p-n junction diode offers a flow of heavy currents during forward bias, a small amount of current (which is due to presence of minority charge carriers across the junction) opposes the flow of forward biased current (I_D). In otherwords the p-n junction diode offers a small amount of forward resistance (r_f) which opposes the flow forward biased current (I_D).

Similarly when the diode is reverse biased the switch is assumed to be open but in practical the p-n junction diode allows the flow small amounts of currents through the junction which can be refered as leakage current (or) minority charge carrier current.

IDEAL VS PRACTICAL

The following figure shows super imposition of practical characteristics in ideal characteristics. From the figure we can observe that the ideal diode does not offer any forward resistance and hence it does not require any additional voltage to make I_D to flow through the diode. In otherwords the forward bias current I_D starts flowing through the diode even when applied forward bias voltage is at 0V.

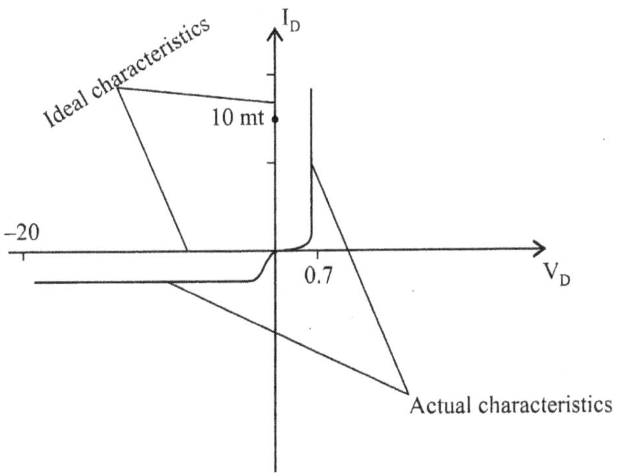

Where as in case of practical diode as it offers a small amount of forward resistance (r_f) it requires a knee voltage (or) threshold voltage of 0.6 in case of silicon (0.3 of germanium) to make forward current I_D to flow through the junction.

ANALOGY: CLOSED SWITCH

Ideal: When the switch is closed, the resistance between the contacts is assumed to be zero. In the plot a point chosen in vertical axis where the diode current 10 mA and voltage across diode is 0V. According to Ohm's law.

$$R_F = \frac{V_D}{I_D} = \frac{0V}{10\ mA} = 0\ \Omega$$

Practical: In practical diode the minimum voltage required to obtain I_D is 0.6 (for Si). Hence

$$R_F = \frac{V_D}{I_D} = \frac{0.6}{10\ mA} = 60\ \Omega$$

ANALOGY: OPEN SWITCH

Ideal: For any value of voltage on horizontal axis (any point on X-axis has zero coordinate) the current is zero. Applying ohm's law. In plot V_D is taken as 20.

$$R_r = \frac{V_D}{I_D} = \frac{20}{0 \text{ mA}} = \infty$$

Practical: For practical diode a small level of current (usually in order of ηA for si and μA for Ge) flows when the diode is reverse biased. Hence

$$R_r = \frac{20}{10 \text{ mA (assume)}} = \frac{20}{10 \times 10^{-9}} = 2000 \text{ M}\,\Omega \text{ for si}$$

	Ideal	Practical
When switch is closed	$R_f = 0$	$R_f = R_f$ (in order of ohms)
When switch is open	$R_r = \infty$	$R_r = R_r$ (in order of $M\Omega$)

1.11 DIODE EQUIVALENT CIRCUITS

An equivalent circuit is a combination of elements that are chosen to represent actual characteristics of device for particular operating region.

***Piecewise Linear Equivalent Circuit*:** The best method of obtaining equivalent circuit is to approximate its characteristics with straight line segments. The forward bias characteristics of p-n junction diode is shown in following Fig. 1.11.

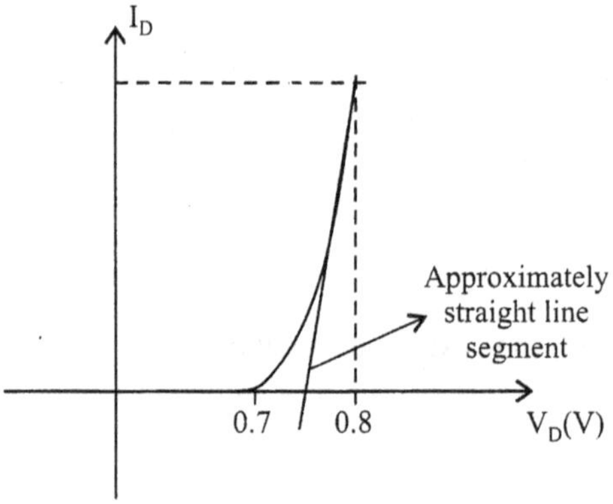

Fig. 1.11 Approximation of diode characteristics

The above Fig. 1.11 represent approximation of exact characteristics of p-n junction diode. From the figure it is clear that this approximation does not yield the exact duplication of actual characteristics of p-n junction diode. However these approximation yields sufficiently close to actual curve to represent the behaviour of the diode.

The actual characteristics have knee voltage of 0.7 V but the approximate model has knee voltage greater than 0.7 V. To achieve this 0.7 V of knee voltage connect a battery of 0.7 V to the p-n junction diode assuring its polarity is in opposition with the applying voltage as show in following Fig.1.12 and assuming the diode to be ideal, the ideal diode does not conduct any current when it is in reverse bias condition hence approximation of reverse bias characteristics of diode can be neglected.

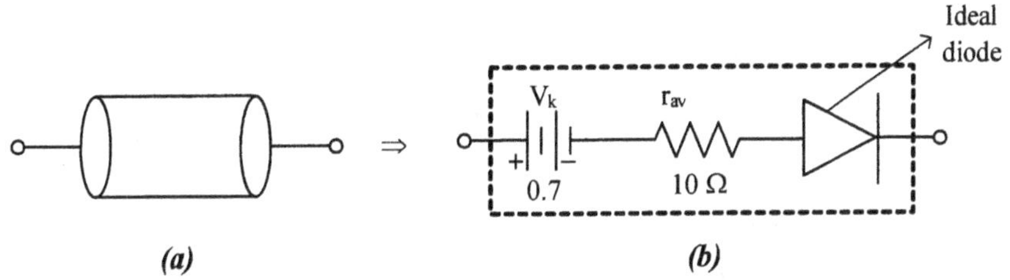

Fig. 1.12 Equivalent circuit of diode

The above Fig. 1.12 assures that the diode is not conducted until voltage across it reaches greater than its knee voltage i.e., 0.7. And once the diode starts conducting it offers a specified value of resistance r_{av}.

The approximate value of r_{av} can be obtained from the operating point of the diode. Assuming its operating point to be (10, 0.8) i.e., the current through the diode is 10 mA when applied voltage is 0.8 (show in Fig. 1.12(a)). We get

$$r_{av} = \left.\frac{\Delta V_d}{\Delta I_d}\right|_{\text{pt to pt}} = \frac{0.8 - 0.7}{10\text{ mA} - 0\text{ mA}} = \frac{0.1}{10\text{ mA}} = 10\ \Omega$$

$$\therefore \qquad r_{av} = 10\ \Omega$$

The 0.7 V is knee voltage i.e., minimum voltage required to make diode ready for conducting and at exact knee point the diode will be ready to conduct, but it will not be conducting hence I_D at knee voltage is 0 mA. When the voltage across the diode just crosses its knee voltage the diode starts conducting and hence I_D starts increasing. But at point when applied voltage is equal to knee voltage the diode will be in equilibrium resulting $I_D = 0$.

Simplified Equivalent Circuit

In general the value of r_{av} is very small in comparison with the other elements of the circuit and hence it can be neglected. Removing r_{av} from diode equivalent circuits give simplified equivalent diode circuit. The following shows the simplified equivalent circuit of the diode and its characteristics.

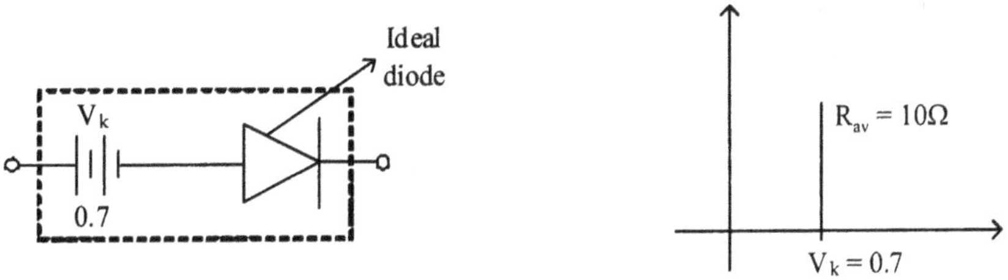

1.12 BREAK DOWN MECHANISMS IN SEMICONDUCTOR DIODES

There are three types of breakdown mechanisms in semiconductor devices.

1. Avalanche Breakdown 2. Zener Breakdown 3. Thermal Breakdown

1.12.1 Avalanche Breakdown

When there is no bias applied to the diode, there are certain number of thermally generated carriers. When bias is applied, electrons and holes acquire sufficient energy from the applied potential to produce new carriers by removing valence electrons from their bonds. These thermally generated carriers acquire additional energy from the applied bias. They strike the lattice and impart some energy to the valence electrons. So the valence electrons will break away from their parent atom and become free carriers. These newly generated additional carriers acquire more energy from the potential (since bias is applied). So they again strike the lattice and create more number of free electrons and holes. This process goes on as long as bias is increased and the number of free carriers gets multiplied. This is known as ***avalanche multiplication***, Since the number of carriers is large, the current flowing through the diode which is proportional to free carriers also increases and when this current is large, avalanche breakdown will occur.

1.12.2 Zener Breakdown

Now if the electric field is very strong to disrupt or break the covalent bonds, there will be sudden increase in the number of free carriers and hence large current and consequent breakdown. Even if thermally generated carriers do not have sufficient energy to break the covalent bonds, the electric field is very high, then covalent bonds are directly broken. This is ***Zener Breakdown***. A junction having narrow depletion layer and hence high field intensity will have zener breakdown effect. ($\simeq 10^6$ V/m). If the doping concentration is high, the depletion region is narrow and will have high field intensity, to cause Zener breakdown.

1.12.3 Thermal Breakdown

If a diode is biased and the bias voltage is well within the breakdown voltage at room temperature, there will be certain amount of current which is less than the breakdown current. Now keeping the bias voltage as it is, if the temperature is increased, due to the thermal energy, more number of carriers will be produced and finally breakdown will occur. This is ***Thermal Breakdown***.

In zener breakdown, the covalent bonds are ruptured. But the covalent bonds of all the atoms will not be ruptured. Only those atoms, which have weak covalent bonds such as an atom at the surface which is not surrounded on all sides by atoms will be broken. But if the field strength is not greater than the critical field, when the applied voltage is removed, normal covalent bond structure will be more or less restored. This is Avalanche Breakdown. But if the field strength is very high, so that the covalent bonds of all the atoms are broken, then normal structure will not be achieved, and there will be large number of free electrons. This is ***Zener Breakdown***.

In Avalanche Breakdown, only the excess electron, loosely bound to the parent atom will become free electron because of the transfer of energy from the electrons possessing higher energy.

1.13 ZENER DIODE CHARACTERISTICS

This is a *p-n junction* device, in which zener breakdown mechanism dominates. ***Zener diode is always used in Reverse Bias.***

Its constructional features are:

1. *Doping concetration is heavy on* p *and* n *regions of the diode, compared to normal* p-n junction *diode.*
2. *Due to heavy doping, depletion region width is narrow.*
3. *Due to narrow depletion region width, electric field intensity* $E = \frac{V}{d} = \frac{V_Z}{W}$ *will be high, near the junction, of the order of* $10^6 V/m$. *So Zener Breakdown mechanism occurs.*

In normal *p-n junction* diode, avalanche breakdown occurs if the applied voltage is very high. The reverse characteristic of a p-n junction diode is shown in Fig. 1.13.

When the Zener diode is reverse biased, the current flowing is only the reverse saturation current I_o which is constant like in a reverse biased diode. At $V = V_Z$, due to high electric field $\left(\frac{V_Z}{W}\right)$, Zener breakdown occurs. Covalent bonds are broken and suddenly the number of free electrons increases. So I_Z increases sharply and V_Z remains constant, since, I_Z increases through Zener resistance R_Z decreases. So the product $V_Z = R_Z \cdot I_Z$ almost remains constant. If the input voltage is decreased, the Zener diode regains its original structure. (But if V_i is increased much beyond V_Z, electrical breakdown of the device will occur. The device looses its semiconducting properties and may become a short circuit or open circuit. ***This is what is meant by device breakdown.***)

Applications

1. *In Voltage Regulator Circuits*
2. *In Clipping and Clamping Circuits*
3. *In Wave Shaping Circuits.*

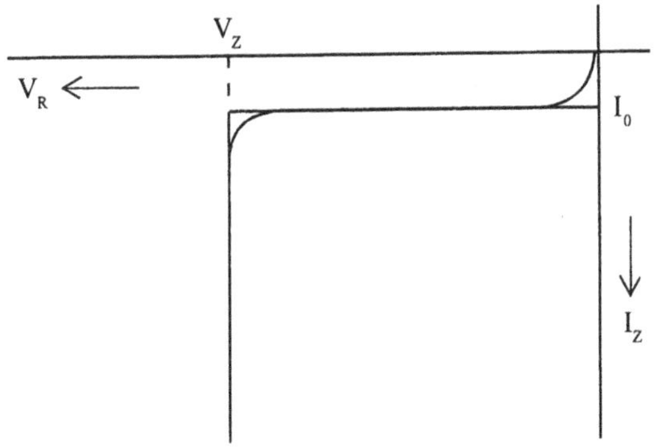

Fig 1.13 Reverse characteristic of a Zener diode.

1.14 p-n JUNCTION DIODE AS RECTIFIER

The electronic circuits require a D.C. source of power. For transistor A.C. amplifier circuit for biasing, D.C supply is required. The input signal can be A.C. and so the output signal will be amplified A.C. signal. But without biasing with D.C. supply, the circuit will not work. So more or less all electronic A.C. instruments require D.C. power. To get this, D.C. batteries can be used. But they will get dried quickly and replacing them every time is a costly affair. Hence it is economical to convert A.C. power into D.C.

Rectifier is a circuit which offers low resistance to the current in one direction and high resistance in the opposite direction.

Rectifier converts sinusoidal signal to unidirectional flow and not pure D.C.

Filter converts unidirectional flow into pure D.C.

If the input to the rectifier is a pure sinusoidal wave, the average value of such a wave is zero, since the positive half cycle and negative half cycle are exactly equal.

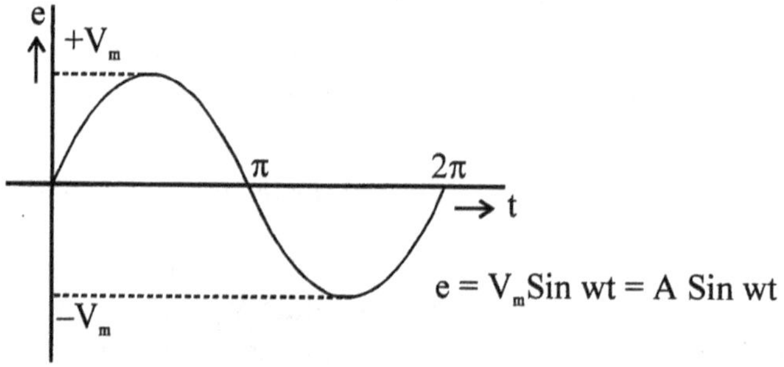

Fig 1.14 AC input wave form.

$$T_{av} = \frac{1}{2\pi}\int_0^{2\pi} A \sin(\omega t)\, dt$$

1.15 HALF-WAVE RECTIFIER

If this signal is given to the rectifier circuit, say Half Wave Rectifier Circuit, the output will be as shown in Fig. 1.15.

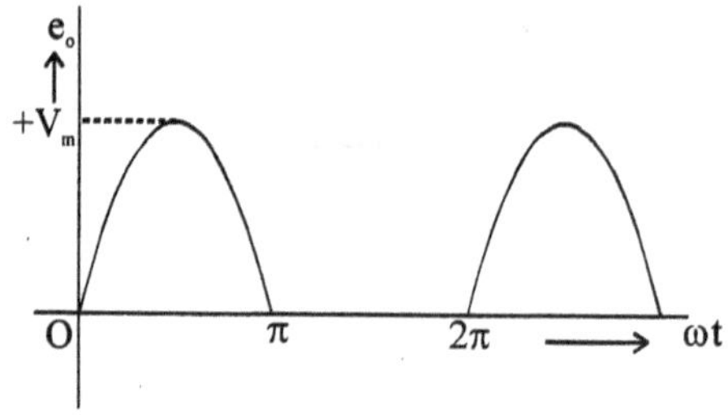

Fig. 1.15 Half wave rectified output (unidirectional flow).

Now the average value of this waveform is *not zero,* since there is no negative half. Hence a rectifier circuit converts A.C. Signal with zero average value to a unidirectional wave form with non zero average value.

The rectifying devices are semiconductor diodes for low voltage signals and vacuum diodes for high voltage circuit. A basic circuit for rectification is as shown in Fig. 1.16.

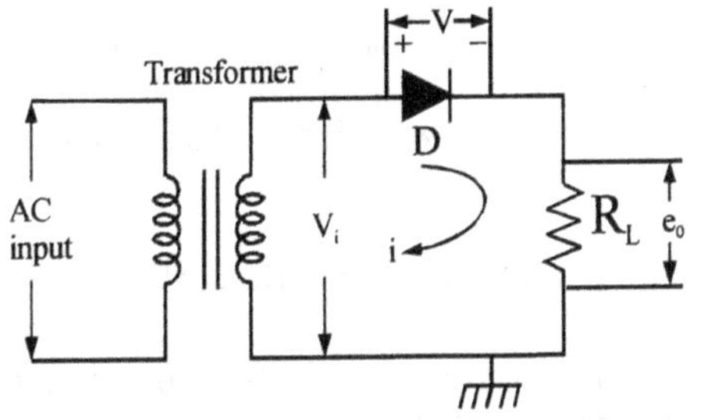

Fig 1.16 Halfwave Rectifier (HWR) circuit.

A.C input is normally the A.C. main supply. Since the voltage is 230V, and such a high voltage cannot be applied to the semiconductor diode, step down transformer should be used. If large D.C. voltage is required vacuum tubes should be used. Output voltage is taken across the load resistor R_L. Since the peak value of A.C. signal is much larger than V_γ, we neglect V_γ, for analysis.

1.15.1 Maximum or Peak Current

The output current waveform for half wave rectification is shown in Fig. 1.17.

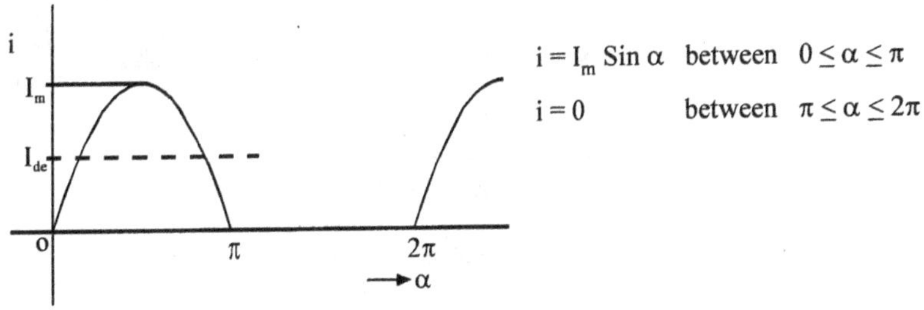

Fig 1.17 Half wave rectified output.

$$\therefore \quad i = I_m \sin\alpha \qquad \alpha = \omega t \qquad 0 \le \alpha \le \pi$$

$$i = 0 \qquad \pi \le \alpha \le 2\pi$$

$$I_m = \frac{V_m}{R_f + R_L}$$

R_f is the forward resistance of the diode. R_L is the load resistance

1.15.2 Reading of D.C. Ammeter

If a D.C. Ammeter is connected in the rectifier output circuit, what reading will it indicate? Is it the peak value or will the needle oscillate from zero to maximum and then to zero and so on, or will it indicate average value? The meter is so constructed that it reads the average value.

$$\text{By definition, average value} = \frac{\text{Area of the curve}}{\text{Base}}$$

$\therefore$ For Half wave rectified output, base value is 2π for one cycle,

$$I_{DC} = \frac{1}{2\pi}\int_0^{\pi} I_m \text{Sin}\alpha \, d\alpha = \frac{I_m}{2\pi} . \left| -\text{Cos}\alpha \right|_0^{\pi}$$

$$I_{DC} = \frac{I_m}{2\pi}[1+1] = \frac{I_m}{\pi}$$

Upper limit is only π, because the signal is zero from π to 2π. The complete cycle is from 0 to 2π.

1.15.3 Reading of A.C. Ammeter

An A.C. ammeter is constructed such that the needle deflection indicates the effective or RMS current passing through it.

Effective or RMS value of an A.C. quantity is the equivalent D.C. value which produces the same heating effect as the alternating component. If some A.C. current is passed through a resistor, during positive and negative half cycles, also because of the current, the resistor gets heated, or there is some equivalent power dissipation. What is the value of D.C. which produces the same heating effects as the A.C. quantity? The magnitude of this equivalent D.C. is called the RMS or Effective Value of A.C.

By definition $I_{rms} = \sqrt{\frac{I}{2\pi}\int_0^{2\pi}(I_m \text{Sin}\alpha)^2 d\alpha} = \sqrt{\frac{I_m^2}{2\pi}\int_0^{2\pi}\text{Sin}^2\alpha . d\alpha}$

$$= \frac{I_m}{\sqrt{2\pi}}\sqrt{\int_0^{2\pi}\left\{\frac{1-\text{Cos}2\alpha}{2}\right\}d\alpha} = \frac{I_m}{\sqrt{2\alpha}}\sqrt{\left.\frac{\alpha}{2}\right|_0^{2\pi} + \left.\frac{\text{Sin}2\alpha}{4}\right|_0^{2\pi}}$$

$\therefore$ RMS value of a sine wave $= \frac{I_m}{\sqrt{2\pi}} \times \sqrt{\frac{2\pi}{2}+0}$

$$= \frac{I_m\sqrt{\pi}}{\sqrt{2}\sqrt{\pi}} = \frac{I_m}{\sqrt{2}} = 0.707\, I_m$$

$$\text{Form Factor} = \frac{\text{RMS Value}}{\text{Average Value}}$$

For a sine wave, $I_{average} = 0.636 I_m = 2 \times \frac{1}{2\pi}\int_0^{\pi}(I_m \ \text{Sin}\alpha)d\alpha$

$$I_{rms} = \frac{I_m}{\sqrt{2}} = 0.707 I_m$$

$$\text{Form factor} = \frac{\left(\frac{I_m}{\sqrt{2}}\right)}{\left(\frac{I_m}{\pi}\right)} = \frac{0.707 I_m}{0.636 I_m} = 1.11$$

1.15.4 Peak Inverse Voltage

For the circuit shown, the input is A.C. signal. Now during the positive half cycle, the diode conducts. The forward resistance of the diode R_f will be small.

$\therefore$ The voltage across the diode $v = i \times R_f$

$i = I_m \text{ Sin } \alpha \qquad 0 \le \alpha \le \pi.$

$\therefore$ $v = I_m R_f \text{ Sin } \alpha \qquad 0 \le \alpha \le \pi.$

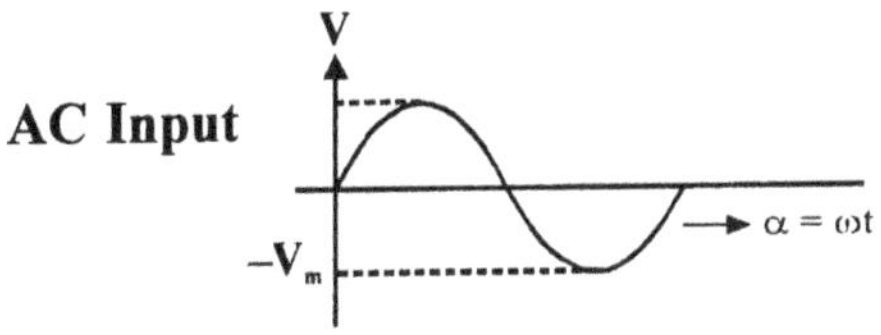

Fig. 1.18

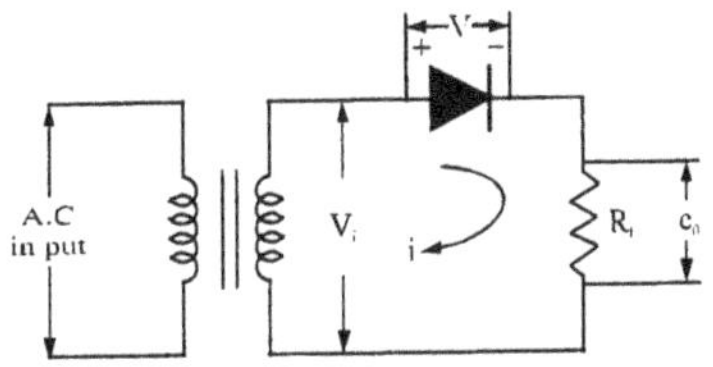

Fig. 1.19 Half wave rectifier circuit

Since R_f is small, the voltage across the diode 'V' during positive half cycle will be small, and the waveform is as shown. But during the negative half cycle, the diode will not conduct. Therefore, the current 'i' through the circuit is zero. So the voltage across the diode is not zero but the voltage of the secondary of the transformer V_i will appear across the diode ($\because$ effectively the diode is across the secondary of the transformer.)

$\therefore \qquad v = V_m \text{ Sin } \alpha \qquad \pi \leq \alpha \leq 2\pi.$

The waveform across the diode is Fig. 1.20.

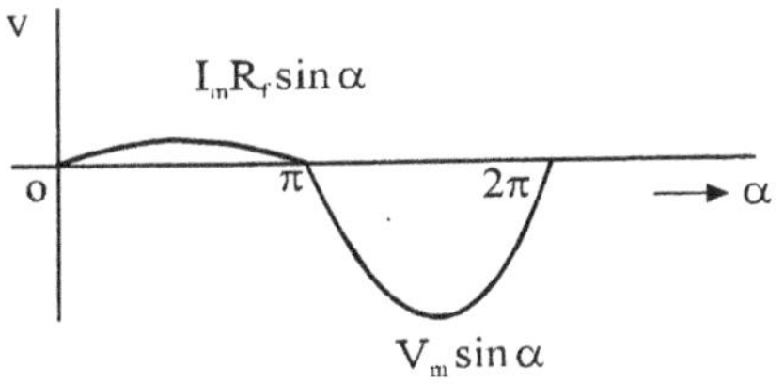

Fig 1.20 Voltage waveform across the diode

So the D.C. Voltage that is read by a D.C. Voltmeter is the average value.

$$V_{DC} = \frac{1}{2\pi}\left[\int_0^{\pi} I_m R_f.\text{Sin}\alpha \, d\alpha + \int_{\pi}^{2\pi} V_m \text{Sin}\alpha \, d\alpha\right]$$

$$= \frac{1}{2\pi}\left[(2)I_m R_f - (2)I_m (R_f + R_L)\right]$$

$$\because \quad V_m = I_m (R_f + R_c)$$

$$\therefore \quad \boxed{V_{DC} = \frac{-I_m R_L}{\pi}} \qquad \text{.......... (1.24)}$$

If we connect a CRO across the diode, this is the waveform that we see is as shown in Fig. 1.20. So in the above circuits the diode is being subjected to a maximum voltage of V_m. It occurs when the diode is not conducting. Hence it is called the ***Peak Inverse Voltage*** (**PIV**)

1.15.5 Regulation

The variation of D.C output voltage as a function of D.C load current is called '*regulation*'.

$$\% \text{ Regulation} = \frac{V_{\text{No Load Voltage}} - V_{\text{Full Load}}}{V_{\text{Full Load}}} \times 100\,\%$$

For an ideal power supply output voltage is independent of the load or the output in voltage remains constant even if the load current varies, like in Zener diode near breakdown. Therefore, ***regulation is zero or it should be low for a given circuit.***

Expression for V_{DC} the Output DC Voltage

For half wave rectifier circuit (Fig. 1.21), I_{DC} the average value is :

$$I_{DC} = \frac{1}{2\pi}\int_0^{2\pi} i.d\alpha$$

$$= \frac{1}{2\pi}\int_0^{\pi} I_m \sin\alpha.d\alpha = \frac{I_m}{\pi}$$

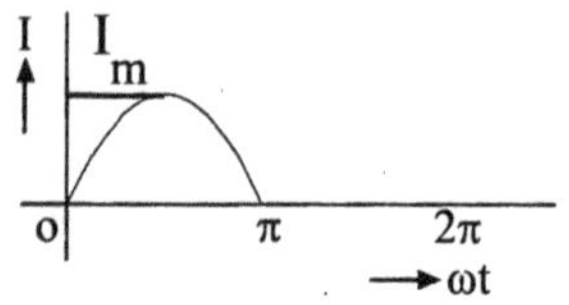

Fig 1.21 HWR current output

$$I_{DC} = \frac{I_m}{\pi} \qquad \text{.......... (1.25)}$$

But $I_m = \dfrac{V_m}{R_f + R_L}$ $\qquad R_L$ = Load Resistance

R_f = Forward Resistance of the Diode.

$$V_{DC} = I_{DC} R_L$$

$$\therefore \quad I_{DC} = \frac{V_m/\pi}{R_f + R_L}$$

But $\quad I_{DC} = \dfrac{V_m}{\pi(R_f + R_L)}$

$$\therefore \quad V_{DC} = \frac{V_m.R_L}{\pi(R_f + R_L)}$$

Adding and Subtracting R_f,

$$V_{DC} = \frac{V_m.(R_L + R_f - R_f)}{\pi(R_f + R_L)} = \frac{V_m}{\pi} - \frac{V_m.R_f}{\pi(R_f + R_L)}$$

$$I_{DC} = \frac{V_m}{\pi(R_f + R_L)}$$

I_{DC} is determined by R_L. Hence V_{DC} depends upon R_L.

$$\boxed{V_{DC} = \frac{V_m}{\pi} - I_{DC} \times R_f} \qquad \text{.......... (1.26)}$$

This expression indicates that V_{DC} is $\frac{V_m}{\pi}$ at ***no load or when the load current is zero***, and it decreases with increase in I_{DC} linearly since R_f is more or less constant for a given diode. The larger the value of R_f, the greater is the decrease in V_{DC} with I_{DC}. But, the series resistance of the secondary winding of the transformer R_s should also be considered.

For a given circuit of half wave rectifier, if a graph is plotted between V_{DC} and I_{DC} the slope of the curve gives ($R_f + R_s$) where R_f is the forward resistance of diode and R_s the series resistance of secondary of transformer.

Regulation For HWR

The regulation indicates how the DC voltage varies as a function of DC load current. In general, the percentage of regulation for ideal power supply is zero. The percentage of regulation is defined as

$$\% \text{ Regulation} = \frac{V_{NL} - V_{FL}}{V_{FL}} \times 100$$

As we know that,

$$V_{DC} = \frac{V_m}{\pi} - I_{DC} \cdot R_f$$

$$V_{NL} = \frac{V_m}{\pi} \text{ and } V_{FL} = \frac{V_m}{\pi} - I_{DC} \cdot R_f$$

$$\therefore \% \text{ Regulation} = \frac{\frac{V_m}{\pi} - \frac{V_m}{\pi} + I_{DC}.R_f}{\frac{V_m}{\pi} - I_{DC}.R_f} \times 100$$

$$= \frac{I_{DC}.R_f}{\frac{V_m}{\pi} - I_{DC}.R_f} = \frac{1}{\frac{V_m}{\pi\,(I_{DC}.R_f)} - 1} \times 100$$

$$= \frac{1}{\frac{V_m}{\pi} \times \frac{\pi}{V_m}\frac{(R_f + R_L)}{R_f} - 1} \times 100 \left[\because I_{DC} = \frac{V_m}{\pi(R_f + R_L)}\right]$$

$$\boxed{\% \text{ Regulation} = \frac{R_f}{R_L} \times 100}$$

Suppose for a given rectifier circuit, the specifications are 15V and 100 mA, i.e., the no load voltage is 15V and max load current that can be drawn is 100 mA. If the value of $(R_f + R_s) = 25\ \Omega$ then the percentage regulation of the circuit is :

No Load Voltage = 15V

Drop across Diode = $I_m\ (R_f + R_s)$

R_S = Transformer Secondary Resistance

Max. Voltage with load = $15V - (I_m \times (R_f + R_s))$

$= 15 - 100\ mA \times 25\Omega$

$= 15 - 2.5$ volts = 12.5V

$$\therefore \quad \text{Percentage Regulation} = \frac{15 - 12.5}{12.5} \times 100 = \frac{2.5}{12.5} \times 100 \simeq 20\%.$$

1.15.6 Ripple Factor

The purpose of a rectifier circuit is to convert A.C. to D.C. But the simple circuit shown before will not achieve this. Rectifier converts A.C. to unidirectional flow and not D.C. So filters are used to get pure D.C. Filters convert unidirectional flow into D.C. Ripple factor is a measure of the fluctuating components present in rectifier circuits.

$$\text{Ripple factor, } \gamma = \frac{\text{RMS Value of alternating components of the waveform}}{\text{Average Value of the waveform}}$$

$$\gamma = \frac{I'_{rms}}{I_{DC}} = \frac{V'_{rms}}{V_{DC}}$$

I'_{rms} and V'_{rms} denote the value of the A.C components of the current and voltage in the output respectively. While determining the ripple factor of a given rectifier system experimentally, a voltmeter or ammeter which can respond to high frequencies (greater than power supply frequency 50 Hz) should be used and a capacitor should be connected in series with the input meter in order to block the D.C. component. Ripple factor should be small. (Total current $i = I_m \sin \omega t$ according to Fourier Series, only A.C. is sum of D.C. and harmonics).

We shall now derive the expression for the ripple factor. The instantaneous current is given by $i' = i - I_{DC}$.

'i' is the total current. In a half wave rectifier, some D.C components are also present. Hence A.C component is,

$$i' = (i - I_{DC})\ [\ (\text{Total current} - I_{DC})\ ]$$

$$\text{RMS value is } I'_{rms} = \sqrt{\frac{1}{2\pi}\int_0^{2\pi} (i - I_{DC})^2\, d\alpha}$$

$$\therefore \quad I'_{rms} = \sqrt{\frac{1}{2\pi}\int_0^{2\pi}\left(i^2 - 2I_{DC}.i + I_{DC}{}^2\right)d\alpha}$$

Now, $\frac{1}{2\pi}\int_0^{2\pi} i^2 d\alpha$ = Square of the rms value of a sine wave by definition.

$$= (I_{rms})^2$$

$\frac{1}{2\pi}\int_0^{2\pi} i.d\alpha$ = The average value or D.C value I_{DC}.

I_{DC} is constant. So taking this term outside,

$$\frac{I_{DC}{}^2}{2\pi}\int_0^{2\pi} d\alpha = \frac{I_{dc}}{2\pi}[2\pi]$$

$$\therefore \quad I'_{rms} = \sqrt{(I_{rms})^2 - 2I_{DC}{}^2 + I_{DC}{}^2}$$

The rms ripple current is

$$I'_{rms} = \sqrt{(I_{rms})^2 - I_{DC}{}^2}$$

Ripple factor, $\gamma = \frac{I'_{rms}}{I_{dc}} = \frac{\sqrt{(I_{rms})^2 - I_{DC}{}^2}}{I_{DC}}$

$$\gamma = \sqrt{\left(\frac{I_{rms}}{I_{DC}}\right)^2 - 1} \qquad \text{.......... (1.27)}$$

This is independent of the current waveshape and is not restricted to a half wave configuration. If a capacitor is used to block D.C. and then I_{rms} or V_{rms} is measured,

Then, $\gamma = \frac{I_{rms}}{I_{DC}}$

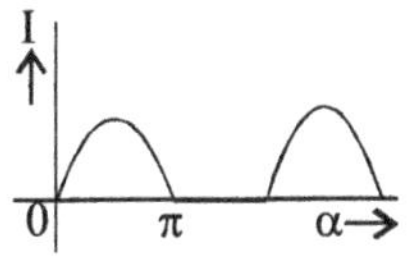

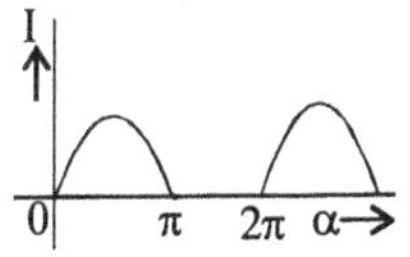

$$\because \quad I'_{rms} = \sqrt{I_{rms}{}^2 - I_{DC}}$$

and $I_{DC} = 0$ (blocked by Capacitor) output waveform

For Half Wave Rectifier Circuit (HWR),

$$I = I_m \sin\alpha$$

$$I_{av} = I_{DC} = \frac{1}{2\pi}\int_0^{\pi} I_m \sin\alpha.d\alpha = \frac{I_m}{\pi}$$

$$I_{rms} = \sqrt{\frac{1}{2\pi}\int_0^{\pi} I_m^2 Sin^2(\alpha)\, d\alpha} = \frac{I_m}{2}$$

$$I_{DC} = \frac{I_m}{\pi};\ V_{DC} = \frac{V_m}{\pi} \quad \text{Peak Inverse Voltage (PIV)} = V_m$$

Ripple Factor $\gamma = \dfrac{\text{RMS value of ripple current}}{\text{Average value of the current}} = \dfrac{I'_{rms}}{I_{DC}}$

Total current $I = I_{DC} + I'(\text{ripple})$

$I'(\text{ripple}) = (I - I_{DC})$

$$I'_{rms} = \sqrt{\frac{1}{2\pi}\int_0^{2\pi}\left(I - I_{DC}\right)^2 d\alpha} = \sqrt{\frac{1}{2\pi}\int_0^{2\pi}\left(I^2 - 2I.I_{DC} + I_{DC}^2\right).d\alpha}$$

$$= \sqrt{\left(I_{rms}\right)^2 - 2I_{DC}^2 + I_{DC}^2}$$

$$\therefore \quad I'_{rms} = \sqrt{\left(I_{rms}\right)^2 - I_{DC}^2}$$

$$\gamma = \frac{I'_{rms}}{I_{DC}} = \sqrt{\left(\frac{I_{rms}}{I_{DC}}\right)^2 - 1}$$

for Half Wave Rectifier, $\gamma = \sqrt{\left(\dfrac{I_m \times \pi}{2 \times I_m}\right)^2 - 1} = \sqrt{\left(\dfrac{\pi}{2}\right)^2 - 1} = 1.21$

1.15.7 Ratio of Rectification (η)

$$\text{Ratio of Rectification} = \frac{\text{DC power delivered to the load}}{\text{AC input power from transfer secondary}}$$

$$P_{DC} = I_{DC}^2 \times R_L$$

But I_{DC} for HWR $= \dfrac{I_m}{\pi}$

$$\therefore \quad P_{DC} = \left(\frac{I_m}{\pi}\right)^2 \times R_L$$

P_{AC} is what a Wattmeter would indicate if placed in the HWR circuit with its voltage terminal connected across the secondary of the transformer.

From $\pi - 2\pi$ the diode is not conducting. Hence $I_{ac} = 0$.

$$P_{ac} = (I_{rms})^2\ (R_f + R_L)$$

Q from $\pi - 2\pi$ the AC is not being converted to D.C. So this power is wasted, as heat across diode and transformer.

$\therefore$ $I_{rms}(\pi - 2\pi)$ portion of AC is $= \sqrt{\frac{1}{2\pi}\int_{\pi}^{2\pi} I_m^2 \sin^2 \alpha d\alpha} = \frac{I_m}{2}$

I_{rms} for HWR is $I_m/2$. During negative half cycle, diode is not conducting and current I in the loop is zero even though Voltage V is present.

$\therefore$ $$P_{AC} = \left(\frac{I_m}{2}\right)^2 (R_f + R_L) = \left(\frac{I_m}{2}\right)^2 \times R_L$$

$\therefore$ $$\frac{P_{DC}}{P_{AC}} = \frac{I_m^2 . R_L \times 4}{\pi^2 I_m^2 \times R_L} = \frac{4}{\pi^2} = 0.406$$

$\therefore$ Ratio of rectification for HWR = 0.406

Considering ideal diode. If we consider R_f also, the expression $= \dfrac{0.406}{1 + \left(\dfrac{R_f}{R_L}\right)}$

The AC input power is not converted to D.C. Only part of it is converted to D.C and is dissipated in the load. The balance of power is also dissipated in the load itself as AC power. We have to consider the rating of the secondary of the transformer. In ratio of rectification we have considered only the A.C output as the secondary of the transformer.

1.15.8 Transformer Utilization Factor

$$\text{Transformer Utilization Factor (TUF)} = \frac{\text{D.C power delivered to the load}}{\text{AC rating of transformer secondary}}$$

$$= \frac{P_{DC}}{P_{AC} \text{ rated}}$$

This term TUF is not ratio of rectification, because all the rated current of the secondary is not being drawn by the circuit.

$$P_{ac} = I_{DC}^2 . R_L - \left(\frac{I_m}{\pi}\right)^2 . R_L$$

$$P_{ac} = V_{rms} \times I_{rms}$$

$$V_{rms} = \frac{V_m}{\sqrt{2}} = \text{Rated voltage of (Secondary Transformer)}$$

$$I_{rms} = I_m/2$$

$$P_{ac} = V_{rms} \times I_{rms}$$

But $V_m = I_m (R_f + R_L) \simeq I_m R_L$ (R_f is small)

$$TUF = \frac{I_m^2}{\pi^2} \times R_L \bigg/ \frac{I_m R_L}{\sqrt{2}} \times \frac{I_m}{2} = \frac{2\sqrt{2}}{\pi^2} = 0.287$$

Transformer Utilization Factor for Half Wave Rectifier is 0.287

1.15.9 DISADVANTAGES OF HALF WAVE RECTIFIER

1. High Ripple Factor (1.21)
2. Low ratio of rectification (0.406)
3. Low TUF (0.287)
4. D.C saturation of transformer.

For half wave configuration,

$$I_{rms} = \sqrt{\frac{1}{2\pi}\int_0^{\pi} I_m^2 \sin^2 \alpha.d\alpha} = \frac{I_m}{2}$$

Because during negative half cycle the diode will not conduct hence i = 0 in the loop from $\pi - 2\pi$. So integration is from $0 - \pi$ only. Even though V is present, i = 0 for one half cycle. Therefore, Power is zero ($V \times I = P = 0$, since, $I = 0$)

$$I_{DC} = \frac{1}{2\pi}\int_0^{\pi} I_m \sin\alpha.d\alpha = \frac{I_m}{2\pi}\left|-\cos\alpha\right|_0^{\pi} = \frac{I_m}{2\pi}|1+1| = \frac{I_m}{\pi}$$

$\therefore$ Ripple Factor for Half Wave Rectification $\frac{I_{rms}}{I_{DC}} = \frac{I_m/2}{I_m/\pi} = \frac{\pi}{2}$

$$\gamma = \sqrt{(1.57)^2 - 1} = 1.21$$

$\because \quad \frac{\pi}{2} = 1.57$

$\therefore \quad \gamma > 1$

So the ripple voltage exceeds D.C voltage. Hence HWR is a poor circuit for converting AC to DC.

1.15.10 POWER SUPPLY SPECIFICATIONS

The input characteristics which must be specified for a power supply are :

1. *The required output D.C voltage*
2. *Regulation*
3. *Average and peak currents in each diode*
4. *Peak inverse voltage (**PIV**)*
5. *Ripple factor.*

1.16 RIPPLE FACTOR

The purpose of a rectifier circuit is to convert A.C. to D.C. But the simple circuit shown before will not achieve this. Rectifier converts A.C. to unidirectional flow and not D.C. So filters are used to get pure D.C. Filters convert unidirectional flow into D.C. Ripple factor is a measure of the fluctuating components present in rectifier circuits.

$$\text{Ripple factor, } \gamma = \frac{\text{RMS Value of alternating components of the waveform}}{\text{Average Value of the waveform}}$$

$$\gamma = \frac{I'_{rms}}{I_{DC}} = \frac{V'_{rms}}{V_{DC}}$$

I'_{rms} and V'_{rms} denote the value of the A.C components of the current and voltage in the output respectively. While determining the ripple factor of a given rectifier system experimentally, a voltmeter or ammeter which can respond to high frequencies (greater than power supply frequency

50 Hz) should be used and a capacitor should be connected in series with the input meter in order to block the D.C. component. Ripple factor should be small. (Total current $i = I_m \sin \omega t$ according to Fourier Series, only A.C. is sum of D.C. and harmonics).

We shall now derive the expression for the ripple factor. The instantaneous current is given by $i' = i - I_{DC}$.

i is the total current. In a half wave rectifier, some D.C components are also present. Hence A.C component is,

$$i' = (i - I_{DC}) \, [\, (\text{Total current} - I_{DC})\,]$$

$$\text{RMS value is } I'_{rms} = \sqrt{\frac{1}{2\pi}\int_0^{2\pi}(i - I_{DC})^2 \, d\alpha}$$

$$\therefore \quad I'_{rms} = \sqrt{\frac{1}{2\pi}\int_0^{2\pi}\left(i^2 - 2I_{DC}.i + I_{DC}{}^2\right) d\alpha}$$

Now, $\frac{1}{2\pi}\int_0^{2\pi} i^2 d\alpha$ = Square of the rms value of a sine wave by definition.

$$= (I_{rms})^2$$

$$\frac{1}{2\pi}\int_0^{2\pi} i.d\alpha = \text{The average value or D.C value } I_{DC}.$$

I_{DC} is constant. So taking this term outside,

$$\frac{I_{DC}{}^2}{2\pi}\int_0^{2\pi} d\alpha = \frac{I_{dc}}{2\pi}[2\pi]$$

$$\therefore \quad I'_{rms} = \sqrt{(I_{rms})^2 - 2I_{DC}{}^2 + I_{DC}{}^2}$$

The rms ripple current is

$$I'_{rms} = \sqrt{(I_{rms})^2 - I_{DC}{}^2}$$

$$\text{Ripple factor,} \quad \gamma = \frac{I'_{rms}}{I_{dc}} = \frac{\sqrt{(I_{rms})^2 - I_{DC}{}^2}}{I_{DC}}$$

$$\gamma = \sqrt{\left(\frac{I_{rms}}{I_{DC}}\right)^2 - 1} \qquad \text{.......... (1.28)}$$

This is independent of the current waveshape and is not restricted to a half wave configuration. If a capacitor is used to block D.C. and then I_{rms} or V_{rms} is measured,

Then, $\gamma = \dfrac{I_{rms}}{I_{DC}}$

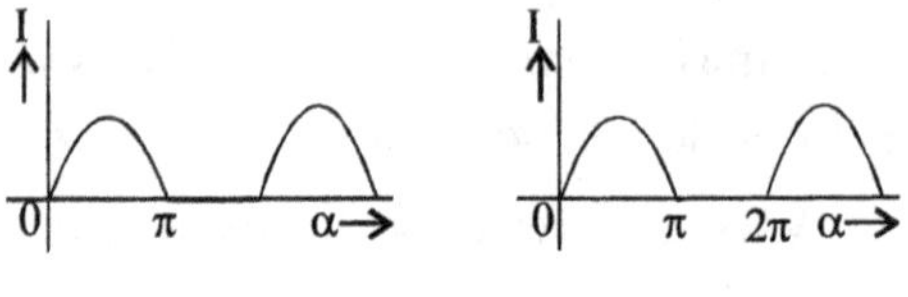

$\because \quad I'_{rms} = \sqrt{I_{rms}{}^{2} - I_{DC}}$

and $\quad I_{DC} = 0$ (blocked by Capacitor) output waveform

For Half Wave Rectifier Circuit (HWR),

$$I = I_m \sin \alpha$$

$$I_{av} = I_{DC} = \frac{1}{2\pi}\int_0^{\pi} I_m \sin\alpha . d\alpha = \frac{I_m}{\pi}$$

$$I_{rms} = \sqrt{\frac{1}{2\pi}\int_0^{\pi} I_m^2 \text{Sin}^2(\alpha)\, d\alpha} = \frac{I_m}{2}$$

$$I_{DC} = \frac{I_m}{\pi}\ ;\ V_{DC} = \frac{V_m}{\pi} \quad \text{Peak Inverse Voltage (PIV)} = V_m$$

Ripple Factor $\quad \gamma = \dfrac{\text{RMS value of ripple current}}{\text{Average value of the current}} = \dfrac{I'_{rms}}{I_{DC}}$

Total current $\quad I = I_{DC} + I'(\text{ripple})$

$$I'(\text{ripple}) = (I - I_{DC})$$

$$I'_{rms} = \sqrt{\frac{1}{2\pi}\int_0^{2\pi}\left(I - I_{DC}\right)^2 d\alpha} = \sqrt{\frac{1}{2\pi}\int_0^{2\pi}\left(I^2 - 2I.I_{DC} + I_{DC}{}^2\right).d\alpha}$$

$$= \sqrt{(I_{rms})^2 - 2I_{DC}{}^2 + I_{DC}{}^2}$$

$$\therefore \quad I'_{rms} = \sqrt{(I_{rms})^2 - I_{DC}{}^2}$$

$$\gamma = \frac{I'_{rms}}{I_{DC}} = \sqrt{\left(\frac{I_{rms}}{I_{DC}}\right)^2 - 1}$$

for Half Wave Rectifier, $\gamma = \sqrt{\left(\dfrac{I_m \times \pi}{2 \times I_m}\right)^2 - 1} = \sqrt{\left(\dfrac{\pi}{2}\right)^2 - 1} = 1.21$

1.17 FULL WAVE RECTIFIER (FWR)

Since half wave rectifier circuit has poor ripple factor, for ripple voltage is greater than DC voltage, it cannot be used. So now analyze a full wave rectifier circuit.

The circuit is as shown in Fig. 1.22.

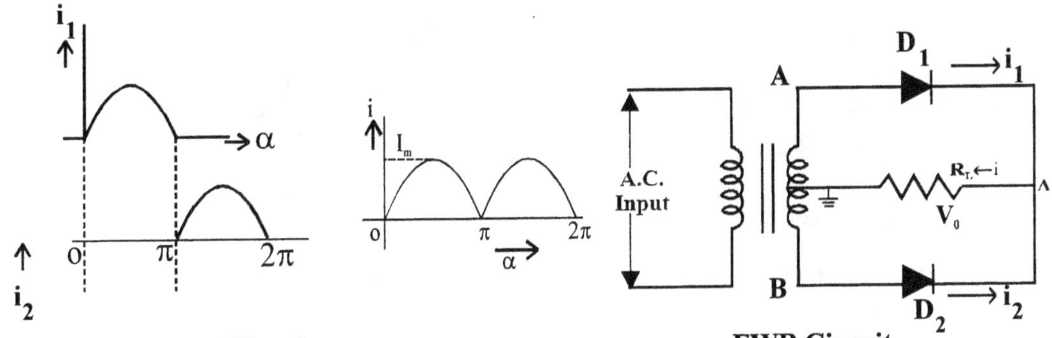

Waveforms **FWR Circuit**

Fig 1.22

During the +ve half cycle, D_1 conducts and the current through D_2 is zero

During the – ve half cycle, D_2 conducts and the current through D_1 is zero

A centre tapped transformer is used.

The total current i flows through the load resistor R_L and the output voltage V_0 is taken across R_L.

For half wave rectifier, circuit, $I_{DC} = \frac{1}{2\pi}\int_0^{2\pi} I_m \sin\alpha d\alpha = \frac{I_m}{\pi}$

For full wave rectifier, circuit, I_{DC} = Twice that of half wave rectifier circuit

$\therefore \quad I_{DC} = 2\, I_m/\pi$

For half wave rectifier circuit , $I_{rms} = \sqrt{\frac{1}{2\pi}\int_0^{\pi} I_m^2 \sin\alpha d\alpha} = \frac{I_m}{2}$

For full wave rectifier circuit, $I_{rms} = \sqrt{2\times\frac{1}{2\pi}\int_0^{\pi} I_m^2 \sin^2\alpha d\alpha}$

$$I_{rms} = \sqrt{2} \times \frac{I_m}{2} = \frac{I_m}{\sqrt{2}}$$

$$I_m = \frac{V_m}{R_f + R_L}$$

R_f is the forward resistance of each diode.

A centre tapped transformer is essential to get full wave rectification. So there is a phase shift of 180^0, because of centre tapping. So D_1 is forward biased during the input cycle of 0 to π, D_2 is forward biased during the period π to 2π since the input to D_2 has a phase shift of 180^0 compared to the input to D_1. So positive half cycle for D_2 starts at a full wave rectifier circuit, while the D.C. current starts through the load resistance R_L is twice that of the Half Wave Rectifier Circuit.

Therefore, for Half Wave Rectifier Circuit,

$$I_{DC} = \frac{I_m}{\pi}$$

For Full Wave Rectifier Circuit,

$$I_{DC} = \frac{2I_m}{\pi}$$

Hence, Ripple Factor is improved.

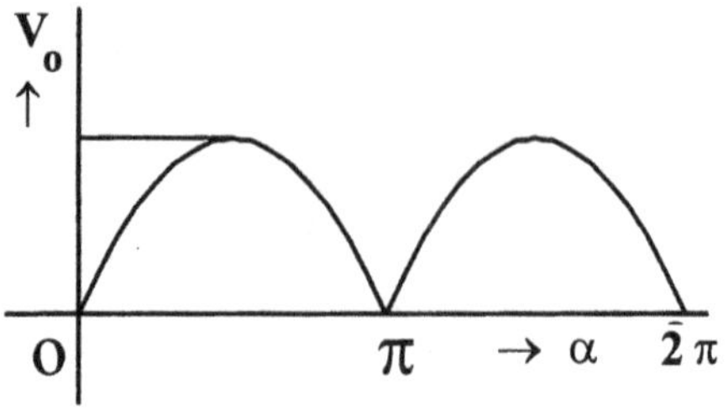

Fig 1.23 FWR voltage output.

1.17.1 Ripple Factor

$$\frac{I'_{rms}}{I_{DC}} = \frac{I_m/\sqrt{2}}{2I_m/\pi} = \frac{\pi}{2\sqrt{2}} = 1.11$$

$$I'_{rms} = \sqrt{I_{rms}^2 - I_{DC}^2}\ ; \quad \gamma = \frac{\sqrt{I_{rms}^2 - I_{DC}^2}}{I_{DC}}$$

Therefore, $\gamma = 1.21$ for HWR, and it is is 0.482 for FWR.

1.17.2 Regulation for FWR

The regulation indicates how D.C voltage varies as a function of load current. The percentage of regulation is defined as

$$\%\ \text{Regulation} = \frac{V_{NL} - V_{FL}}{V_{FL}} \times 100$$

The variation of DC voltage as a function of load current is as follows

$$V_{DC} = I_{DC}\,R_L$$

$$= \frac{2I_m}{\pi} \cdot R_L \quad \left[\because I_m = \frac{V_m}{R_f + R_L}\right]$$

$$= \frac{2V_m}{\pi(R_f + R_L)} \cdot R_L \quad \left[\because I_m = \frac{V_m}{R_f + R_L}\right]$$

$$= \frac{2V_m}{\pi}\left[1 - \frac{R_f}{R_f + R_L}\right]$$

$$= \frac{2V_m}{\pi} - \frac{2V_m}{\pi(R_f + R_L)} \times R_f$$

$$V_{DC} = \frac{2V_m}{\pi} - I_{DC} \cdot R_f$$

$$V_{NL} = \frac{V_m}{\pi}, \qquad V_{FL} = \frac{V_m}{\pi} - I_{DC} \cdot R_f$$

$$\% \text{Regulation} = \frac{\frac{2V_m}{\pi} - \frac{2V_m}{\pi} + I_{DC} \cdot R_f}{\frac{2V_m}{\pi} - I_{DC} \cdot R_f} \times 100$$

$$= \frac{I_{DC} R_f}{\frac{2V_m}{\pi} - I_{DC} . R_f} \times 100$$

$$= \frac{I_{DC} R_f}{I_{DC} R_f \left(\frac{2V_m}{\pi I_{DC} . R_f} \right) - 1} \times 100$$

$$= \frac{1}{\left(\frac{2V_m}{\pi} \times \frac{(R_f + R_L)}{\frac{2V_m}{\pi} . R_f} - 1 \right)} \times 100$$

$$\% \text{Regulation} = \frac{R_f}{R_L} \times 100$$

1.17.3 Ratio of Rectification

$$\frac{\text{D.C. power delivered to Load}}{\text{A.C. input power from transformer secondary}}$$

$p_{DC} = I_{DC}^2 \times R_L$ for FWR,

$$I_{DC} = \frac{2I_m}{\pi} = \frac{4.I_m^2}{\pi^2} R_L.$$

P_{AC} is what a Wattmeter would indicate if placed in the FWR Circuit with its voltage terminals connected across the transformer secondary.

$$P_{ac} = (I_{rms})^2 . (R_F + R_L); \; I_{rms} \text{ for FWR} = \frac{I_m}{\sqrt{2}}$$

$$= \left(\frac{I_m}{\sqrt{2}} \right)^2 .(R_F + R_L).$$

$$= \frac{I_m^2}{2} (R_F + R_L).$$

If we assume that R_f is the forward resistance of the diode is very small, compared to R_L.

$$R_f + R_L \cong R_L.$$

$$\therefore \quad P_{AC} = \frac{I_m^2}{2} (R_L).$$

$$\text{Ratio of Rectification} = \frac{p_{DC}}{p_{AC}} = \frac{4 I_m^2 R_L \times 2}{\pi^2 I_m^2 \times R_L} = \frac{8}{\pi^2} = 0.812$$

For HWR it is 0.406. Therefore, for FWR the ratio of rectification is twice that of HWR.

1.17.4 Transformer Utilization Factor

In fullwave rectifier using center-tapped transformer, the secondary current flows through each half separately in every half cycle. While the primary of transformer carries current continuously. Hence TUF is calculated for primary and secondary windings separately and then the average TUF is determined.

$$\text{Secondary TUF} = \frac{\text{DC power delevered to load}}{\text{AC rating of transformer secondary}}$$

$$= \frac{(I_{DC})^2 R_L}{V_{rms} I_{rms}} = \frac{\left(\frac{2 I_m}{\pi}\right)^2 . R_L}{\frac{V_m}{\sqrt{2}} \times \frac{I_m}{\sqrt{2}}}$$

$$= \frac{4I_m^2}{\pi^2} \times \frac{2R_L}{I_m^2 (R_f + R_L)} = \frac{8}{\pi^2} \times \frac{1}{1 + \frac{R_f}{R_L}}$$

if $R_f << R_L$, then secondary TUF = 0.812.

The primary of the transformer in feeding two HWR's separately. These two HWR's work independently of each other but feed a common load. Therefore,

$$\text{TUF for primary} = 2 \times \text{TUF of HWR} = 2 \times \frac{0.287}{1 + \frac{R_f}{R_L}}$$

if $R_f << R_L$, then TUF for primary = 0.574

The average TUF for FWR using center tapped transformer

$$= \frac{\text{TUF of primary} + \text{TUF of secondary}}{2} = \frac{0.812 + 0.574}{2} = 0.693$$

Therefore, the transformer is utilized 69.3% in FWR using center tapped transformer.

1.17.5 Peak Inverse Voltage for Full Wave Rectifier

With reference to the FWR circuit, during the –ve half cycle, D_1 is not conducting and D_2 is conducting. Hence maximum voltage across R_L is V_m, since voltage is also present between A and O of the transformer, the total voltage across $D_1 = V_m + V_m = 2V_m$ (Fig. 1.22).

1.17.6 D.C. Saturation

In a FWR, the D.C. currents I_1 and I_2 flowing through the diodes D_1 and D_2 are in opposite direction and hence cancel each other. So there is no problem of D.C. current flowing through the core of the transformer and causing saturation of the magnetic flux in the core.

PIV is 2 V_m for each diode in FWR circuit because, when D_1 is conducting, the drop across it is zero. Voltage delivered to R_L is V_m. D_2 is across R_L. During the same half cycle, D_2 is not conducting. Therefore, peak voltage across it is V_m from the second half of the transformer. The total voltage across D_2 is $V_m + V_m = 2V_m$.

1.18 BRIDGE RECTIFIERS

The circuit is shown in Fig. 1.24. During the positive half cycle, D_1 and D_2 are forward biased. D_3 and D_4 are open. So current will flow through D_1 first and then through R_L and then through D_2 back to the ground. During the –ve half cycle D_4 and D_3 are forward biased and they conduct. The current flows from D_3 through R_L to D_4. Hence the direction of current is the same. So we get full wave rectified output.

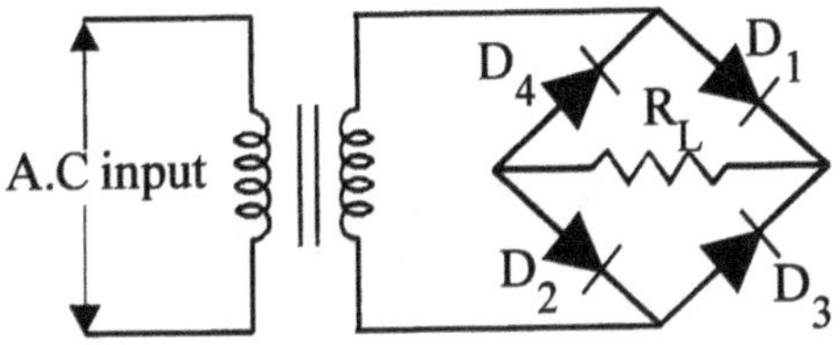

Fig 1.24 Bridge rectifier circuit.

In Bridge rectifier circuit, there is no need for centre tapped transformer. So the transformer secondary line to line voltage should be one half of that, used for the FWR circuit, employing two diodes.

$$I_{DC} = \frac{2I_m}{\pi}$$

where

$$I_m = \frac{V_m}{R_S + 2R_f + R_L}$$

$$V_{DC} = \frac{2V_m}{\pi} - I_{DC}(R_S + 2R_F)$$

R_S = Resistance of Transformer Secondary.

R_f = Forward Resistance of the Diode.

R_L = Load Resistance.

$2R_f$ should be used since two diodes in series are conducting at the same time. The ripple factor and ratio of rectification are the same as for Full Wave Rectifier.

1.18.1 ADVANTAGES OF BRIDGE RECTIFIER

1. *The peak inverse voltage (PIV) across each diode is V_m and not $2V_m$ as in the case of FWR. Hence the Voltage rating of the diodes can be less.*
2. *Centre tapped transformer is not required.*
3. *There is no D.C. Current flowing through the transformer since there is no centre tapping and the return path is to the ground. So the transformer utilization factor is high.*

1.18.2 DISADVANTAGES

1. *Four diodes are to be used.*
2. *There is some voltage drop across each diode and so output voltage will be slightly less compared to FWR. But these factors are minor compared to the advantages.*

Bridge rectifiers are available in a package with all the 4 diodes incorporated in one unit. It will have two terminals for A.C. Input and two terminals for DC output. Selenium rectifiers are also available as a package.

1.19 THE HARMONIC COMPONENTS IN RECTIFIER CIRCUIT

The analytical representation of output current wave of halfwave rectifier using fourier series is given by

$$i = I_m \left[\frac{1}{\pi} + \frac{1}{2}\sin\omega t - \frac{2}{\pi} \sum_{k=\text{even}} \frac{\cos\omega t}{(k+1)(k-1)} \right] \quad .....(1.29)$$

The lowest angular frequency present in eq. (1.29) is of a.c source. And all other terms present in above equation are of even harmonic of power source.

Similarly in case of full wave rectifier in which one diode conducts for positive half cycle and the other diode conducting for negative half cycle and the net output current is difference of the currents conducted in positive half cycle ($i_1(\alpha)$) and the current conducted in negative half cycle ($i_2(\alpha)$). i.e.,

$$\text{Net current } i = i_1(\alpha) + i_2(\alpha) \qquad (i_2(\alpha) = i_2(\alpha + \pi)) \quad ....(1.30)$$

Here both $i_1(\alpha)$ and $i_2(\alpha)$ are same in magnitude but opposite in polarity. The difference $i_1(\alpha) + i_2(\alpha)$ results in cancellation of lowest angular frequency of power source present in eq. (1.29). And hence the total current.

$i = i_1 + i_2$ attains of form

$$i = I_m \left[\frac{2}{\pi} - \frac{4}{\pi} \sum_{\substack{k \text{ even} \\ k \neq 0}} \frac{\cos k\omega t}{(k+1)(k-1)} \right] \quad .....(1.31)$$

Here in eq. (1.31) the fundamental frequency 'ω' has been eliminated from the equation. And hence lowest frequency in the output being 2ω, i.e., the second hormonic term. This offers an definite advantage in the effectiveness of filtering the output.

The other advantage of fullwave rectifier is that the direction of the current during two halfs of cycle is of opposite in nature, thus the magnetic cycle through which iron core is take is essentially that of the altering current and this eliminates d.c. saturation of the transformer, which would yield additional hormonics in the output.

1.20 INDUCTOR FILTER

FILTERS

A power supply must provide ripple free source of power from an A.C. line But the output of a rectifier circuit contains ripple components in addition to a D.C. term. It is necessary to include a filter between the rectifier and the load in order to eliminate these ripple components. Ripple components are high frequency A.C. Signals in the D.C output of the rectifier. These are not desirable, so they must be filtered. So filter circuits are used.

Flux linkages per ampere of current $L = \frac{N\phi}{I}$. The ability of a component to develop induced voltage when alternating current is flowing through the element is the property of the inductor. Types of Inductors are :

1. *Iron Cored* 2. *Air Cored*

An inductor opposes any change of current in the circuit. So any sudden change that might occur in a circuit without an inductor are smoothed out with the presence of an inductor. In the case of AC, there is change in the magnitude of current with time.

Inductor is a short circuit for DC and offers some impedance for A.C. ($X_L = j\omega L$). So it can be used as a filter. AC voltage is dropped across the inductors, where as D.C. passes through it. Therefore, A.C is minimized in the output.

Inductor filter is used with FWR circuit. Therefore, HWR gives 121% ripple and using filter circuit for such high ripple factor has no meaning . FWR gives 48% ripple and by using filter circuit we can improve it. According to Fourier Analysis, current, 'i' in HWR Circuit is

$$i = \frac{I_m}{\pi} + \frac{I_m}{2}\text{Sin}\,\omega t - \frac{2I_m}{\pi}\frac{\text{Cos}\,4\omega t}{3}$$

Current in the case of FWR is $i \cong \frac{2I_m}{\pi} - \frac{4I_m}{3\pi}\cos 2\omega t - \frac{4I_m \cos 4\omega t}{15\pi}$

The reactance offered by inductor to higher order frequencies like $4\omega t$ can be neglected. So the output current is

$$i \cong \frac{2I_m}{\pi} - \frac{4I_m}{3\pi}\cos 2\omega t$$

the fundamental harmonic ω is eliminated.

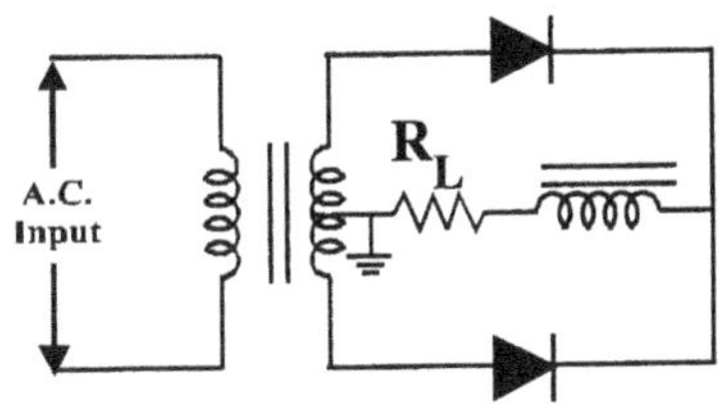

Fig 1.25 Inductor filter circuit.

The circuit for FWR with inductor filter is as shown in Fig. 1.25.

One winding of transformer can be used as 'L'. For simplification, diode and choke (inductor) resistances can be neglected, compared with R_L. The D.C. component of the current is

$$I_m = \frac{V_m}{R_L}$$

Impedance due to L and R_L in Series $|Z| = \sqrt{R_L^2 + (2\omega L)^2}$

For second Harmonic, the frequency is 2ω.

Therefore, A.C. Component of current $I_m = V_m / \sqrt{R_L^2 + 4\omega^2 L^2}$. So substituting these values in the expression, for current,

$$I = \frac{2I_m}{\pi} - \frac{4I_m}{3\pi}\text{Cos } 2\omega t$$

by inductor to higher order frequencies like $4\omega t$ etc., we get,

$$i = \frac{2V_m}{\pi R_L} - \frac{4V_m}{3\pi\sqrt{R_L^2 + 4\omega L^2}}\cos(2\omega t - \theta)$$

where θ is the angle by which the load current lags behind the voltage and is given by

$$\theta = \text{Tan}^{-1}\frac{2\omega L}{R_L}.$$

1.20.1 Ripple Factor

$$\gamma = I_{r'rms}/I_{D.C.} = I'_{rms}/I_{DC} \quad I_{DC} = 2V_m/\pi R_L$$

$$I_{rrms} = \frac{I_m}{\sqrt{2}} = \frac{4V_m}{3\sqrt{2}\pi\sqrt{R_L^2 + 4\omega^2 L^2}}$$

$$\therefore \quad \text{Ripple factor} = \frac{4v_m \times \pi R_L}{3\sqrt{2}\pi\sqrt{R_L^2 + 4\omega^2 L^2} \times 2V_m}$$

$$= \frac{2}{3\sqrt{2}} \times \frac{1}{\sqrt{1 + \frac{4\omega^2 L^2}{R_L^2}}}$$

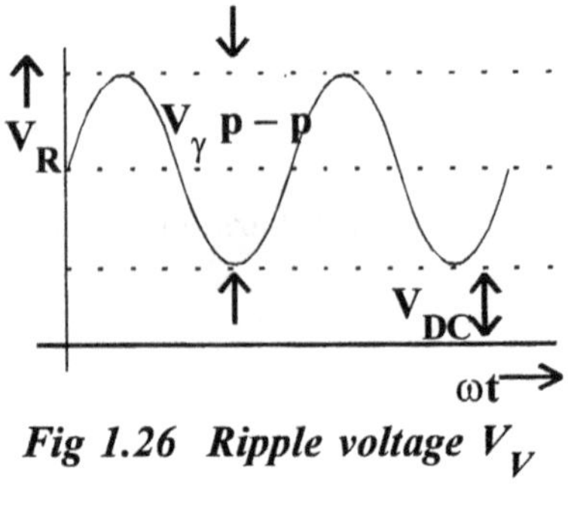

Fig 1.26 Ripple voltage V_V

If $\frac{4\omega^2 L^2}{R_L^2} >> 1$, then

$$\text{Ripple Factor } r = \frac{R_L \times 2}{3\sqrt{2} \times 2\omega L}$$

$$\therefore \quad \text{Ripple factor} = \frac{R_L}{3\sqrt{2}\omega L} \qquad \text{.......... (1.32)}$$

Therefore, for higher values of L, the ripple factor is low. If R_L is large, then also γ is high. Hence inductor filter should be used where the value of R_L is low. Suppose the output wave form from FWR supply is as shown in Fig. 1.26, then,

$V_{\gamma p-p}$ is the peak to peak value of the ripple voltage. Suppose

$V_{DC} = 300V$, and $V_{\gamma p-p}$ is 10V.

then V_R maximum $= \frac{10}{2} = 5V$. Therefore, $V_\gamma rms = \frac{V_\gamma max}{\sqrt{2}} = \frac{5}{\sqrt{2}} = 3.54V$

$$\therefore \quad \text{Ripple Factor } \gamma = \frac{V_r rms}{V_x} = \frac{3.54}{300} = 0.0118.$$

$$\% \text{ ripple} = \text{Ripple factor} \times 100\,\% = 1.18$$

1.20.2 REGULATION

$$V_{DC} = I_{DC}.R_L; \qquad I_{DC} \text{ for F W R is} \frac{2I_m}{\pi}$$

$$V_{DC} = \frac{2I_m R_L}{\pi} = \frac{2V_m}{\pi}$$

$$I_m.R_L = V_m$$

Therefore, V_{DC} is constant irrespective of R_L. But this is true if L is ideal. In practice

$$V_{DC} = \frac{2V_m}{\pi} - I_{DC}.R_f$$

where R_f is the resistance of diode.

An inductor stores magnetic energy when the current flowing through it is greater than the average value and releases this energy when the current is less than the average value.

Another formula that is used for Inductor Filter for

$$\text{Ripple Factor} = 0.236 \frac{R_C + R_L}{\omega_L} \qquad \text{.......... (1.33)}$$

where R_L is load resistance R_c is the series resistance of the inductors. This is the same as the above formula $\frac{R_L}{3\sqrt{2}\omega_L}$

$$\because \quad \frac{1}{3\sqrt{2}} = 0.236.$$

Problem 1.18

A FWR is used to supply power to a 2000Ω Load. Choke Inductors of 20 H inductance and capacitors of 16μf are available. Compute the ripple factor using **1**. One Inductor filter **2**. One capacitor filter **3**. Single L type filter.

Solution

1. One Inductor Filter :

$$I = \frac{2I_m}{\pi} - \frac{4I_m}{3\pi} \cos(2\omega t - \phi) \text{ where } \phi = \text{Tan}^{-1}\left(\frac{2\omega L}{R}\right)$$

$$I_{DC} = \frac{2V_m}{\pi R_L}$$

$$V = \frac{I \text{ ripple}}{I_{DC}} = \frac{R_L}{3\sqrt{2}\omega_L} = \frac{100}{3 \times 1.414 \times 2} = 0.074$$

2. Capacitor filter :

The ripple voltage for a capacitor filter is of Triangular waveform approximately. The rms value for Triangular wave is $\frac{E_d}{2\sqrt{3}}$ or $\frac{V_{rms}}{2\sqrt{3}}$.

$$\therefore \quad E_{rms} = \frac{E_d}{2\sqrt{3}}$$

Suppose the discharging of the capacitor will cut one for one full half cycle .then the change last by the capacitor

$$Q = \omega t = I_{DC} \times \frac{1}{2f}.$$

$$\therefore \quad \text{voltage} = \frac{Q}{c}$$

$$\therefore \quad E_d = \frac{I_{DC}}{2fc}$$

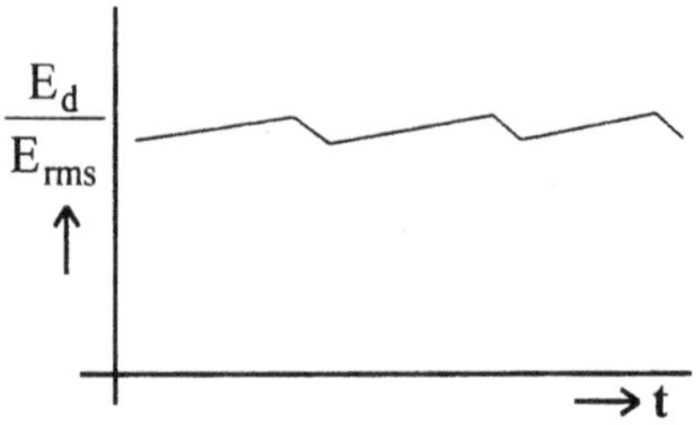

Fig 1.27 V_0 of capacitor filter.

$$\gamma = \frac{I_{DC}}{2fc \times 2\sqrt{3} \times I_{DC}} = \frac{1}{4fc\sqrt{3}R_L} = \frac{1}{4 \times \sqrt{3} \times 50 \times 16 \times 10^{-6} \times 2000} = 0.09$$

3. L Type filter :

$$\gamma = \frac{\sqrt{2}}{3} \times \frac{1}{4\omega^2 LC} = 0.0037$$

Problem 1.19

A diode whose internal resistance is 20 Ω is to supply power to a 1000Ω load from a 110V ((rms) source of supply. Calculate **(a)** The peak load current. **(b)** The DC load current **(c)** AC Load Current **(d)** The DC diode voltage. **(e)** The total input power to the circuit. **(f)** % regulation from no load to the given load.

Solution

Since only one diode is being used, it is for HWR Circuit.

(a) $$I_m = \frac{V_m}{R_f + R_L} = \frac{110\sqrt{2}}{1020} = 152.5\text{mA}.$$

(b) $$I_{DC} = \frac{I_m}{\pi} = \frac{152.5}{\pi} = 48.5\text{mA}.$$

(c) $$I_{rms} = \frac{I_m}{2} = \frac{152.5}{2} = 76.2\text{mA}.$$

(d) $$V_{dc} = \frac{-I_m.R_L}{\pi} = -48.5 \times 1 = -48.5\text{V}.$$

(e) $$P_i = I_{rms} \times (R_f + R_L) = 5.92\Omega.$$

(f) $$\% \text{ regulation} = \frac{V_{NL} - V_{FL}}{V_{FL}} \times 100\% = \frac{\frac{V_m}{\pi} - I_{DC}.R_L}{I_{DC}.R_L} = 2.06\%$$

Problem 1.20

Show that the maximum DC output power $P_{DC} = V_{DC}\, I_{DC}$ in a half wave single phase circuit occurs when the load resistance equals the diode resistance R_f.

Solution

$$P_{DC} = I_{DC}^2.R_L = \frac{V_m^2.R_L}{\pi^2(R_F + R_L)}$$

when R_L is very large, $V_{DC} \cong \frac{V_m}{\pi}$; $I_{DC} = \frac{V_{DC}}{R_L} = \frac{V_m}{\pi.R_L}$

for P_{DC} to be maximum, $\frac{dp_{DC}}{dR_L} = 0.$

or $$\frac{V_m^2}{\pi^2}\left(\frac{(R_F + R_L)^2 - (R_F + R_L).R_L}{(R + R_L)^4}\right) = 0.$$

$$R_F + R_L = 2R_L \text{ or } R_L = R_F.$$

Therefore, P_{DC} is maximum when R_L is equal to R_F.

Problem 1.21

A 1mA meter whose resistance is 10Ω is calibrated to read rms volts when used in a bridge circuit with semiconductor diodes. The effective resistance of each element may be considered to be zero in the forward direction and ∞ in the reverse direction. The sinusoidal input voltage is applied in series with a 5 - KΩ resistance. What is the full scale reading of this meter?

Solution

$$I_{DC} = \frac{2I_m}{\pi} \text{ for bridge rectifier circuit.}$$

$$I_m = \frac{V_m}{R_L};$$

$$V_m = \sqrt{2}V_{rms}$$

$$\therefore \quad I_{DC} = \frac{2\sqrt{2} \times V_{rms}}{\pi R_L};$$

$$R_L = 5K\Omega + 10\,\Omega = (5000 + 10)\Omega = 5010\,\Omega$$

$$1mA = \frac{2\sqrt{2} \times V_{rms}}{\pi \times 5010}$$

$$\therefore \quad V_{rms} = \frac{1 \times 10^{-3} \times \pi \times 5010}{2\sqrt{2}} = 5.56V.$$

1.21 L-SECTION FILTER

In an inductor filter, ripple decreases with increase in R_L. In a capacitor filter, ripple increases with increase in R_L. A combination of these two into a choke ***input or L-Section Filter*** should then make the ripple independent of load resistance.

If L is small, the capacitor will be charged to V_m, the diodes will be cut off allowing a short pulse of current. As L is increases, the pulses of current are smoothed and made to flow for a larger period, but at reduced amplitude. But for a critical value of inductance L_c, either one diode

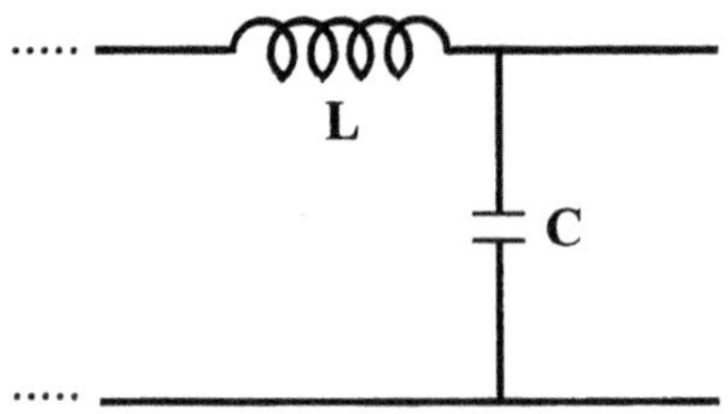

Fig 1.28 (a) Circuit.

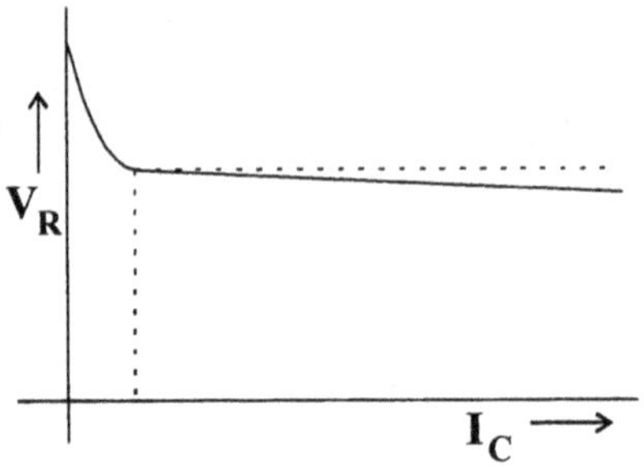

Fig 1.28 (b) LC filter characteristic.

or the other will be always conducting, with the result that the input voltage V and I to the filter are full wave rectified sine waves.

The circuit and graph of DC output voltage for LC Filter is as shown in Fig. 1.28(a) and (b) respectively.

$$V_{DC} = \frac{2V_m}{\pi}$$

for Full Wave Rectifier Circuit. Considering ideal elements, conduction angle is increased when inductor is placed because there is some drop across 'L'. So C cannot charge to V_m. Therefore, the diode will be forward biased for a longer period.

The ripple circuit which passes through 'L', is not allowed to develop much ripple voltage across R_L, if X_c at ripple frequency is small compared to R_L. Because the current will pass through C only. Since X_c is small and Capacitor will get charged to a constant voltage. So V_o across R_L will not vary or ripple will not be there. Since for a properly designed LC Filter,

$$X_c << R_L$$

and $$X_L >> X_c \quad \text{at } \omega = 2\pi f$$

X_L should be greater than X_c because, all the AC should be dropped across X_L itself so that AC Voltage across C is nil and hence ripple is low.

1.21.1 Ripple Factor in LC Filter

AC Component of current through L is determined by X_L.

$$X_L = 2\omega L$$

RMS value of ripple current for Full Wave Rectifier with L Filter is,

$$I'rms = \frac{4V_m}{3\pi\sqrt{2}X_L} = \text{RMS Value of Ripple Current for L Filter}$$

$$I'rms = \frac{2}{3\sqrt{2}X_L} = \frac{2V_m}{\pi}$$

$$\therefore \quad I'rms = \frac{2}{3\sqrt{2}X_L} \cdot V_{DC} = \frac{\sqrt{2}V_{DC}}{3X_L}$$

The ripple voltage in the output is developed by the ripple current flowing through X_C.

$$\therefore \quad V'rms = I'rms \cdot X_C = \frac{\sqrt{2}}{3} \cdot \frac{X_C}{X_L} \cdot V_{DC}$$

$$\text{Ripple factor} = \frac{V'rms}{V_{DC}}$$

$$\gamma = \frac{\sqrt{2}.X_C}{3X_L}$$

$$X_C = \frac{1}{j2\omega fc} \;.\; X_L = jL\omega L = 2\pi fL$$

$$\therefore \quad \gamma = \frac{\sqrt{2}}{3}.\frac{1}{2\omega C}.\frac{1}{2\omega L}$$

$X_L = 2wL$ since ripple is being considered at a frequency, twice the line frequency.

If $\quad f = 60\text{ Hz}, \quad \gamma = \frac{0.83}{LC}$

1.21.2 Bleeder Resistance

For the LC filter, the graph between I_{DC} and V_{DC} is as shown in Fig. 1.29. For light loads, i.e., when the load current is small, the capacitor gets charged to the peak value and so the no load voltage is $2V_m/\pi$. As the load resistance is decreased, I_{DC} increases so the drop across other elements V_{iz} diodes and choke increases and so the average voltage across the capacitor will be less than the peak value of $2V_m/\pi$. The output voltage remains constant beyond a certain point I_B and so the regulation will be good. The voltage V_{DC} remains constant even if I_{DC} increases, because, the capacitor gets charged every time to a value just below the peak voltage, even though the drop across diode and choke increases. So for currents above I_B, the filter acts more like an inductor filter than C filter and so the regulation is good. Therefore, for LC filters, the load is chosen such that, the $I_{DC} \geq I_B$. The corresponding resistance is known as ***Bleeder Resistance*** (R_B). So Bleeder Resistance is the value for which $I_{DC} \geq I_B$ and good regulation is obtained, and conduction angle is 180^0. When $R_L = R_B$, the conduction angle = 180^0, and the current is continuous. So just as we have determined the critical inductance L_C, when $R_L = R_B$, the Bleeder Resistance, I_{DC} = the negative peak of the second harmonic term.

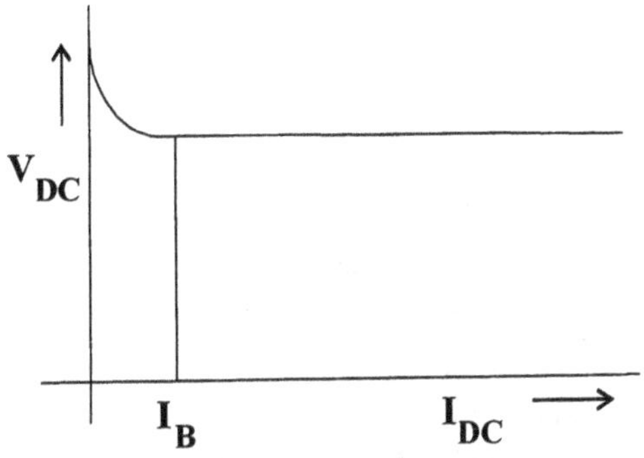

Fig 1.29 LC filter characteristic.

$$\therefore \quad I_{DC} = \frac{2V_m}{\pi R_B}$$

$$I'\text{ peak} = \frac{4V_m}{3\pi}.\frac{1}{XL}$$

$$\therefore \quad \frac{2V_m}{\pi R_B} = \frac{4V_m}{3\pi}.\frac{1}{X_L}$$

$$R_B = \frac{3.X_L}{2};$$ R_B is Bleeder Resistance.

I_{DC} should be equal to or greater than 'I' peak, since, 'X_L' determines the peak value of the ripple component. If 'X_L' is large 'I' peak can be $\leq I_{DC}$ and so the ripple is negligible or pure D.C. is obtained, where $X_L = 2\omega L$ corresponding to the second harmonic. Therefore, R should be at least equal to this value of R_B, the Bleeder Resistor.

1.21.3 Swinging Choke

The value of the inductance in the LC circuits should be > L_C the minimum value so that conduction angle of the diodes is 180^0 and ripple is reduced. But if the current I_{DC} is large, then the inductance for air cored inductors as $L = N\phi/I$ (flux linkages per ampere of current), as 'I' increases, 'L' decreases and may become below the critical inductance value L_C. Therefore, Iron cored inductors or chokes are chosen for filters such that the value of L varies within certain limits and when 'I' is large, the core saturates, and the inductance value will not be < L_C. Such chokes, whose inductance varies with current within permissible limits, are called ***Swinging Chokes***. (In general for inductors ϕ/I remains constant so that L is constant for any value of current I) This can be avoided by choosing very large value of L so that, even if current is large, Inductance is large enough > L_C But this increases the cost of the choke. Therefore, swinging choke are used.

$$L_C \geq R_L / 3\omega$$

R_L = Load Resistance

L_C = Critical Inductance

$\omega = 2\pi f$

At no load $R_L = \infty$. Therefore, L should be ∞ which is not possible. Therefore, Bleeder Resistance of value = $3X_L / 2$ is connected in parallel with R_L, so that even when R_L is ∞, the conduction angle is 180^0 for each diode.

The inductance of an iron - cored inductor depends up on the D.C. current flowing through it, L is high at low currents and low at high currents. Thus its L varies or swings within certain limits. This is known as ***Swinging Choke***.

Typical values are

$$L = 30 \text{ Henrys at } I = 20 \text{ mA and}$$

$$L = 4 \text{ Henrys at } I = 100 \text{ mA.}$$

Problem 1.22

Design a full wave rectifier with an LC filter (single section) to provide 9V DC at 100mA with a maximum ripple of 2%. Line frequency f = 60 HZ.

Solution

$$\text{Ripple factor } \gamma = \frac{\sqrt{3}}{2} \times \frac{1}{2\omega C} \times \frac{1}{2\omega L} = \frac{0.83}{LC}\ \mu$$

$$\therefore \quad 0.02 = \frac{0.83}{LC} \quad \text{or} \quad LC = \frac{0.83}{0.02} = 42.\mu$$

$$R_L = \frac{V_{DC}}{I_{DC}} = \frac{9V}{0.1} = 90\Omega$$

$$LC \geq \frac{R_L}{3\omega} \geq \frac{R_L}{1130}$$

But LC should be 25% larger. ∴ for f = 60 Hz,the value of LC should be $\geq \frac{R_L}{900}$.

$$LC \geq \frac{R_L}{900} \geq \frac{90}{900} = 0.1 \text{ Henry.}$$

If L = 0.1H, then $C = \frac{42}{0.1} = 420\mu f$. This is high value

If L = 1H, then $C = 42\ \mu f$.

If the series resistance of L is assumed to be 50 Ω, the drop across L is

$$I_{DC} \times R = 0.1 \times 50 = 5V.$$

Transformer Rating :

$$V_{DC} = 9V + 5V = 14V$$

$$\therefore \quad V_m = \frac{\pi}{2}(9+5) = 22V$$

$$\therefore \quad \text{rms value is } \frac{22}{\sqrt{2}} = 15.5V$$

Therefore, a 15.5 – 0 –15.5 V, 100mA transformer is required. PIV of the diodes is $2V_m$ Because it is FWR Circuit.

$$\therefore \quad PIV = 44V$$

So, diodes with 44V, 100mA ratings are required.

Problem 1.23

In a FWR with C filter circuit, V_rp-p = 0.8v and the maximum voltage is 8.8 volts (Fig. 1.30). $R_L = 100\Omega$ and $C = 1050\ \mu f$. Power line frequency = 60 Hz. Determine the Ripple Factor and DC output voltage from the graph and compare with calculated values.

Solution

$$V_r' \text{ p-p} = 0.8V.$$

$$\therefore \quad V_{rms}' = \frac{0.8}{2\sqrt{3}} = 0.231V$$

$$V_{DC} = Vm - \frac{V_r'p-p}{2} = 8.8 - \frac{0.8}{2} = 8.4V.$$

$$\therefore \quad \gamma = \frac{V'rms}{V_{DC}} = \frac{0.2}{8.4} = 0.0238 \text{ or } 2.38\%$$

Theoretical values, $\gamma = \frac{1}{4\sqrt{3}fR_L \cdot C}$

$$= \frac{1}{4\sqrt{3} \times 60 \times 100 \times 1050\mu f}$$

$= 0.057$ or 5.75%

$$V_{DC} = \frac{4fR_L C}{1 + 4fR_L C} \times V_m = 8.46V$$

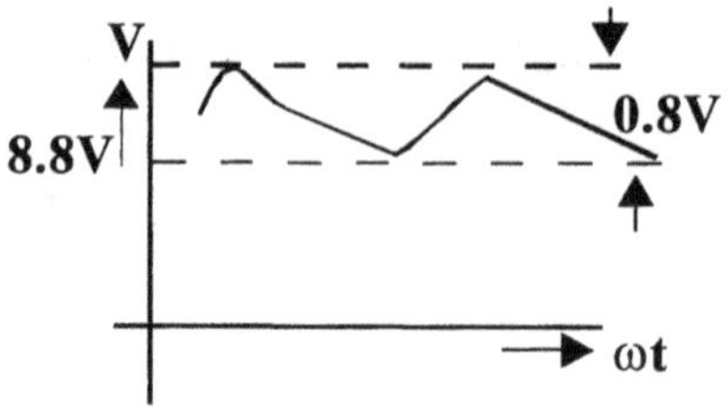

Fig 1.30 For problem 23

1.22 π-SECTION FILTER

In the LC filter or L section filter, there is some voltage drop across L. If this cannot be tolerated and more D.C. output voltage V_0 is desired, pi or π filter or CLC filter is to be used. The ripple factor will be the same as that of L section filter, but the regulation will be poor. It can be regarded as a L section filter with L_1 and C_1, before which there is a capacitor C, i.e. the capacitor filter. The input to the L section filter is the output of the capacitor filter C. The output of capacitor filter will be a triangular wave superimposed over D.C.

Now the output V_0 is the voltage across the input capacitor C less by the drop across L_1. The ripple contained in the output of 'C' filter is reduced by the L section filter L_1C_1.

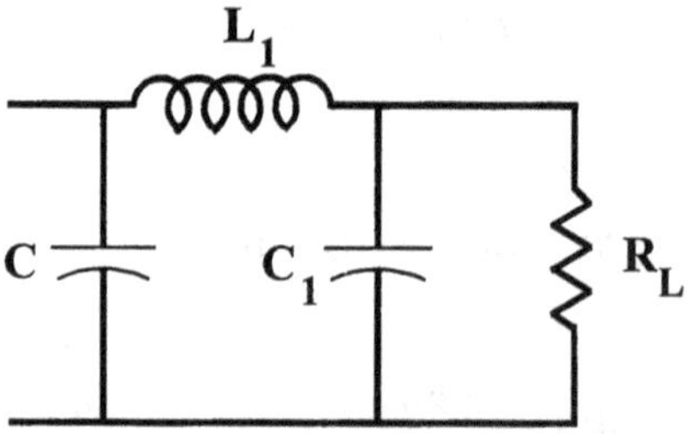

Fig 1.31 CLC or π section filter

1.22.1 π-Section Filter with a Resistor Replacing the Inductor

Consider the circuit shown here. It is a π filter with L replaced by R. If the resistor R is chosen equal to the reactance of L (the impedance in Ω should be the same), the ripple remains unchanged.

V_{DC} for a single capacitor filter = ($V_m - I_{DC}/4fC$) If you consider C_1 and R, the output across C_1 is ($V_m - I_{DC}/4fC$). There is some drop across R. Therefore, the net output is

$$V_m - \frac{I_{DC}}{4fC} - I_{DC} \times R$$

The Ripple Factor for π sector is

$$\gamma = \frac{\sqrt{2}X_c X_{c_1}}{R_L . X_{L_1}} \qquad \text{.......... (1.34)}$$

So by this, saving in the cost, weight and space of the choke are made . But this is practical only for low current power supplies.

Suppose a FWR output current is 100 mA and a 20 H choke is being used in π section filter. If this curve is to be replaced by a resistor, $X_L = R$. Taking the ripple frequency as 2f = 100 Hz,

$$X_L = \omega L = 2\pi(2f).L = 4\pi fL$$
$$= 4 \times 3.14 \times 50 \times 20 = 12560\Omega$$
$$\simeq 12\ K\Omega$$

Voltage drop across R = 12,000 × 0.1 = 1200V, which is very large.

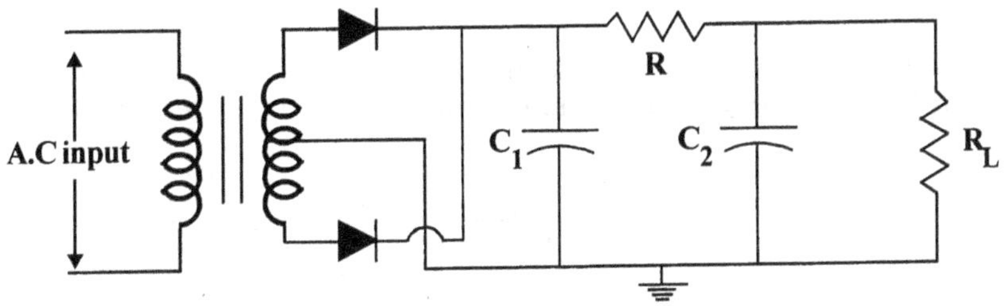

Fig 1.32 **FWR circuit π section / CRC filter.**

1.22.2 Expression for Ripple Factor of π-Section Filter

The output of the capacitor in the case of a capacitor filter is a Triangular wave. So assuming the output V_o across C to be a Triangular wave, it can be represented by Fourier series as

$$V = V_{DC} - \frac{V'P-P}{\pi}(\text{Sin } 2\omega t - \frac{\sin 4\omega t}{2} + \frac{\sin 6\omega t}{3} ...)$$

But the ripple voltage peak to peak $V'\ p\text{-}p = \frac{I_{DC}}{2fc}$. (for Δ wave)

Neglecting 4th and higher harmonic, the rms voltage of the Second Harmonic Ripple is

$$\frac{v'p-p}{\pi\sqrt{2}} = \frac{I_{DC}}{2\sqrt{2}\pi fc} = \frac{I_{DC} \times 2}{4\sqrt{2}\pi fc} = \frac{\sqrt{2}I_{DC}}{4\pi fc}$$

$$\frac{1}{4\pi fc} = X_c$$

It is the capacitive reactance corresponding to the second harmonic.

$$V'_{rms} = \sqrt{2}I_{DC}X_C \qquad \text{(across 'C' filter only)}$$

The output of C filter is the input for the L section filter of L_1C_1 . Therefore, as we have done in the case of L section filter,

Current through the inductor (harmonic component) = $\frac{\sqrt{2}I_{DC}X_C}{X_{L_1}}$

Output Voltage = Current (X_C)   ($V_o = I.X_C$)

$$V'_{rms} = \frac{\sqrt{2}I_{DC}X_C}{X_{L_1}}.X_{C_1} \qquad \text{(after L section)}$$

Ripple factor, $\gamma = \frac{v'_{rms}}{V_{DC}} = \frac{\left(\sqrt{2}.I_{DC}.X_C\right)}{V_{DC}} \frac{X_{C_1}}{X_{L_1}}$

$$V_{DC} = I_{DC}.R_L$$

$$\therefore \quad \gamma = \frac{\left(\sqrt{2}.X_C\right)}{R_L}\left(\frac{X_{C_1}}{X_{L_1}}\right)$$

where all reactances are calculated at the 2nd harmonic frequency $\omega = 2\pi f$.

DC output voltage = (The DC voltage for a capacitor filter) – (The drop across L_1).

Problem 1.24

Design a power supply using a π-filter to give DC output of 25V at 100 mA with a ripple factor not to exceed 0.01%. Design of the circuit means, we have to determine L,C, diodes and transformers.

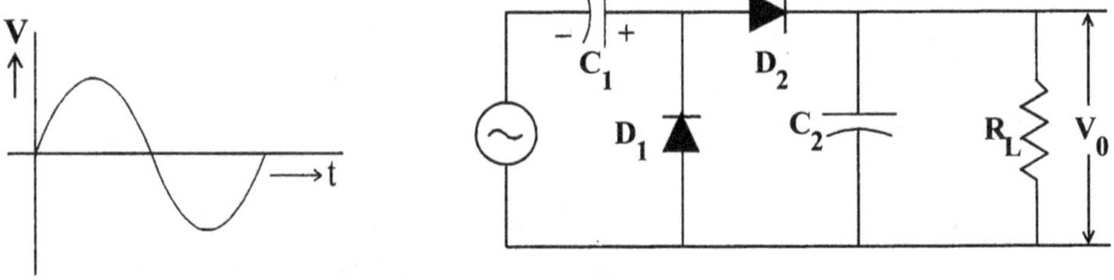

Fig 1.33 Peak to peak detector.

Solution

Design of the circuit means, we have to determine L, C, Diodes and Transformer

$$R_L = \frac{V_{DC}}{I_{DC}} = \frac{25V}{I_{DC}} = \frac{25V}{100\ mA} = 250\Omega$$

Ripple factor $\gamma = \sqrt{2}.\frac{X_c}{R_L}.\frac{X_{C1}}{X_{L1}}$

X_C can be chosen to be = Xc_1

$$\therefore \quad \gamma = \sqrt{2}.\frac{X_C^{\ 2}}{R_L.X_{L1}}$$

This gives a relation between C and L.

$$C^2 L = y$$

There is no unique solution to this.

Assume a reasonable value of L which is commercially available and determine the corresponding value of capacitor. Suppose L is chosen as 20 H at 100 mA with a D.C. Resistance of 375Ω (of Inductor).

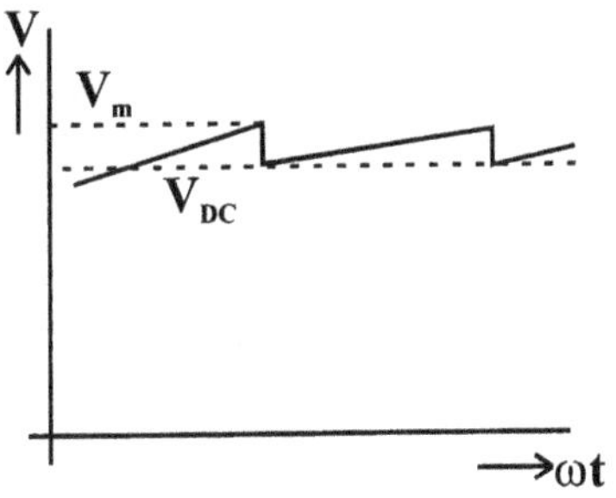

Fig 1.34 For problem 1.7.

$$\therefore \quad c^2 = \frac{y}{L}$$

or $$c = \sqrt{\frac{y}{L}}$$

$$V_{DC} = Vm - \frac{V_\gamma}{2}$$

$$V_\gamma = \frac{I_{DC}}{2fc}$$

Now the transformer voltage ratings are to be chosen.

The voltage drop across the choke = choke resistance × $I_{DC} = 375 \times 100 \times 10^{-3} = 37.5V$.

$$V_{DC} = 25V.$$

Therefore, voltage across the first capacitor C in the π-filter is

$$V_c = 25 + 37.5 = 62.5V.$$

The peak transformer voltage, to centre tap is

$$V_m = (V_C) + \frac{V_\gamma}{2} \text{ (for C filter)}$$

$$V_\gamma = \frac{I_{DC}}{2fc}$$

$$\therefore \quad V_m = 62.5 \text{ V} + \frac{0.1}{4 \times 50 \times c}$$

$$V_{rms} = \frac{V_m}{\sqrt{2}} \cong 60v$$

Therefore, a transformer with 60-0-60V is chosen . The ratings of the diode should be, current of 125 mA and voltage = PIV = $2V_m = 2 \times 84.6V = 169.2V$.

Problem 1.25

A full wave rectifier with LC filter is to supply 250 V at 100 mA. D.C. Determine the ratings of the needed diodes and transformer, the value of the bleeder resistor and the ripple, if R_C of the choke = 400Ω. L = 10H and C = 20 μF.

Solution

$$R_L = \frac{V_{DC}}{I_{DC}}$$

$$R_L = \frac{250v}{0.1} = 2{,}500\Omega.$$

For the choke input resistor,

$$E_{DC} = \frac{2E_m}{\pi(1+\frac{R_C}{R_L})}$$

and $I_{rms} \cong I_{DC}$

$$E_m = \frac{\pi E_{DC}}{2}(1+\frac{R_C}{R_L}) = \frac{\pi\times 250}{2}(1+\frac{400}{2500}) = 455V$$

$$\therefore \quad E_{rms} = \frac{455}{\sqrt{2}} = 322V$$

Therefore, the transformer should supply 322V rms on each side of the centre tap. This includes no allowance for transformer impedance, so that the transformer should be rated at about 340 volts 100 mA, D.C

The Bleeder Resistance $R_B = \frac{3X_L}{2}$

$$L = 10000\Omega.$$

$$I_B = \frac{2E_m}{3\pi\omega L} = \frac{2\times 455}{3\pi\times 377\times 10} = 0.0256A$$

Ripple factor $\gamma = \frac{0.47}{4\omega^2 LC-1} = \frac{0.47}{4\times 377^2\times 20\times 10^{-6}}$

The current ratings of each diode = 0.00413. Should be 50mA.

1.23 USE OF ZENER DIODE AS REGULATOR

Voltage Regulator Circuits are electronic circuits which give constant DC output voltage, irrespective of variations in *Input Voltage* V_i, *current drawn by the load* I_L from output terminals, and *Temperature T*. Voltage Regulator circuits are available in discrete form using BJTs, Diodes etc and in IC (Integrated Circuit) form. The term voltage regulator is used when the output delivered is DC voltage. The input can be DC which is not constant and fluctuating. If the input is AC, it is converted to DC by Rectifier and Filter Circuits and given to I.C. Voltage Regulator circuit, to get constant DC output voltage. If the input is A.C 230 V from mains, and the output desired is constant DC, a stepdown transformer is used and then Rectifier and filter circuits are used, before the electronic regulator circuit. The block diagrams are shown in Fig. 1.35 and 1.36.

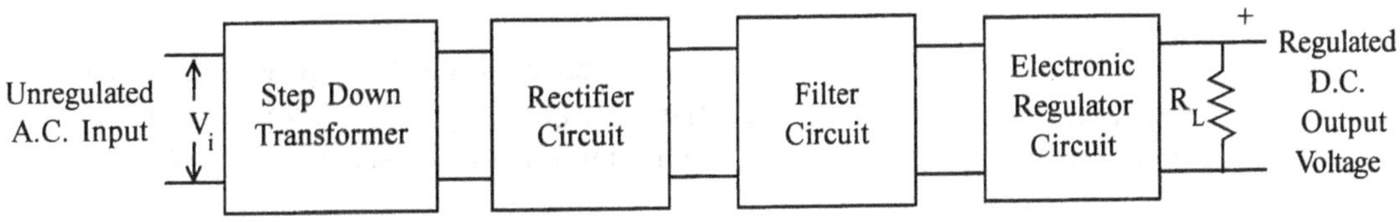

Fig. 1.35 Block diagram of voltage regulator with A.C input.

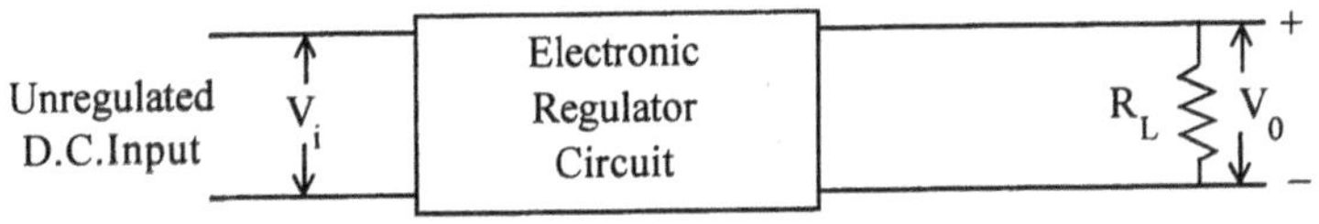

Fig. 1.36 Block diagram of voltage regulator with D.C input.

The term *Voltage Stabilizer* is used, if the output voltage is AC and not DC. The circuits used for voltage stabilizers are different. The voltage regulator circuits are available in IC form also. Some of the commonly used ICs are, μA 723, LM 309, LM 105, CA 3085 A.

7805, 7806, 7808, 7812, 7815 : Three terminal positive Voltage Regulators.

7905, 7906, 7908, 7912, 7915 : Three terminal negative Voltage Regulators.

The Voltage Regulator Circuits are used for electronic systems, electronic circuits, IC circuits, etc.

The specifications and Ideal Values of Voltage Regulators are :

Specifications		Ideal Values
1. Regulation (S_V)	:	0 %
2. Input Resistance (R_i)	:	∞ ohms
3. Output Resistance (R_0)	:	0 ohms
4. Temperature Coefficient (S_T)	:	0 mv/oc.
5. Output Voltage V_0	:	-
6. Output current range (I_L)	:	-
7. Ripple Rejection	:	0 %

Different types of Voltage Regulators are

1. Zener regulator
2. Shunt regulator
3. Series regulator
4. Negative voltage regulator
5. Voltage regulator with foldback current limiting
6. Switching regulators
7. High Current regulator

1.23.1 Zener Voltage Regulator Circuit

A simple circuit without using any transistor is with a zener diode Voltage Regulator Circuit. In the reverse characteristic voltage remains constant irrespective of the current that is flowing through Zener diode. The voltage in the break down region remains constant (Fig. 1.37).

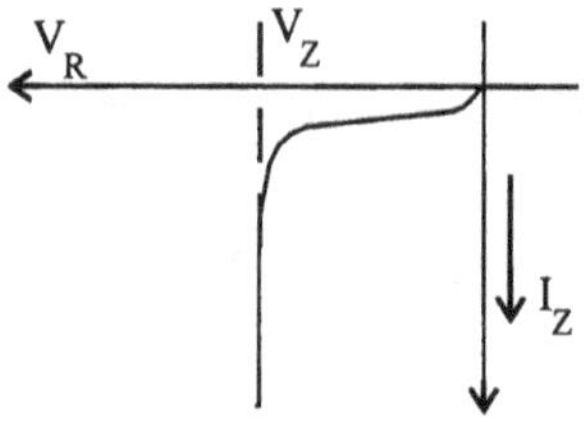

Fig. 1.37 Zener diode reverse characteristic.

Therefore in this region the zener diode can be used as a voltage regulator. If the output voltage is taken across the zener, even if the input voltage increases, the output voltage remains constant. The circuit as shown in Fig. 1.38.

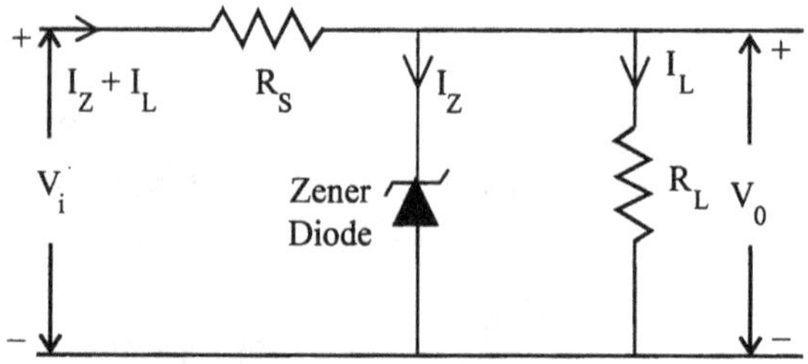

Fig. 1.38 Zener regulator circuit.

The input V_i is DC. Zener diode is reverse biased.

If the input voltage V_i increases, the current through R_s increases. This extra current flows, through the zener diode and not through R_L. Therefore zener diode resistance is much smaller than R_L when it is conducting. Therefore I_L remains constant and so V_0 remains constant.

The limitations of this circuits are

1. The output voltage remains constant only when the input voltage is sufficiently large so that the voltage across the zener is V_Z.
2. There is limit to the maximum current that we can pass through the zener. If V_i is increased enormously, I_Z increases and hence breakdown will occur.
3. Voltage regulation is maintained only between these limits, the minimum current and the maximum permissible current through the zener diode. Typical values are from 10 mA to 1 ampere.

1.23.2 Shunt Regulator

The shunt regulator uses a transistor to amplify the zener diode current and thus extending the Zener's current range by a factor equal to transistor h_{FE}. (Fig. 1.39)

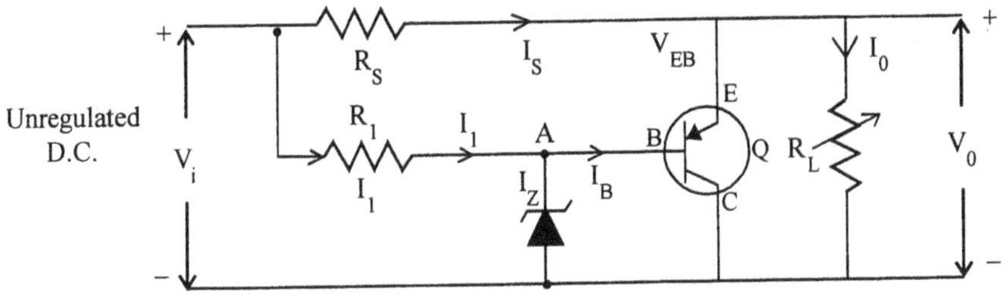

Fig. 1.39 Shunt regulator circuit

Zener current, passes through R_1

Nominal output voltage

$$= V_Z + V_{EB}$$

The current that gets branched as I_B is amplified by the transistor. Therefore the total current $I_0 = (\beta + 1)\, I_B$, flows through the load resistance R_L. Therefore for a small current through the zener, large current flows through R_L and voltage remains constant. In otherwords, for large current through R_L, V_0 remains constant. Voltage V_0 does not change with current.

Problem 1.26

For the shunt regulator shown, determine

1. The nominal voltage
2. Value of R_1,
3. Load current range
4. Maximum transistor power dissipation.
5. The value of R_S and its power dissipation.

V_i = Constant. Zener diode 6.3V, 200mW, requires 5mA minimum current.

Transistor Specifications :

$$V_{EB} = 0.2V,\ h_{FE} = 49,\ I_{CBO} = 0.$$

1. The nominal output voltage is the sum of the transistor V_{EB} and zener voltage.

$$V_0 = 0.2 + 6.3 = 6.5V = V_{Eb} + V_Z$$

2. R_1 must supply 5mA to the zener diode :

$$\therefore \quad R_1 = \frac{8V - 6.3}{5 \times 10^{-3}} = \frac{1.7}{5 \times 10^{-3}} = 340\ \Omega$$

3. The maximum allowable zener current is

$$\frac{\text{Power rating}}{\text{Voltage rating}} = \frac{0.2}{6.3} = 31.8\ \text{mA}$$

The load current range is the difference between minimum and maximum current through the shunt path provided by the transistor. At junction A, we can write,

$$I_B = I_Z - I_1$$

I_1 is constant at 5 mA

$$\therefore \quad I_B = I_Z - I_1$$

I_1 is constant at 5 mA

$$\therefore \quad I_B = 5 \times 10^{-3} - 5 \times 10^{-3} = 0$$

$$I_B\,(\text{min}) = I_{Z\,Max} - I_2$$
$$= 31.8 \times 10^{-3} - 5 \times 10^{-3} = 26.8 \text{ mA}$$

The transistor emitter current $I_E = I_B + I_C$

$$I_C = \beta\, I_B = h_{FE}\, I_B$$

$$\therefore \quad I_E = (\beta + 1)\, I_B = (h_{FE} + 1)\, I_B$$

I_B ranges from a minimum of 0 to maximum of 26.8 mA

$\therefore$ Total load current range is $(h_{FE} + 1)\, I_B$

$$= 50\,(26.8 \times 10^{-3}) = \mathbf{1.34\ A}$$

4. The maximum transistor power dissipation occurs when the current is maximum $I_E \simeq I_C$

$$P_D = V_o\, I_E = 6.5\,(1.34) = \mathbf{8.7\ W}$$

5. R_S must pass 1.34 A to supply current to the transistor and R_L.

$$R_S = \frac{V_i - V_0}{1.34} = \frac{8 - 6.5}{1.34} = 1.12\ \Omega$$

The power dissipated by R_S,

$$= I_S^2\, R_S$$
$$= (1.34)^2 \,.\, (1.12) = 3\text{W}$$

REGULATED POWER SUPPLY

An unregulated power supply consists of a transformer, a rectifier, and a filter. For such a circuit regulation will be very poor i.e. as the load varies (*load means load current*) [*No load means no load current or 0 current. Full load means full load current or short circuit*], we want the output voltage to remain constant. But this will not be so for unregulated power supply. The short comings of the circuits are :

1. Poor regulation
2. DC output voltage varies directly as the a.c. input voltage varies
3. In simple rectifiers and filter circuits, the d.c. output voltage varies with temperature also, if semiconductors devices are used.

An electronic feedback control circuit is used in conjuction with an unregulated power supply to overcome the above three short comings. Such a system is called a "*regulated power supply*".

Stabilization

The output voltage depends upon the following factors in a power supply.

1. Input voltage V_i
2. Load current I_L
3. Temperature

$\therefore$ Change in the output voltage ΔV_0 can be expressed as

$$\Delta V_0 = \frac{\partial V_0}{\partial V_i}.\Delta V_i + \frac{\partial V_0}{\partial I_L}.\Delta I_L + \frac{\partial V_0}{\partial T}.\Delta T$$

$$\Delta V_0 = S_1\, \Delta V_i + R_0\, \Delta I_L + S_T\, \Delta T$$

Where the three coefficients are defined as,

(i) **Stability factors.**

$$S_V = \frac{\Delta V_0}{\Delta V_i}\Bigg|_{\Delta I_L = 0,\ \Delta_T = 0}$$

This should be as small as possible. Ideally 0 since V_0 should not change even if V_i changes.

(ii) ***Output Resistance***

$$R_0 = \frac{\Delta V_0}{\Delta I_L}\Bigg|_{\Delta V_i = 0,\ \Delta_T = 0}$$

(iii) ***Temperature Coefficient***

$$S_T = \frac{\Delta V_0}{\Delta_T}\Bigg|_{\Delta V_i = 0,\ \Delta I_L = 0}$$

The smaller the values of the three coefficients, the better the circuit.

SUMMARY

- Energy possessed by an electron rotating in an orbit with radius r;

$$E = -\frac{e^2}{8\pi \epsilon_0 r}$$

- Expression for the radius of orbit, $r = \frac{n^2 h^2 \epsilon_0}{\pi m e^2}$
- Expression for wavelength of emitted radiation, $\lambda = \frac{12,400}{(E_2 - E_1)}$
- Types of Electronic Emissions from the surface.
  1. Thermionic Emission
  2. Secondary Emission
  3. Photo electric Emission
  4. High Field Emission
- Expression for Threshold Frequency f_T in photoelectric emission,

$$f_T = \frac{e\phi}{h}$$

- Fermi Level lies close to E_C in *n-type* semiconductor
- Fermi Level lies close to E_V in *p-type* semiconductor
- $\sigma = ne\mu_n + pe\mu_p$
- Hall Voltage, $V_H = \frac{BI}{\rho\omega}$
- Expression for current through a *p-n junction.*

$$\text{Diode is } I = I_O \left(e^{\frac{V}{\eta V_T}} - 1 \right)$$

- The three types of breakdown mechanisms in semiconductor diodes are
  1. Avalanche Breakdown
  2. Zener Breakdown
  3. Thermal Breakdown.
- Rectifier circuit converts AC to Unidirectional Flow.
- Filter Circuit converts Unidirectional flow to DC. It minimises Ripple.
- The different types of filter circuits are
  1. Capacitor Filter
  2. Inductor Filter
  3. L-Section (LC) Filter.
  4. π-Section Filter
  5. CRC and CLC Filters

- Ripple Factor $= \frac{I'_{rms}}{I_{DC}}$. For Half Wave Rectifier, $\gamma = 1.21$, for Full Wave Rectifier $\gamma = 0.482$
- Expression for Ripple Factor for C Filter,

$$\gamma = \frac{1}{4\sqrt{3} f CR_L}$$

- Expression for Ripple Factor for L Filter,

$$\gamma = \frac{R_L}{4\sqrt{3}\omega L}$$

- Expression for Ripple Factor for LC Filter,

$$\gamma = \frac{\sqrt{2}}{3} \times \frac{1}{2\omega C} \times \frac{1}{2\omega L}$$

- Expression for Ripple Factor for π Filter,

$$\gamma = \frac{\sqrt{2} X_C}{R_L}\left(\frac{X_{C_1}}{X_{L_1}}\right)$$

- Critical Inductance $L_C \geq \frac{R_L}{3\omega}$
- Bleeder Resistance $R_B = \frac{3X_L}{2}$

OBJECTIVE TYPE QUESTIONS

1. Free electron concentration in semiconductors is of the order of ....
2. Insulators will have resistivity of the order of ....
3. Expression for J in the case of a semiconductor with concentrations n and p is ....
4. According to Law of Mass Action in semiconductors, ...
5. Intrinsic concentration depends on temperature T as ....
6. Depletion region width varies with reverse bias voltages as ....
7. In *p-type* semiconductor, Fermi Level lies close to ....
8. In *n-type* semiconductor, Fermi Level lies close to ....
9. The rate at with I_o changes with temperature in Silicon diode is ....
10. Value of Volt equivalent of Temperature at 25^oC is ....
11. Einstein's relationship in semiconductors is ....
12. A very poor conductor of electricity is called ....

13. When donor impurities are added allowable energy levels are introduced a little .... the .... band.
14. Under thermal equilibrium, the product of the free negative and positive concentration is a constant independent of the amount of .... doping and this relationship, called the mass action low and is given by np = ...
15. The unneutralized ions in the neighbourhood of the junction are referred to as .... charges and this region depleted of mobile charges is called .... charge region.
16. General expression for E_0, the contact difference of potential in an open circuited *p-n junction* in terms of N_C, N_A and n_i is ....
17. Typical value of E_o = ...eV
18. The equation governing the law of the junction is ....
19. The expression for the current in a forward biased diode is ....
20. The value of cut in voltage in the case of Germanium diode .... and Silicon Diode .... at room temperature.
21. Expression for V_B the barrier potential in terms of depletion region width W is V_B = ...
22. Expression for I_o in terms of temperature T and V_T is ....
23. Zener breakdown mechanism needs relatively .... electric field compared to Avalanche Breakdown.
24. Rectifier Converts .... to ....
25. Filter Circuit Converts .... to ....
26. Ripple Factor in the case of Full Wave Rectifier Circuit is ....
27. The characteristics of a swinging choke is ....
28. In the case of LC Filter Circuits, Bleeder Resistance ensures
    (1) ....
    (2) ....

ESSAY TYPE QUESTIONS

1. Explain the concept of 'hole'. How *n-type* and *p-type* semiconductors are formed? Explain.
2. Derive the expression for E_G in the case of intrinsic semiconductor.
3. Derive the expression for E_G in the case of of *p-type* and *n-type* semiconductors.
4. With the help of necessary equations, Explain the terms Drift Current and Diffusion Current.
5. Explain about Hall Effect. Derive the expression for Hall Voltage. What are the applications of Hall Effect?
6. Distinguish between Thermistors and Sensistors.

7. Derive the expression for contact difference of potential V_o in an open circuited p-n junction.
8. Draw the forward and reverse characteristics of a p-n junction diode and explain them qualitatively.
9. Derive the expression for Transistor Capacitance C_T in the case of an abrupt p-n junction.
10. Compare Avalanche, Zener and Thermal Breakdown Mechanisms.
11. Derive the expression for E_o in the case of open circuited *p-n junction* diode.
12. Qualitatively explain the forward and reverse characteristic of *p-n junction* diode.
13. How junction capacitances come into existance in *p-n junction* diode.
14. Distinguish between Avalanche, Zener and Thermal Breakdown Mechanisms
15. Obtain the expression for ripple factor in the case of Full Wave Rectifier Circuit with Capacitor Filter.
16. Explain the terms Swinging Choke and Bleeder Resistor.
17. Compare C, L, L-Section, π-Section (CLC and CRC) Filters in all respects.

MULTIPLE CHOICE QUESTIONS

1. The force of electron 'F' between nucleus and electron with charge 'e' and radius 'r' is, proportional to, F α

(a) $\frac{e^2}{r^2}$ (b) $\frac{r^2}{e^2}$ (c) $e^2 r^2$ (d) $\frac{e}{r}$

2. The energy possesed by the electron, W orbitting round the nucleus is,

(a) $\frac{-e}{8\pi \varepsilon_0 r}$ (b) $\frac{-e^2}{8\pi \varepsilon_0 r}$ (c) $\frac{-e.r}{8\pi \varepsilon_0}$ (d) $\frac{-e^2}{8\pi \varepsilon_0 r^2}$

3. The expression for the Kinetic Energy E of free electron in terms of momentum of the electron 'p' and mass of the electron m is,

(a) $\frac{p}{2m}$ (b) $\frac{p}{2m^2}$ (c) $\frac{p^2}{2m^2}$ (d) $\frac{p^2}{2m}$

4. The expression for radius of stable state 'e' r, is

(a) $\frac{n^2h^2\varepsilon_o}{\pi me^2}$ (b) $\frac{nh\varepsilon_o}{\pi^2 m^2e^2}$ (c) $\frac{n^2 h\varepsilon_o^2}{\pi m e^2}$ (d) $\frac{n^2 h^2 \varepsilon_o^2}{\pi m e^2}$

5. The value of the radius of the lowest state or ground state is, given, h = 6.626 × 10^{-34} J - Secs ε_o = $10^{-9}/36\pi$

(a) 0.28 A° (b) 0.98 A° (c) 0.18 A° (d) 0.58 A°

6. The energy required to detach an electron from its parent atom is called,

(a) Ionization potential (b) Eletric potential
(c) Kinetic energy (d) Threshold potential

7. **Electron collision without transfer of energy in collision is called,**
   (a) Null collision (b) Impact collision
   (c) Elastic collision (d) Stiff collision
8. **Wave mechanics in electron theory is also known as**
   (a) Eienstein theory (b) Quantum mechanics
   (c) Bohr mechanics (d) Classical theory
9. **The expression for threshold frequency f_T to cause photoelectric emission is, with usual notation is,**
   (a) $\frac{e\phi^2}{h}$ (b) $\frac{e\phi}{h}$ (c) $\frac{eh}{\phi}$ (d) $\frac{e^2\phi^2}{h^2}$
10. **The conductivity of a good conductor is typically**
   (a) 10^{-3} Ω/cm (b) 10^3 Ω/cm (c) 10^3 ℧/cm (d) 10 ℧/m
11. **The free electron concentration of a good conductor is of the order of**
   (a) 10^{10} electrons/m3 (b) 10^{28} electron/m^3
   (c) 10^2 electrons/cm^3 (d) 10^{10} electrons/m^3
12. **Fermi level in Intrinsic semiconductor lies**
   (a) close to conduction band (b) close to valence band
   (c) In the middle (d) None of the these
13. **The potential which exists in a p-n junction to cause drift of charge carriers is called**
   (a) contact potential (b) diffusion potential
   (c) ionisation potential (d) threshold potential
14. **The rate of increase of reverse saturation current for Germanium diode is,**
   (a) 5% (b) 4% (c) 1% (d) 7%
15. **Special types of diodes in which transition time and storage time are made small are called ...**
   (a) Snap diodes (b) Rectifier diodes (c) Storage diodes (d) Memory diodes
16. **The Circuit which converts undirectional flow to D.C. is called**
   (a) Rectifier circuit (b) Converter circuit
   (c) filter circuit (d) Eliminator
17. **For ideal Rectifier and filter circuits, % regulations must be ...**
   (a) 1% (b) 0.1 % (c) 5% (d) 0%
18. **The value of current that flows through R_L in a 'π' section filter circuit at no load is**
   (a) ∞ (b) 0.1 mA (c) 0 (d) few mA

CONCEPTUAL QUESTIONS (Interview Questions)

1. When it is said that an electron moves from valence band to conduction band, will there be physical movement of the electron ?

Ans : No, The energy band diagram represents the energy levels of the electrons. When an electron "moves" from valence band to conduction band, it only means that the energy possessed by the electron has increased from Ev electron volts to Ec+ electron volts. The energy diagram is only Pictorial representation of the energy possessed by the electrons. The energy levels of different electrons will be different.

2. What does the arrow in the symbol of a P-n junction diode (—▷|—) indicate?

Ans : The arrow mark A▷|K (p) (n) indicates the direction of current flow when the P - n juction is forward biased.

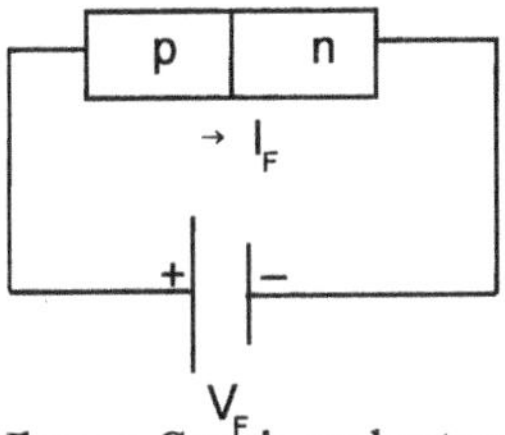

Direction of I_F is the same as that of the direction of arrow.

3. When a Semiconductor device 'breaks down'; what does it mean?

Ans : It does not mean physical breakdown. The shape of the device remains the same. But the device looses its Semiconductor properties. The device may become a short or open circuited.

4. What is the difference between Avalanche and zener breakdown mechanisms?

Ans : In Zener breakdown mechanism, due to sudden increase in the electric field strength ε, when it reaches a critical value, covalent bonds break and due to this there will be sudden increase in the number of free electrons and hence, the reverse current which was small till that time increases suddenly. But in Avalanche breakdown the field strength is not that high. A free electron with lot of energy collides with another electron, transfers its energy and makes it a free electron. This in turn generates some more free electrons. Thus there is multiplication of carriers. This will occur usually in a forward biased device. Avalanche breakdown precedes Zener breakdown, for comparision, though both cannot occur in the same device at a time.

5. How for a low voltage like 6.8 V, in the case of Fz 6.8 Zener doide zener breakdown occurs, which requires high electric field strength?

Ans : Electric field strength $\varepsilon = \frac{V}{d}$. Even if V is small, if 'd' is very small, ε can be very high. This is what happens in a Zener diode. Due to heavy doping depletion, region width near the junction becomes very small. So, when the voltage V_z reaches the required value, electric field 'ε' reaches the critical value, covalent bonds will be broken and Zener breakdown occurs.

Specifications for Silicon Diode

Sl.No.	Parameter	Symbol	Typical Value	Units
1.	Reverse breakdown voltage	V_{br}	75	V
2.	Static Reverse Current	I_R	5	μA
3.	Static Forward voltage (for silicon)	V_F	0.5	V
4.	Total Capacitance	C_T	2	pf
5.	Reverse Recovery Time	t_{rr}	4	ns
6.	Continuous Power Dissipation	P	500	mw
7.	Max. For Ward Current	I_F	50	mA

Some type numbers of Diodes

1. 0A79 : 0 → Semiconductor Device.
   A → Denotes Germanium device.
2. BY 127 : B → Denotes Silicon Device.
   Y → Rectifying Diode.
   Number 127 is owing a type no and has significance.
3. 1N4153 : N → Bipolar Device.
   1 → Single Polar function.
4. BY 100:800V, 1 A Silicon Diode.

Specifications of a junction diode

Sl.No.	Parameter	Symbol	Typical Value	Units
1.	Working Inverse voltage	WIV	80	V
2.	Average Rectifield current	I_U	100	A
3.	Continious forward current	I_F	300	mA
4.	Peak repetetive forward current	i_f	400	mA
5.	Forward Voltage	V_f	0.6	V
6.	Reverse current	I_R	500	nA
7.	Breakdown voltage	BV	100	V

Specifications of a Germanium Tunnel Diode IN 2939

Sl.No.	Parameter	Symbol	Typical Value	Units
1.	Forward current	I_F	5	mA
2.	Reverse current	I_R	10	mA
3.	Peak current	I_p	1	mA
4.	Valley current	I_v	0.1	mA
5.	Peak voltage	V_P	50	V
6.	Valley voltage	V_V	30	V
7.	Ratio of I_p to I_V	I_P/I_V	10	-

2 Transistor and FET Characteristics

In this Chapter,

- The principle of working of Bipolar Junctions Transistors, (which are also simply referred as Transistors) and their characteristics are explained.
- The operation of Transistor in the three configurations, namely Common Emitter, Common Base and Common Collector Configurations is explained.
- The variation of current with voltage, in the three configurations is given.
- The structure of Junction Field Effect Transistor (JFET) and its V-I characteristics are explained.
- The structure of Metal Oxide Semiconductor Field Effect Transistor (MOSFET) and its V-I characteristics are explained.
- JFET amplifier circuits in Common Source (CS) Common Drain (CD) and Common Gate (CG) configurations are given.

2.1 TRANSISTOR CONSTRUCTION

Two commonly employed methods for the fabrication of diodes, transistors and other semiconductor devices are :

1. Grown type **2.** Alloy type

2.1.1 Grown Type

It is made by drawing a single crystal from a melt of Silicon or Germanium whose impurity concentration is changed during the crystal drawing operation.

2.1.2 Alloy Type

A thin wafer of ***n-type*** Germanium is taken. Two small dots of indium are attached to the wafer on both sides. The temperature is raised to a high value where indium melts, dissolves Germanium beneath it and forms a saturator solution. On cooling, indium crystallizes and changes Germanium to ***p-type***. So a PNP transistor formed. The doping concentration depends upon the amount of indium placed and diffusion length on the temperature raised.

2.1.3 Manufacture of Grown Junction Transistor

Grown Junction Transistors are manufactured through growing a single crystal which is slowly pulled from the melt in the crystal growing furnace Fig. 2.1. The purified polycrystalline semiconductor is kept in the chamber and heated in an atmosphere of N_2 and H_2 to prevent oxidation. A seed crystal attached to the vertical shaft which is slowly pulled at the rate of (1mm / 4hrs) or even less. The seed crystal initially makes contact with the molten semiconductors. The crystal starts growing as the seed crystal is pulled. To get ***n-type*** impurities, to the molten polycrystalline solution, impurities are added. Now the crystal that grows is ***n-type*** semiconductor. Then ***p-type*** impurities are added. So the crystal now is ***p-type***. Like this NPN transistor can be fabricated.

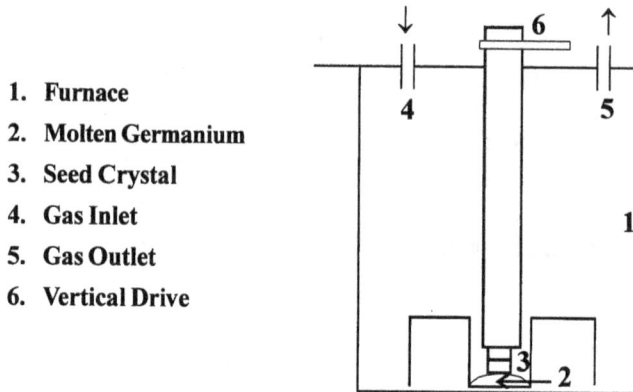

1. Furnace
2. Molten Germanium
3. Seed Crystal
4. Gas Inlet
5. Gas Outlet
6. Vertical Drive

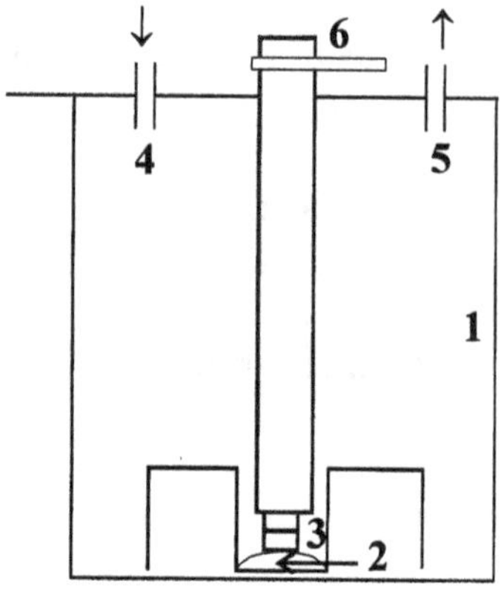

Fig 2.1 Crystal growth.

2.1.4 Diffusion Type of Transistor Construction

To fabricate NPN transistor, ***n-type*** 'Si' (or 'Ge') wafer is taken. This forms the collector. Base-collector junction areas is determined by a mask and diffusion length. ***p-type*** impurity (Boron) is diffused into the wafer by diffusion process. The wafer to be diffused and the Boron wafer which acts as ***p-type*** impurity, are placed side by side. Nitrogen gas will be flowing at a particular rate, before the Boron is converted to Boron Oxide and this gets deposited over Siliconwafer. Baron is driven inside by drive in diffusion process. So the base-collector junction will be formed. Now again ***n-type*** impurity (Phosphorus) is diffused into the Base Area in a similar way. The different steps in process are :

1. *Oxidation*
2. *Photolithography*
3. *Diffusion*
4. *Metallization*
5. *Encapsulation*

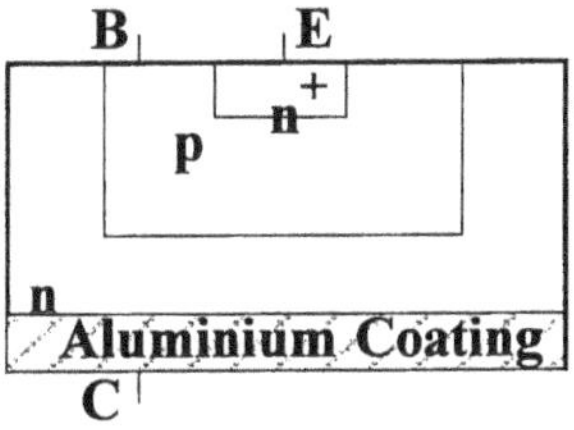

2.1.5 Epitaxial Type

In this process, a very thick, high purity (high resistance), single crystal layer of Silicon or Germanium is grown as a heavily doped substrate of the same material. (n on n^+) or (p on p^+). This forms the collector. Over this base and emitter are diffused. ***Epitaxi*** is a greek word. *Epi* mean 'ON' *Taxi* means 'ARRANGEMENT'. This technique implies the growth of a crystal on a surface, with properties identical to that of the surface. By this technique abrupt step type p-n junction can be formed. The structure is shown in Fig 2.2.

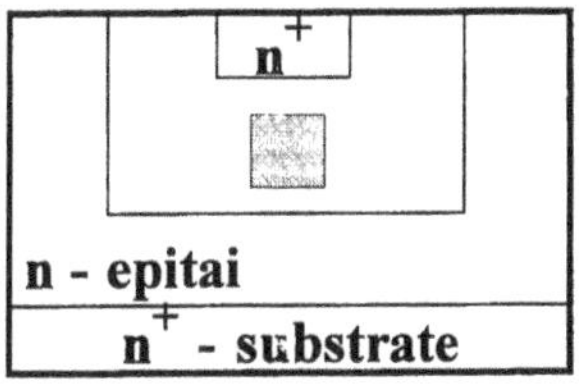

Fig 2.2 Epitaxial type transistor.

2.2 BIPOLAR JUNCTION TRANSISTOR (BJT)

The device in which conduction takes place due to two types of carriers, electrons and holes is called a Bipolar Device. As *p-n junctions* exist in the construction of the device, it is a junction device. When there is transfer of resistance from input side which is Forward Biased (low resistance) to output side which is Reverse Biased (high resistance), it is a **Trans Resistor or Transistor Device**. There are two types of transistors NPN and PNP. In NPN Transistor, a ***p-type*** Silicon (Si) or Germanium (Ge) is sandwiched between two layers of ***n-type*** silicon. The symbol for PNP transistor is as shown in Fig. 2.3(a) and for NPN transistor, is as shown in Fig 2.3 (b).

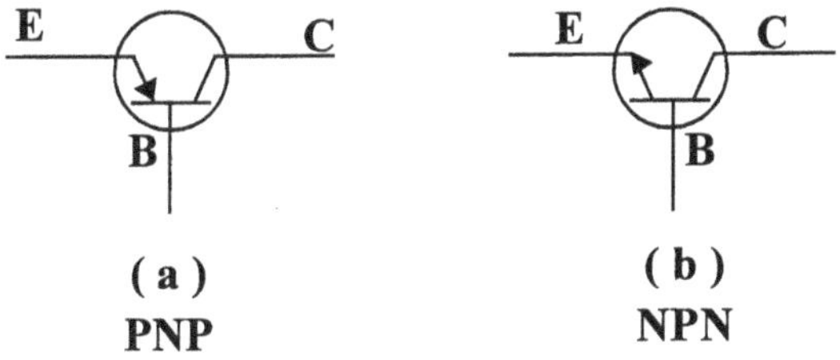

Fig. 2.3 Transistor symbols.

The three sections of a transistor are Emitter, Base and Collector. If the ***arrow mark is towards the base,*** it is PNP transistor. If it is away then it is NPN transistor. The arrow mark on the emitter specifies the direction of current when the emitter base junction is forward biased. When the PNP transistor is forward biased, holes are injected into the base. So the holes move from emitter to base. The conventional current flows in the same direction as holes. So arrow mark is towards the base for PNP transistor. Similarly for NPN transistor, it is away. DC Emitter Current is represented as I_E, Base Current as I_B and Collector Current as I_C. These currents are assumed to be positive when the currents flow into the transistor. V_{EB} refers to Emitter - Base Voltage. Emitter (E) Voltage being measured with reference to base B (Fig.2.4). Similarly V_{CB} and V_{CE}.

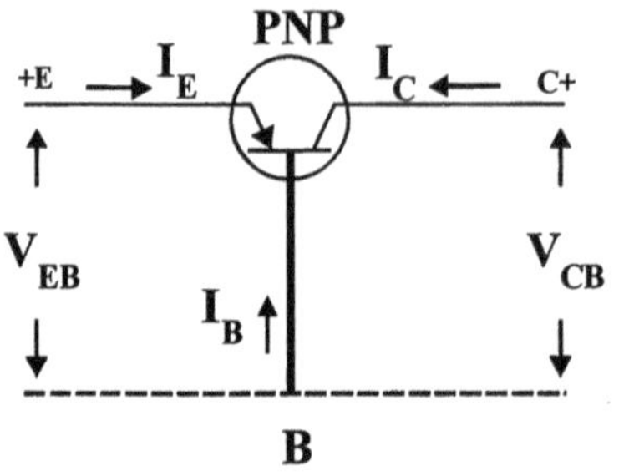

Fig. 2.4 Current components in a pnp transistor.

2.3 TRANSISTOR AS AN AMPLIFIER

A small change ΔV_i of Emitter - Base Voltage causes relatively large Emitter - Current Change ΔI_E. α' is defined as the ratio of the change in collector current to the change in emitter current. Because of change in emitter current ΔI_E and consequent change in collector current, there will be change in output voltage ΔV_o.

$$\Delta V_0 = \alpha' \times R_L \times \Delta I_E \qquad \text{Since } \alpha' = \frac{\Delta I_C}{\Delta I_E}$$

$$\Delta V_i = r_e' \times \Delta I_E$$

R_L is load resistance. Therefore, the voltage amplification is,

$$A_V = \frac{\Delta V_0}{\Delta V_i} = \frac{\alpha' R_L . \Delta I_E}{r_e' \Delta I_E}$$

$$A_V = \frac{\alpha' . R_L}{r_e'} ; \alpha' = \left. \frac{\Delta I_C}{\Delta I_E} \right|_{V_{CB}=K}$$

$R_L > r_e'$

$A_V > 1$

α' = Small Signal Forward - Circuit Current Gain.

r_e' = Emitter Junction Resistance.

The term, r'_e is used since this resistance does not include contact or lead resistances. r'_e value is small, because emitter - base Junction is forward biased.

So a small change in emitter base voltage produces large change in collector base voltage. Hence, the transistor acts as an amplifier. (This is for common base configuration. Voltages are measured with respect to base).

The input resistance of the circuit is low (typical value 40Ω) and output resistance is high (3,000Ω). So current from low resistance input circuit is transferred to high resistance output circuit (R_L). ***So it is a transfer resistor device and abbreviated as transistor.*** The magnitude of current remains the same ($\alpha' \approx 1$).

2.4 TYPES OF TRANSISTOR CONFIGURATIONS

Transistors (BJTs) are operated in three configuration namely,

1. Common Base Configuration (C.B)
2. Common Emitter Configuration (C.E)
3. Common Collector Configuration (C.C)

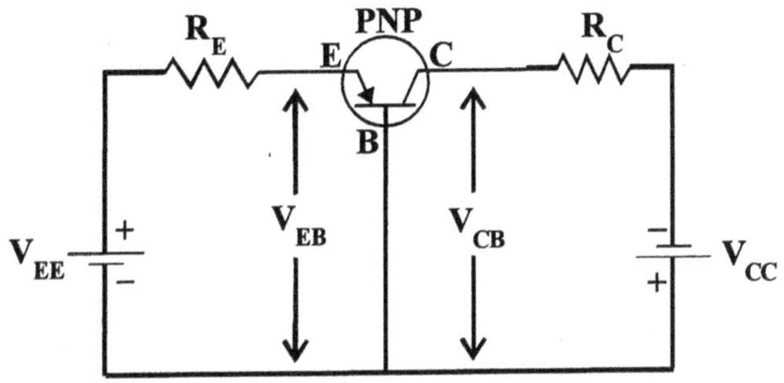

Fig. 2.5 Common base amplifier circuit.

2.4.1 Common Base Configuration (C.B)

The base is at ground potential. So this is known as ***Common Base Configuration*** or ***Grounded Base Configuration.*** Emitter and Collector Voltages are measured with respect to the base. The convention is currents, entering the transistor are taken as positive and those leaving the transistor as negative. For a PNP transistor, holes are the majority carriers. Emitter Base Junction is Forward Biased. Since holes are entering the base or I_E is positive. Holes through the base reach collector. I_C is flowing out of transistor. Hence I_C is negative; similarly I_B is negative. The characteristics of the transistor can be described as,

$$V_{EB} = f_1(V_{CB}, I_E)$$

Dependent Variables are Input Voltage and Output Current.

$$I_C = f_2(V_{CB}, I_E)$$

Independent Variables are Input Current and Output Voltage. For a Forward Biased Junction V_{EB} is positive, for reverse bias collector junction V_{CB} is negative.

Transistor Characteristics in Common Base Configuration

The three regions in the output characteristics of a transistor are

1. *Active region*
2. *Saturation region*
3. *Cut off region*

The input and output characteristics for a transistor in C.B configuration are as shown in Fig. 2.6 (a) and (b). When $I_E = 0$, the emitter is open circuited, so the transistor is not conducting and the I_{CBO} is not considered. Generally the transistor is not used in this mode and it is regarded as cut off ; Even though I_{CBO} is present it is very small for operation, for a small variation of V_{BC} there is very large, change in current I_C with I_E. So current is independent of voltage. There is no control over the current. Hence the transistor can not be operated in this region. So it is named as Saturation region. The literal meaning of saturation should not be taken. The region to the right is called active region. For a given V_{CB}, with I_E increased, there is no appreciable change in I_C. It is in saturation. When transistor is being used as a switch, it is operated between cut off and saturation regions.

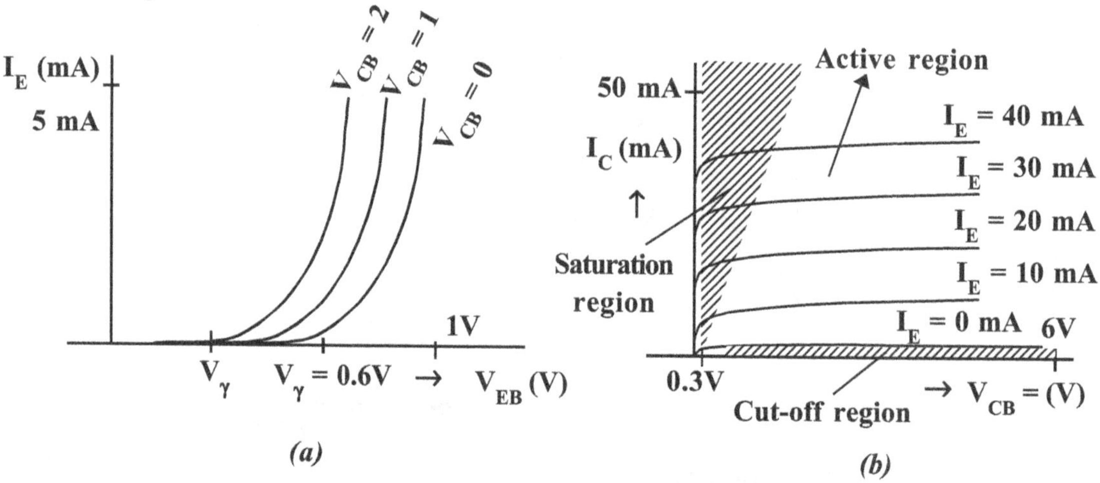

Fig. 2.6 (a) Input characteristics in C.B configuration

(b) Output characterisitics in C.B configuration.

EARLY EFFECT

The collector base junction is reverse biased. If this reverse bias voltage is increased, the space charge width at the *n-p region* of Base - Collector Junction of PNP Transistor increases. So effective base width decreases (Fig. 2.7)

$$V_B = \frac{e.N_A}{2\epsilon}.W^2$$

W = Space Charge Region Width.

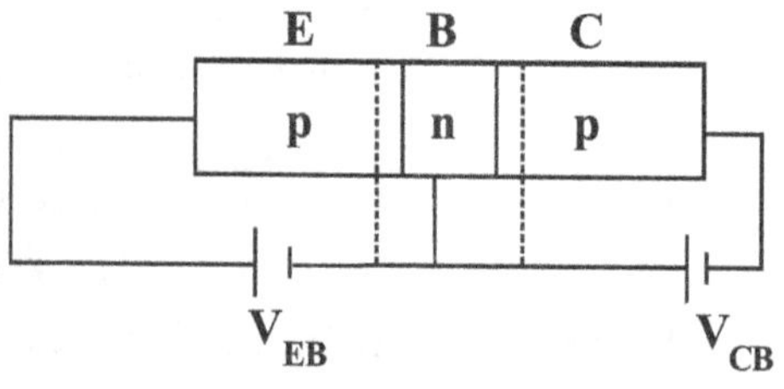

Fig. 2.7 Early effect.

The decrease in base width with increase in collector reverse bias voltage is known as Early Effect. When the base width decreases, the probability of recombination of holes and electrons in the base region is less. So the transportation factors β^* increases.

2.4.2 Common Emitter Configuration (C.E)

In this circuit Fig 2.8, emitter is common to both base and collector. So this is known as CE configuration or grounded emitter configuration. The input voltage V_{BE}, and output current I_C are taken as the dependent variables. These depend upon the output voltage V_{CE} and input current I_B.

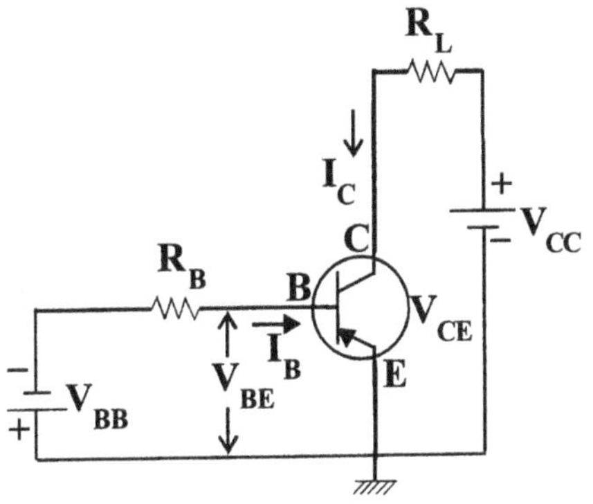

Fig. 2.8 Common Emitter pnp transistor amplifier circuit.

$$\therefore \quad V_{BE} = f_1 (V_{CE}, I_B)$$
$$I_C = f_2(V_{CE}, I_B)$$

Input Characteristics

If the collector is shorted to the emitter, the transistor is similar to a Forward Biased Diode. So I_B increases as V_{BE} increases. The input characteristics are as shown in Fig. 2.9. I_B increases with V_{BE} exponentially, beyond cut-in voltage V_γ. The variation is similar to that of a Forward Biased Diode. If $V_{BE} = 0$, $I_B = 0$, since emitter and collector junctions are shorted. If V_{CE} increases, base width decreases (by Early effect), and results in decreased recombination. As V_{CE} is increased, from –1 to –3 I_b decreases. V_γ is the cut in voltage.

$$I_E + I_B + I_C = 0$$
$$I_E = -(I_B + I_C)$$

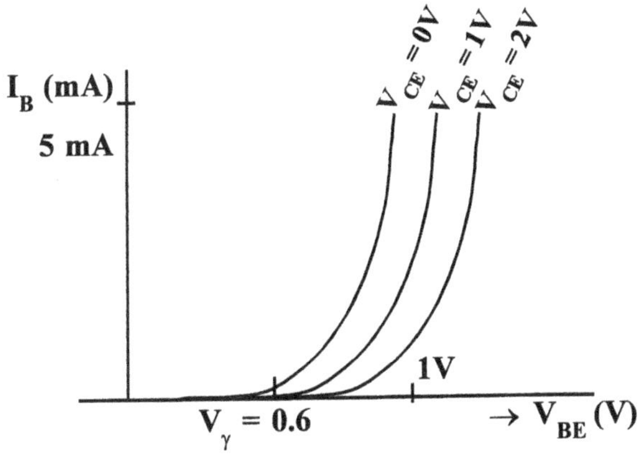

Fig. 2.9 Input characteristics in C.E configuration.

As V_{CE} increases, I_C increases, therefore I_B decreases. Therefore, base recombination will be less. The input characteristics are similar to a forward biased p-n junction diode.

$$I_B = I_o\left(e^{\frac{V_{BE}}{nV_T}} - 1\right)$$

The input characteristics I_B vs V_{BE} follow the equation,

$$I_B = I_o\left(e^{\frac{V_{BE}}{nV_T}} - 1\right)$$

OUTPUT CHARACTERISTICS

The transistor is similar to a Collector Junction Reverse Biased Diode. The output characteristics is divided into three regions namely,

1. *Active Region.*
2. *Saturation Region*
3. *Cut-Off Region.*

These are shown in Fig 2.10. If I_B increases, there is more injection into the collector. So I_C increases. Hence characteristics are as shown.

The output characteristics are similar to a reverse biased *p-n junction.* (Collector Base junction is reverse biased). They follow the equation.

$$I_C = I_o\left(e^{\frac{V_{CE}}{nV_T}} - 1\right)$$

The three different regions are

1. Active Region 2. Saturation Region 3. Cut-off Region.

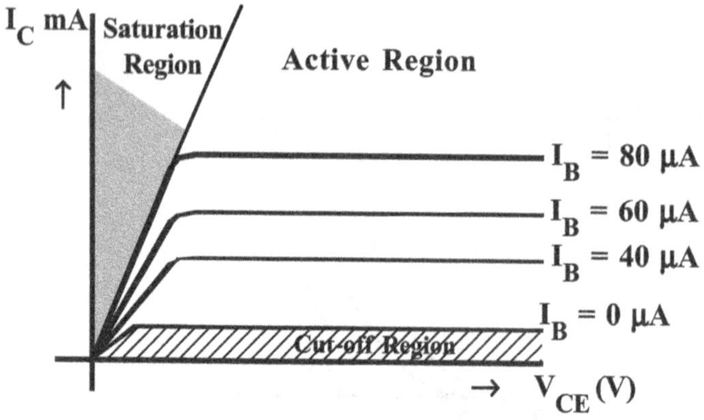

Fig. 2.10 Output characteristics in C.E configuration.

$$I_B + I_C + I_E = 0$$

$$\therefore \quad I_E = -(I_B + I_C)$$

$$-I_C = +I_{CO} - \alpha \cdot I_E$$

But $\quad I_E = -(I_B + I_C)$

$\therefore \quad I_C = -I_{CO} + \alpha \cdot I_B + \alpha \cdot I_C$

$I_C(1-\alpha) = -I_{CO} + \alpha \cdot I_B$

If the transistor were to be at cut off, I_B must be equal to zero.

TEST FOR SATURATION

1. $$|I_B| \geq \left|\frac{I_C}{\beta}\right|$$

2. If V_{CB} is positive for PNP transistor and negative for NPN transistor, the transistor is in saturation.

COMMON EMITTER CUT OFF REGION

If a transistor was to be at cut off, $I_C = 0$. To achieve this, if the emitter-base junction is open circuited, there cannot be injection of carriers from E to C, through B. Hence $I_C = 0$. (Any reverse saturation current flowing because of thermal agitation is very small. So the transistor can not be operated). But in the case of C.E configuration, even if I_B is made 0, the transistor is not cut off. I_C will have a considerable value even when $I_B = 0$. The circuit is shown in Fig. 2.11.

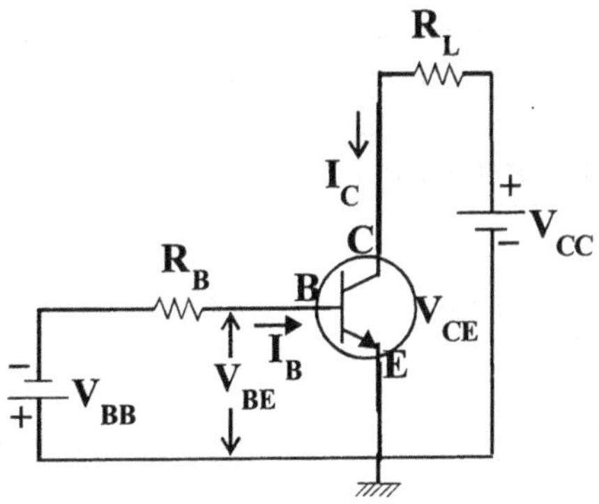

Fig. 2.11 Common Emitter Cut-off region.

Because, $\quad I_C = -\alpha I_E + I_{CO}$

α = Current gain;

I_{CO} = Reverse saturation current in the collector.

But $\quad I_E = -(I_B + I_C)$

If $\quad I_B = 0, I_E = -I_C$

$\therefore \quad I_C = -\alpha(-I_C) + I_{CO}$

or $\quad I_C(1-\alpha) = I_{CO}$

or $$I_C = \frac{I_{CO}}{1-\alpha} = I_{CBO}$$

In order to achieve cut-off in the Common Emitter Configuration, Emitter Base Junction should be reverse biased. α is constant for a given transistor. So the actual collector current with collector junction reverse biased and base open circuited is designated by I_{COB}. If $\alpha \approx 0.9$, then

$I_{CBO} \approx 10\, I_{CO}$. So the transistor is not cut off. Therefore, in the C.E configuration to cut of the transistor, it is not sufficient, if $I_B = 0$. The Emitter Base Junction must be reverse biased, so that I_E is equl to 0. Still I_C will not be zero, but will be I_{CO}. Since I_{CO} is small, the transistor can be regarded as at cut off.

REVERSE COLLECTOR SATURATION CURRENT I_{CBO}

I_{CBO} is due to leakage current flowing not through the junction, but around it. The collector current in a transistor, when the emitter current is zero, is designated by I_{CBO}. $|I_{CBO}|$ can be $> I_{CO}$. I_{CO} is collector reverse saturation current (when collector-base junction is reverse biased).

SATURATION RESISTANCE

For a transistor operating in the saturation region, saturation resistance for Common Emitter configuration is of importance. It is denoted by R_{CES} or R_{CE}'s (saturation) or R_{CS}

$$R_{CS} = \frac{V_{CE}(sat)}{I_C}$$

The operating point has to be specified for R_{CS}.

Saturation Voltage :

Manufacturers specify this for a given transistor. This is done in number of ways. R_{CS} value may be given for different values of I_B or they may supply the output and input characteristic for the transistor itself. V_{CE}(Sat) depends upon the operating point and also the semiconductor material, and on the type of transistor construction. Alloy junction and epitaxial transistors give the lowest values of V_{CE}. Grown junction transistor yield the highest V_{CE}.

DC CURRENT GAIN β_{dc}

β_{dc} is also designated as h_{FE} (DC forward current transfer ratio).

$$\beta = \frac{I_C}{I_B}$$

$$I_C = \frac{V_{CC}}{R_L}$$

So if β is known, $I_B = \left(\frac{I_C}{\beta}\right)$

Therefore, $\frac{I_C}{\beta}$ gives the value of base current to operate the transistor in saturation region.

β varies with I_C for a given transistor.

TEST FOR SATURATION

To know whether a transistor is in saturation or not,

1. $|I_B| \geq \frac{I_C}{h_{fe}\,(\text{or } \beta)}$

2. V_{CB} should be positive for PNP transistor and negative for NPN transistor. In the C.E configuration,

$$I_C = -(I_B + I_E)$$
$$I_C = -\alpha I_E + I_{CO}$$

$$\therefore \qquad I_E = \frac{I_{CO} - I_C}{\alpha}$$

$$\therefore \qquad I_C = \frac{I_{CBO}}{1-\alpha} + \frac{\alpha . I_B}{1-\alpha}$$

But, $\qquad \beta = \dfrac{\alpha}{1-\alpha}$

$$\therefore \qquad (1-\alpha) = \frac{\alpha}{\beta}$$

If we replace I_{CO} by I_{CBO},

$$I_C = I_{CBO} \times (\beta + 1) + \frac{\alpha . I_B . \beta}{\alpha}$$
$$I_C = (1+\beta)\, I_{CBO} + \beta\, I_B$$

Transistor is cut off when $I_C = 0$, $I_C = I_{CBO}$ and $I_B = -I_{CBO}$. So at cut-off $\beta = 0$.

THE RELATIONSHIP BETWEEN α AND β

I_E = Emitter Current
I_B = Base Current
I_C = Collector Current

Since in a PNP transistor, I_E flows into the transistor. I_B and I_C flow out of the transistor. The convention is, currents leaving the transistor are taken as negative.

$$\therefore \qquad I_E = -(I_B + I_C)$$

or $\qquad I_E + I_B + I_C = 0$

or $\qquad I_C = -I_E - I_B$

But $\qquad \alpha = \dfrac{-I_C}{I_E}$

or for PNP transistor taking the convention,

$$I_E = \frac{-I_C}{\alpha}$$

$$\therefore \qquad I_C = +\frac{I_C}{\alpha} - I_B$$

$$I_C\left(1 - \frac{1}{\alpha}\right) = -I_B$$

$$I_C \frac{(\alpha - 1)}{\alpha} = -I_B$$

or $\qquad I_C \dfrac{(1-\alpha)}{\alpha} = +I_B$

$$\frac{I_C}{I_B} = \frac{\alpha}{1-\alpha}$$

$$\frac{I_C}{I_B} = \beta = \text{Large Signal Current Gain} = h_{FE}$$

$$\therefore \quad \beta = \frac{\alpha}{1-\alpha}.$$

The Expression for Collector Current, I_C :

The current flowing in the circuit, when E - B junction is left open i.e., $I_B = 0$ and collector is reverse biased. The magnitude of I_{CEO} is large. Hence the transistor is not considered to be cut off.

The general expression for I_C in the active region is given by,

$$I_C = -\alpha I_E + I_{CBO} \quad .....(2.1)$$

$$I_E = -(I_C + I_B) \quad .....(2.2)$$

Substitute eq. (2.2) in eq. (2.1)

$$I_C = +\alpha (I_C + I_B) + I_{CBO}$$

$$I_C [1 - \alpha] = \alpha I_B + I_{CBO}$$

$$I_C = \frac{\alpha}{1-\alpha} I_B + \frac{I_{CBO}}{1-\alpha}$$

$$I_C = \beta I_B + (\beta + 1) I_{CBO}$$

$$I_C = \beta I_B + I_{CEO}$$

where $\quad I_{CEO} = \dfrac{I_{CBO}}{1-\alpha}$

I_{CEO} is the reverse collector to emitter current when base is open.

RELATIONSHIP BETWEEN I_{CBO} AND I_{CEO}

$$I_C = -\alpha I_E + I_{CBO}$$

This is the general expression for I_C in term of a I_E and I_{CBO}.

when $\quad I_B = 0$, in Common-Emitter (C.E.) Configuration,

$$I_C = -I_E = I_{CEO}$$

$$\therefore \quad I_C + I_B + I_E = 0$$

$$I_B = 0 \quad \text{(Since E-B Junction is left open)}$$

$$\therefore \quad I_C = -I_E$$

$$\therefore \quad I_{CEO} = +\alpha I_{CEO} + I_{CBO}$$

or $\quad I_{CEO} (1 - \alpha) = I_{CBO}$

$$\therefore \quad I_{CEO} = \frac{I_{CBO}}{(1-\alpha)}; \quad \frac{1}{(1-\alpha)} = (\beta + 1)$$

$$\therefore \quad I_{CEO} = (\beta + 1) I_{CBO}$$

In Common Emitter Configuration, for NPN transistor, holes in the collector and electrons in the base are the minority carriers. When one hole from the collector is injected into the base, to neutralize this, the emitter has to inject many more electrons. These excess electrons which do not combine with the hole, travel into the collector resulting in large current. Hence I_{CEO} is large compared to I_{CBO}. When E B junction is reverse biased, the emitter junction is reduced and hence I_{CEO} is reduced.

Therefore, the expression for β gives the ratio of change in collector current $(I_C - I_{CBO})$ to the increase in base current from cut off value I_{CBO} to the I_B. So β is ***large signal current gain of a Transistor in Common Emitter Configuration.***

2.4.3 Common Collector Configuration (C.C)

Here the load resistor R_L is connected in the emitter circuit and not in the collector circuit. Input is given between base and ground. The drop across R_L itself acts as the bias for emitter base junction. The operation of the circuit similar to that of Common Emitter Configuration. When the base current is I_{CO}, emitter current will be zero. So no current flows through the load. Base current I_B should be increased so that emitter current is some finite value and the transistor comes out of cut-off region.

Input characteristics I_B vs V_{BC} Output characteristics I_C vs V_{EC}

The circuit diagram is shown in Fig. 2.12 (a) and the characteristics are shown in Fig. 2.12 (b) and (c).

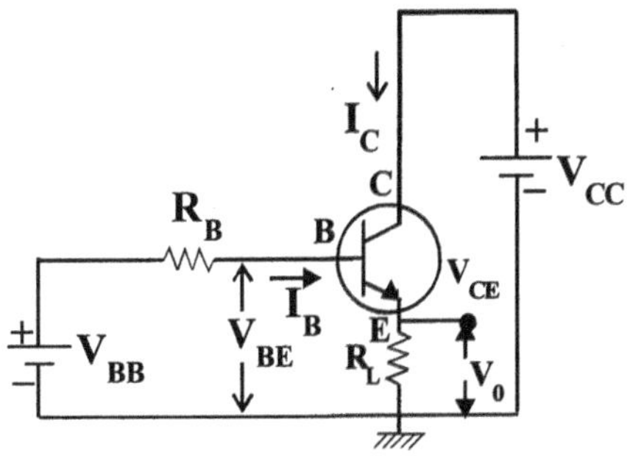

(a) Cirucuit for C.C configuration

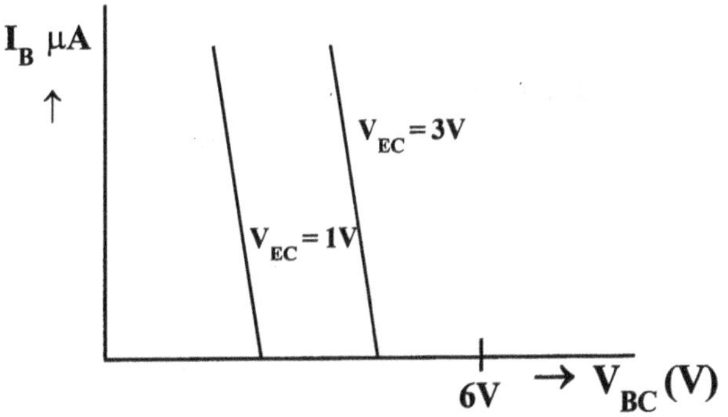

(b) Input characteristics in C.C configuration

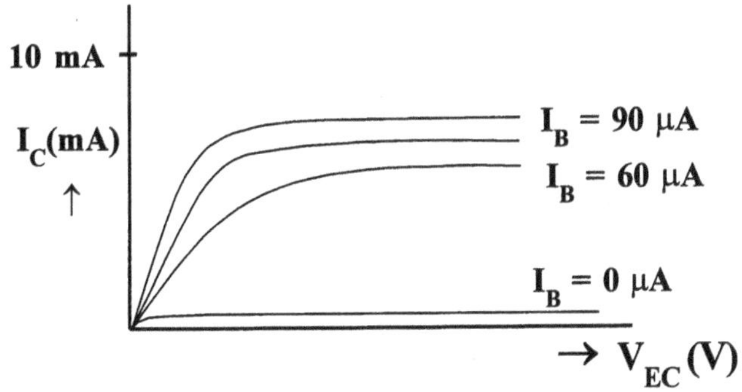

(c) Output characteristics in C.C configuration

Fig. 2.12

The current gain in CC, 'γ' :

It is defined as emitter current to the base current

$$\gamma = -\frac{I_E}{I_B} \quad .....(2.3)$$

Relationship between 'α' and 'γ' :

$$\alpha = -\frac{I_C}{I_B} \quad .....(2.4)$$

$$I_E + I_B + I_C = 0, \ \therefore I_B = (I_C + I_E) \quad .....(2.5)$$

Substitute eq. (2.5) in eq. (2.3)

$$\gamma = \frac{I_E}{+(I_C + I_E)} = \frac{1}{+\frac{I_C}{I_E} + \frac{I_E}{I_E}}$$

$$\gamma = \frac{1}{1-\alpha} \qquad \left[\because \alpha = \frac{-I_C}{I_E}\right]$$

Expression for emitter current, I_E :

$$I_C = -\alpha\, I_E + I_{CBO} \quad .....(2.6)$$

$$I_E = -(I_B + I_C) \quad .....(2.7)$$

Substitute eq. (2.6) in eq. (2.7)

$$I_E = -(I_B - \alpha I_E + I_{CBO})$$

$$I_E\,[1 - \alpha] = -I_B - I_{CBO}$$

$$I_E = \frac{-I_B}{1-\alpha} - \frac{I_{CBO}}{1-\alpha}$$

$$I_E = -\gamma\, I_B + I_{ECO}$$

$$I_{ECO} = -\frac{I_{CBO}}{1-\alpha}$$

Where I_{ECO} is reverse emitter to collector current when base is open.

PARAMETERS

γ = Emitter Efficiency

= I_{pE}/I_E for PNP transistor.

β^* = Transportation factor

$$= \frac{I_{pC}}{I_{pE}} = \frac{\text{No. of holes reaching collector}}{\text{No. of holes emitter}}$$

α = Small signal current gain

$$= \frac{I_C}{I_E} \; ; \alpha = \beta^* \gamma \qquad \alpha < 1$$

α' = Small signal current gain

$$= \partial I_C / \partial I_E = \frac{\Delta I_C}{\Delta I_E} \qquad \alpha' < 1$$

$h_{FE} = \beta$ = Large signal current gain

$$= \frac{I_C}{I_B} \qquad \beta >> 1$$

$h_{FE} = \beta'$ = Small signal current gain

$$= \frac{\partial I_C}{\partial I_B} = \frac{\Delta I_C}{\Delta I_B} \qquad \beta' >> 1$$

I_{CBO} = Reverse saturation current when E-B junction is open circuited.

I_{CO} = Normal Reverse Saturation Current

$$\alpha = \frac{\beta}{1+\beta}$$

$$\beta = \frac{\alpha}{1-\alpha}$$

2.5 LIMITS OF OPERATION

For any transistor there will a region of operation on its characteristics which will ensure that its maximum ratings are not exceeded and its output is minimum distorted.

Let us consider a transistor voltage characteristics are shown in Fig 2.13. From Fig 2.13. it is clear that $I_{C\,(max)} = 50$ mA, $V_{CEO} = 20$ V.

The vertical line 'M' on the characteristics define $V_{CE\,sat}$ which is the minimum value of V_{CE} that can be employed without falling the transistor into nonlinear region.

For the device in Fig. 2.13. the power dissipation is 300 mV. But

$$P_{C_{max}} = V_{CE} I_C = 300 \text{ mW} \qquad .....(2.8)$$

The significance of equation (2.8) is that, we can choose any point 'M' the characteristics of the device whose maximum power should be equal to 300 mW i.e., we can choose different values of V_{CE} and I_C but their product should be equal to 300 mW to prevent the device being get damaged and to give minimum output distortion.

e.g.,

$$P_{C_{max}} = V_{CE} I_C = 300 \text{ mW}$$

$$= V_{CE} (50 \text{ mA}) = 300 \text{ mW}$$

$$\Rightarrow V_{CE} \frac{300 \text{ mW}}{50 \text{ mA}} = 6 \text{ V} \qquad .....(2.9)$$

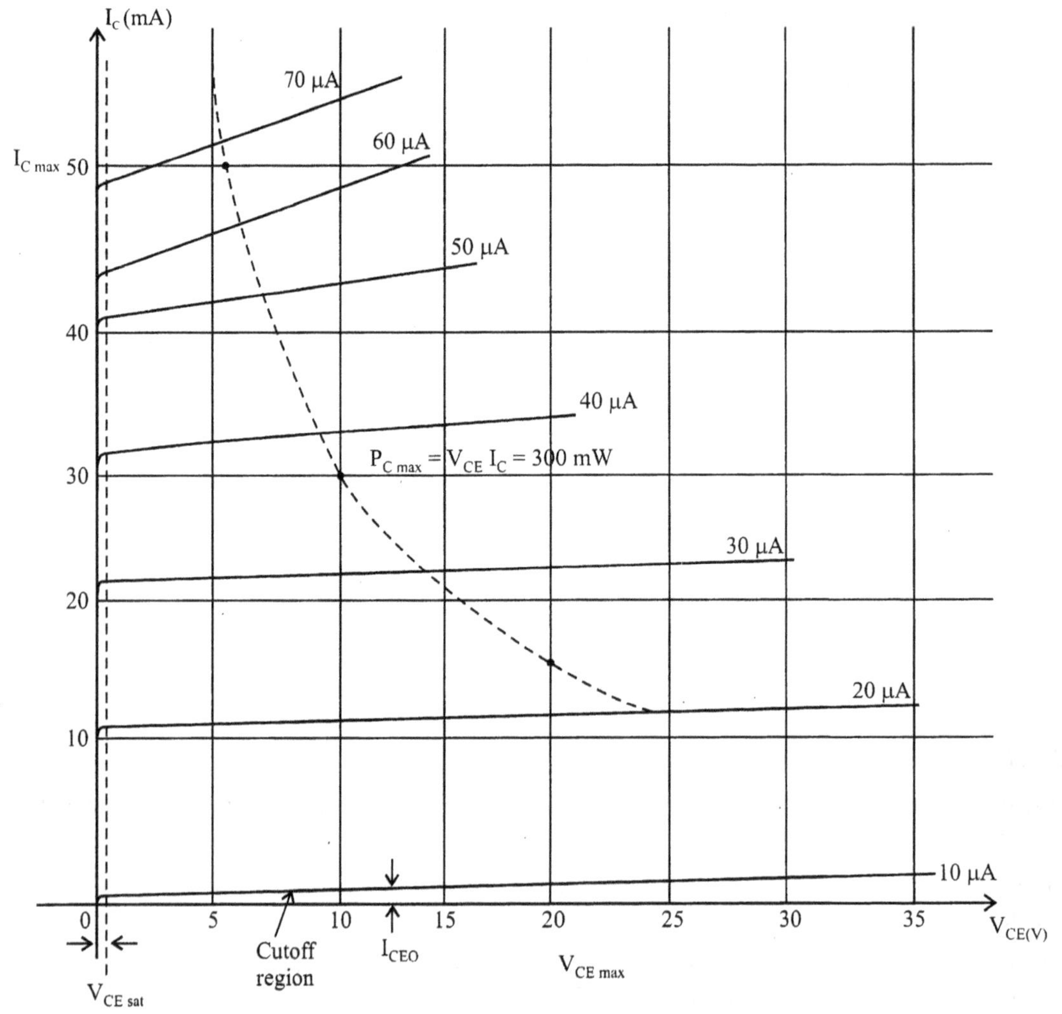

Fig. 2.13 Illustration of limits of operation

Similarly if we choose $V_{CE} = 20$ V we get 20 $I_C = 30$ mW

$$\Rightarrow I_C = 30 \text{ mV}/20 = 15 \text{ mA} \qquad .....(2.10)$$

The values of V_{CE} and I_C in eq. (2.9) gives the operation of transistor is closer to saturation region and similarly the values if V_{CE} and I_C in eq. (2.10) gives the operation of transistor near cutt off region. In order to get a mimum distorted output we should choose a operation of transistor that falls exactly at midway of values given by eqs. (2.9) and (2.10).

i.e., $V_{CE}\,(25) = 300$ mW

$$V_{CE} = \frac{300 \text{ mW}}{25 \text{ mA}} = 12 \text{ V} \qquad .....(2.11)$$

Hence in order to get minimum distorted output the transistor should be operated at $V_{CE} = 12$ and $I_C = 25$ mA. These values also will not damage the transistor.

The cutt off region is defined as region between $I_C = I_{CEO}$ this region must also be avoided if the distortion in the output signal is to have minimum distortion.

If the characteristic curves are unavailable or do not appear on the specification sheet, one most simply be sure that I_C, V_{CE} and their product should fall into following range

$$I_{CEO} \leqq I_C \leqq I_{C\ max}$$

$$V_{CE(sat)} \leqq V_{CE} \leqq V_{CE\ (max)}$$

$$V_{CE}\ I_C \leqq P_{C(max)}$$

2.6 BJT SPECIFICATIONS

Specifications of a Bipolar Junction Transistor (BJT)

Sl.No.	Parameter	Symbol	Typical value	Units
1.	Collector-Emitter Voltage	V_{CE}	40	V
2.	Collector-Base Voltage	V_{CB}	60	V
3.	Emitter-Base Voltage	V_{EB}	6	V
4.	Collector-Current	I_C	500	mA
5.	Total Power Dissipation	P_D	0.4	W
6.	Operating Temperature	T_{op}	25-80	^oC
7.	Collector- Emitter Break down voltage	BV_{CEO}	40	V
8.	Emitter-Base Break down voltage	BV_{EBO}	50	V
9.	Current - Gain Bandwidth product	f_T	250	MHz
10.	Output Capacitance	C_0	8	p_f
11.	Input Capacitance	C_{in}	20	p_f
12.	Small Signal Current gain	h_{fe}	100	-
13.	Output admittance	h_{oe}	10	m℧
14.	Voltage feedback ratio	h_{re}	8×10^{-4}	-
15.	Input impedance	h_{ie}	3	kΩ
16.	Delay time	t_d	10	nSec.
17.	Rise time	t_r	20	nSec.
18.	Fall time	t_f	50	nSec.
19.	Storage time	t_s	200	nSec.
20.	Noise Figure	NF	4	dB

2.7 THE JUNCTION FIELD EFFECT TRANSISTOR (FET OR JFET)

The Field Effect Transistor is a semiconductor device which depends for its operation on the control of current by an electric field. Hence, the name FET.

There are two types (FET) :

1. Junction Field Effect Transistor (JFET) or simply FET.
2. Insulated Gate Field Effect Transistor IGFET. It is also called as Metal Oxide Semiconductor (MOS) transistor or MOST or MOSFET.

The principle of operations of these devices, their characteristic are given in this chapter.

Advantages of FET over BJT (Transistor)

1. JFET is unipolar device. Its operation depends upon the flow of majority carriers only. Vacuum tube is another example of unipolar device. Because the current depends upon the flow of electrons emitted from cathode. Transistor is a bipolar device. So recombination noise is more.
2. JFET has high input resistance (MΩ). BJT (Transistor) has less input resistance. So loading effect will be there.
3. It has good thermal stability.
4. Less noisy than a tube or transistor. Because JFET is unipolar, recombination noise is less.
5. It is relatively immune to radiation.
6. JFET can be used as a symmetrical bilateral switch.
7. By means of small charge stored or internal capacitance, it acts as a memory device.
8. It is simpler to fabricate and occupies less space in IC form. So packing density is high.

Disadvantage

1. Small gain - bandwidth product.

2.7.1 JFET

There are two types;

1. n-channel JFET.
2. p-channel JFET.

If n - type semiconductor bar is used it is n - channel JFET.

If p - type semiconductor bar is used it is p - channel JFET.

Ohmic contacts are made to the two ends of a semiconductors bar of n-type or p-type semiconductor. Current is caused to flow along the length of the bar, because of the voltage supply connected between the ends. If it is n - channel JFET, the current is due to electrons only and if it is p-channel JFET, the current is due to holes only.

The three leads of the device are **1.** Gate **2.** Drain **3.** Source. They are similar to **1.** Base **2.** Collector **3.** Emitter of BJT respectively.

2.7.2 n-CHANNEL JFET

The arrow mark at the gate indicates the direction of current if the Gate source junction is forward biased.

On both sides of the n - type bar heavily doped (p^+) region of acceptor impurities have been formed by alloying on diffusion. These regions are called as Gates. Between the gate and source a voltage V_{GS} is applied in the direction to reverse bias the p-n junction.

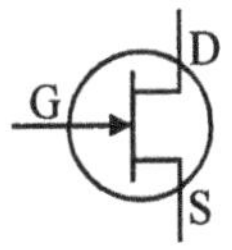

Fig. 2.14 (n-channel JFET).

2.7.3 p-CHANNEL JFET

The symbol for p-channel FET is given in Fig. 2.15. The semiconductor bar is p-type or source is p-type. Gate is heavily doped n-region. If G - S junction is forward biased, electrons from gate will travel towards source. Hence, the conventional current flows outside. So, the arrow mark is as shown in Fig. 2.15. indicating the direction of current, if G-S junction is forward biased.

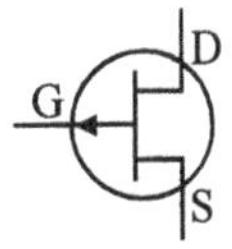

Fig.2.15 (p-channel JFET).

The FET has three terminals, Source, drain and Gate.

Source

The source S is the terminal through which the majority carriers enter the bar (Fig. 2.16). Conventional current entering the bar at S is denoted as I_S.

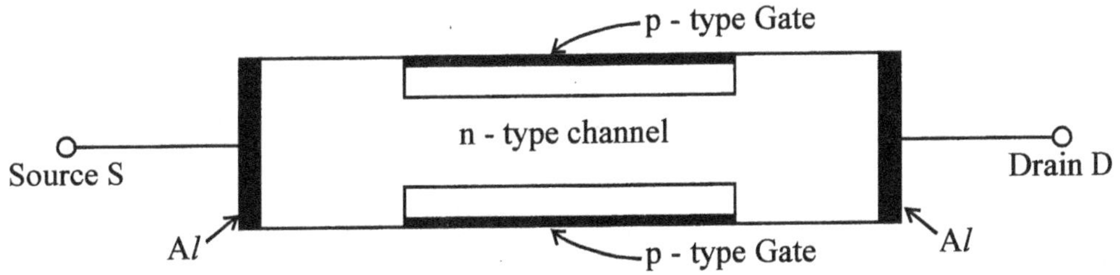

Fig. 2.16 Structure of JFET.

Drain

The drain D is the terminal through which the majority carriers leave the bar. Current entering the bar at D is designated by I_D. V_{DS} is positive if D is more positive than S, source is forward biased.

Gate

In the case of n - channel FET, on both sides of the bar heavily doped (p+) region of acceptor impurities have been formed by diffusion or other techniques. The impurity regions form as Gate. Both these region (p+ regions) are joined and a lead is taken out, which is called as the gate lead of the FET. Between the gate and source, a voltage V_{GS} is applied so as to reverse bias the gate source p-n junction. ***Because of this reason JFET has high input impedance. In BJT Emitter Base junction is forward biased. So it has less Z.*** Current entering the bar at G is designated as I_G.

Channel

The region between the two gate regions is the channel through which the majority carriers move from source to drain. By controlling the reverse bias voltage applied to the gate source junction the channel width and hence the current can be controlled.

2.7.4 FET STRUCTURE

FET will have gate junction on both sides of the silicon bar (as shown in Fig. 2.17).

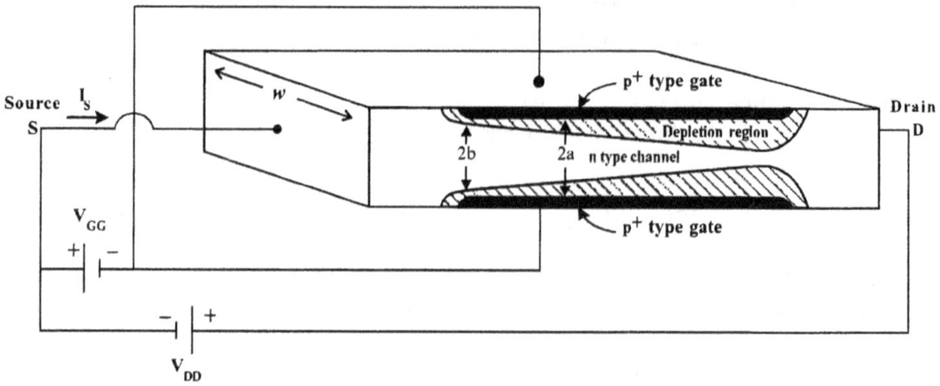

Fig. 2.17 JFET structure.

But it is difficult to diffuse impurities from both sides of the wafer. So the normal structure that is adopted is, a p-type material is taken as substrate. Then n-type channel is epitaxially grown. A p-type gate is then diffused into the n-type channel. The region between the diffused p-type impurity and the source acts as the drain.

2b (x) = Actual channel width, at any point x, the spacing between the depletion regions at any point x from source end because the spacing between the depletion regions is not uniform, it is less towards the drain and more towards the source (see Fig.2.18). 2b depends upon the value of V_{GS}.

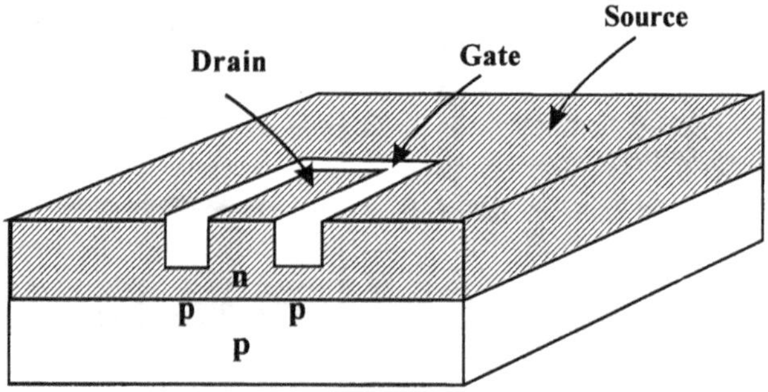

Fig. 2.18 Structure

2a = Channel width, the spacing between the doped regions of the gate from both sides. This is the channel width when $V_{GS} = 0$ (and the maximum value of channel width).

2.7.5 FET OPERATION

If a p-n junction is reverse biased, the majority carriers will move away from the junction. That is holes on the p-side will move away from the junction leaving negative charge or negative ions on the p-side (because each atom is deprived of a hole or an electron fills the hole. So it becomes negative by charge). Similarly, electrons on the n-side will move away from the junction leaving positive ions near the junction. Thus, there exists space charge on both sides of the junction in a reverse biased p-n junction diode. So, the electric field intensity, the lines of force originate from the positive charge region to the negative charge region. This is the source of voltage drop across the junction. As the reverse bias across the junction increases, the space charge region also increases, or the region of immobile uncovered charges increases (i.e., negative region on p-side and positive region on n-side increases as shown in Fig. 2.19). The conductivity of this region is usually zero or very small. Now in a FET, between gate and source, a reverse bias is applied. Therefore the channel width is controlled by the reverse bias applied between gate and source. Space charge region exists near the gate region on both sides. The space between them is the channel. If the reverse bias is increased, the channel width decreases. Therefore, for a fixed drain to source voltage the drain current will be a function of the reverse biasing voltage across the junction. Drain is at positive potential (for n-type FET). Therefore, electrons tend to move towards drain from the source (1) Because, source is at negative potential, they tend to move towards the drain. But because of the reverse bias applied to the gate, there is depletion region or negative charge region near the gate which restricts the number of electrons reaching the drain. Therefore the drain current also depends upon the reverse bias voltage across the gate junction. The term field effect is used to describe this device because the mechanism of current control is the effect of the extensions, with increase reverse bias of the field associated with region of uncovered charges.

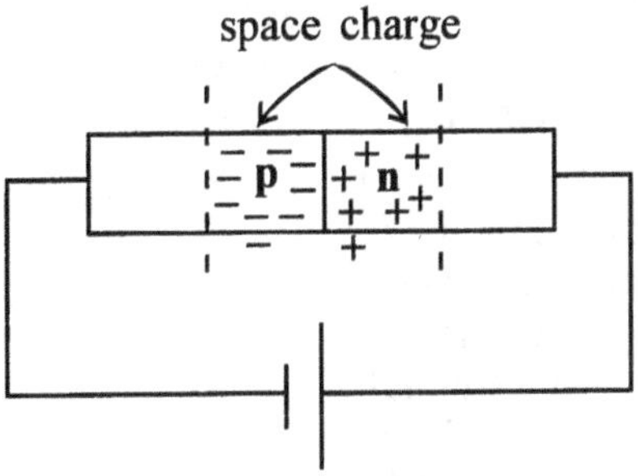

Fig. 2.19 Space charge in p-n junction.

The characteristics of n-channel FET between I_D, the drain current and V_{DS} the drain source voltage are as shown in Fig. 2.20, for different values of V_{GS}.

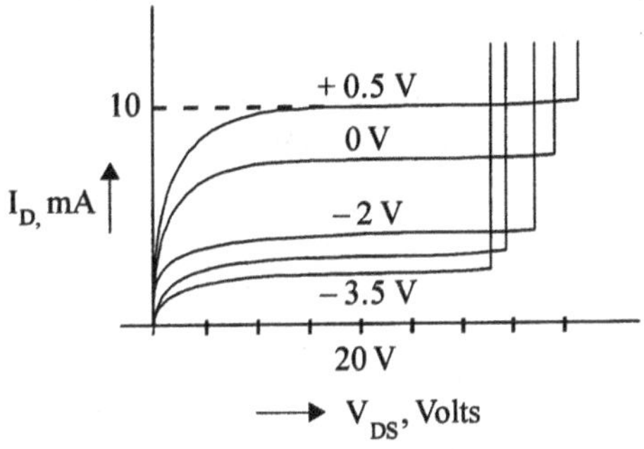

Fig. 2.20 Drain charecteristics

To explain these characteristics, suppose $V_{GS} = 0$. When $V_{DS} = 0$, $I_D = 0$, because the channel is entirely open. When a small V_{DS} is applied (source is forward biased or negative voltage is applied to n-type source), the n-type bar acts as a simple semiconductor resistor and so I_D increases linearly with V_{DS}. With increasing current, the ohmic voltage drop between the source and the channel region reverse biases the junction and the conducting portion of the channel begins to constrict. Because, the source gets reverse biased or the negative potential at the n-type source reduces. Because of the ohmic drop, along the length of the channel it self, the constrictions is not uniform, but is more ***pronounced*** at distances farther from the source. i.e., the channel width is narrow at the ***drain*** and wide at the source. Finally, the current will remain constant at a particular value and the corresponding voltage at which the current begins to level off is called as the ***pinch off voltage***. But the channel cannot be closedown completely and there by reducing the value of I_D to zero, because the required reverse bias will not be there.

Now if a gate voltage V_{GS} is applied in a direction to reverse bias the gate source junction, pinch off will occur for smaller values of (V_{DS}), and the maximum drain current will be smaller compared to when $V_{GS} = 0$. If V_{GS} is made + 0.5V, the gate source junction is forward biased. But at this voltage, the gate current will be very small because for Si FET, 0.5V is just equal to or less than the cut in voltage. Therefore, the characteristic for $V_{GS} = + 0.5V$, I_D value will be comparatively larger, (compared to $V_{GS} = 0V$ or – 0.5V). Pinch off will occur early.

FET characteristics are similar to that of a pentode, in vacuum tubes.

The maximum voltage that can be applied between any two terminals of the FET is the lowest voltage that will cause avalanche breakdown, across the gate junction. From the FET characteristics it can be observed that, as the reverse bias voltage for the gate source junction is increased, avalanche breakdown occurs early or for a lower value of V_{DS}. This is because, the reverse bias gate source voltage adds to the drain voltage because drain though n -type (for n-channel FET), a positive voltage is applied to it. Therefore, the effective voltage across the gate junction is increased.

For n-channel FET, the gate is p-type, source and drain are n-type. The source should be forward-biased, so negative voltage is applied. Positive voltage is applied to the drain. ***Gate source junction should be reverse biased,*** and gate is p-type. Therefore,voltage or negative voltage is applied to the gate. Therefore, n-channel FET is exactly similar to a Vacuum Tube (Triode). Drain is similar to anode (at positive potential), source to cathode and gate to grid, (But the characteristics are similar to pentode).

For p-channel FET, gate is n-type, and positive voltage is applied, drain is at negative potential with respect to source.

Consider n-channel FET. The source and drain are n-type and p-type gate is diffused from both sides of the bar (See Fig. 2.21 below).

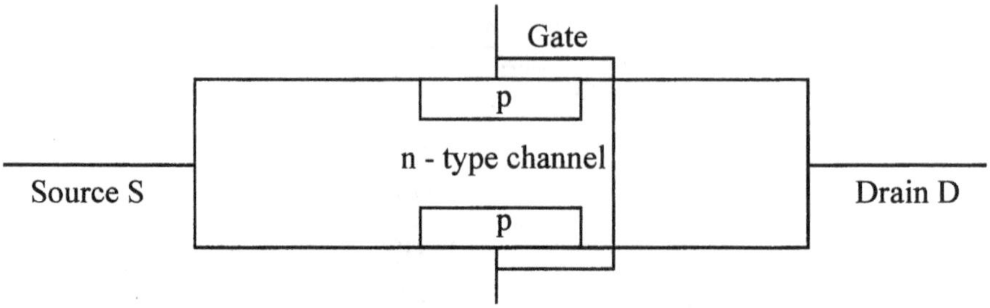

Fig. 2.21 Structure of n-channel JFET.

Suppose source and gate are at ground potential and small positive voltage is applied to the drain. Source is n-type. So it is Forward biased. Because drain is at positive potential, electrons from the source will move towards the drain. Negligible current flows between source and gate or gate and drain, since these p-n junctions are reverse biased and so the current is due to minority carriers only. Because gate is heavily doped, the current between S and G or G and D can be neglected. The current flowing from source to drain I_D depends on the potential between source and drain, V_{DS}, resistance of the n-material in the channel between drain to source. This resistance is a function of the doping of the n-material and the channel width; length and thickness.

If V_{DS} is increased, the reverse bias voltage for the gate drain junction is increased, since, drain is n-type and positive voltage at drain is increased. This problem is similar to that of a reverse biased p-n junction diode.

Consider a p-n junction which is reverse biased. The holes on the p-side remain near the negative terminal and electrons on the n-side reach near the positive terminal as shown in Fig.2.22. There are no mobile charges near the junction. So we call this as the depletion region since it is depleted of mobile carriers or charges. As the reverse voltage is increased, the depletion region width 'l' increases.

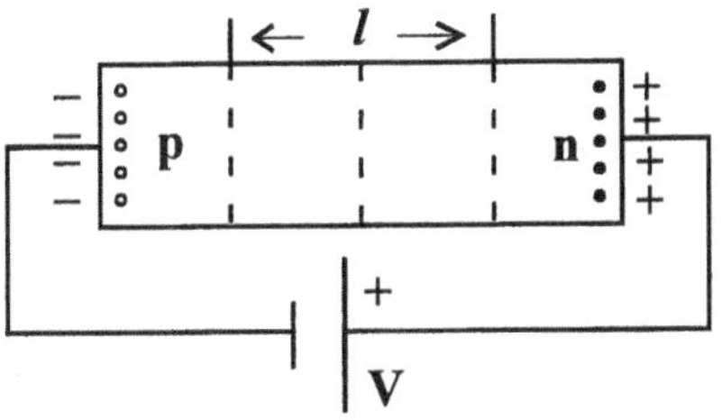

Fig. 2.22 Reverse biased G - D junctions

This result is directly applicable for JFET. The depletion region extends more towards drain because points close to the drain are at higher positive voltage compared to points close to source. So the depletion region is not uniform $V_{DS} < V_{P0}$. But extends more towards drain than source. V_{P0} is the pinch off voltage.

As V_{DS} is increased, depletion region is increased (shaded portion). So channel width decreases and channel resistance increases. Therefore, the rate at which I_D increases reduces, eventhough there is positive potential for the electrons at the drain and hence they tend to move towards the drain and conventional current flows as shown by the arrow mark, in the Fig.2.23.

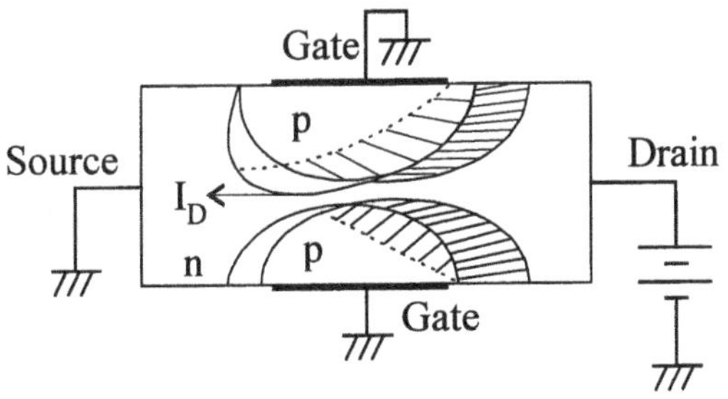

Fig. 2.23 Speed of depletion region

When $V_{DS} = V_{PO}$

When V_{DS} is further increased, the depletion region on each side of the channel join together as shown in Fig.2.24. ***The corresponding V_{DS} is called as V_{PD}, the pinch off voltage,*** because it pinches off the channel connection between drain and source.

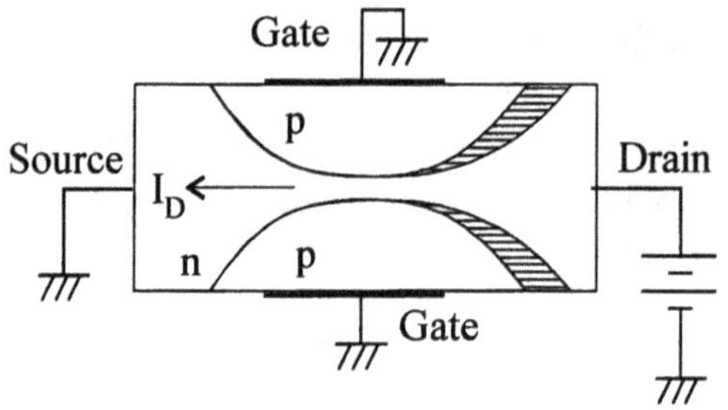

Fig. 2.24

***When* $V_{DS} > V_{P0}$**

If V_{DS} is further increased, the depletion region thickens. So the resistivity of the channel increases. Because drain is at more positive potential, more electrons tend to move towards the drain. Hence I_D should increase. But, because channel resistance increases, I_D decreases. Therefore, the net result is I_D levels off, for any V_{DS} above V_{P0}.

$$I_D = \frac{V_{DS}}{r_{DS}}$$

In the linear region, r_{DS} is almost constant. So as V_{DS} is increased, I_D increases. When the two channels meet, as V_{DS} increases r_{DS} also increases. So I_D remains constant.(See Fig.2.25)

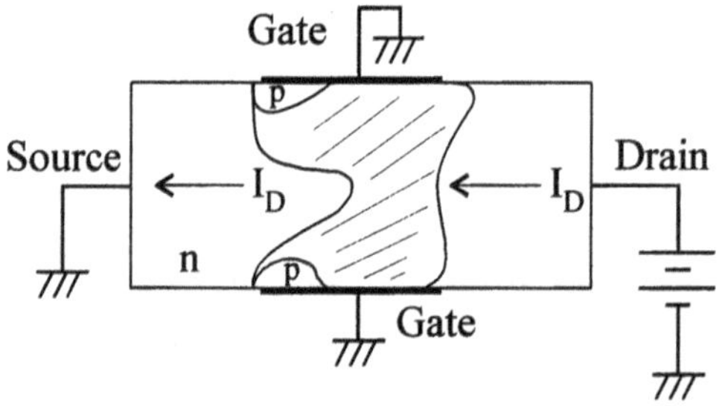

Fig. 2.25 Depletion region width variation in JFET.

2.7.6 PINCH OFF REGION

The voltage V_{DS} at which, the drain current I_D tends to level off is called the ***pinch off voltage***. When the value of V_{DS} is large, the electric field that appears along the x-axis, i.e., along the channel ε_x will also be more. When the value of I_D is more, the drain end of the gate i.e., the gate region near the drain is more reverse biased than the source end. Because, the drain is at reverse potential, for n-channel FET, source is n-type, drain is n-type and positive voltage is applied for the drain, gate is also reverse biased. Therefore, the drop across the channel adds to the reverse bias voltage of the gate. Therefore, channel narrows more near the drain region and less near the source region. So, the boundaries of the depletion region are not parallel when V_{DS} is large and hence ε_x is large.

As V_{DS} increases, ε_x and I_D increase, whereas the channel width $b(x)$ narrows ($\because$ depletion region increases). Therefore the current density $J = I_D / 2b(x)\, w$ increases. If complete pinch off were to take place b = 0. But this cannot happen, because if b were to be 0, $J \to \infty$ which cannot physically happen.

Mobility μ is a function of electric field intensity μ remains constant for low electric fields when $\varepsilon_x < 10^3$V/cm.

$$\mu \ \alpha \ \frac{1}{\sqrt{\varepsilon_x}} \text{ when } \varepsilon_x \text{ is } 10^3 \text{ to } 10^4 \text{ V/cm.(for moderate fields).}$$

$$\mu \ \alpha \ \frac{1}{\varepsilon_x} \text{ for very high field strength} > 10^4 \text{ V/cm.}$$

$$I_D = 2bwe\, N_D\, \mu_n\, \varepsilon_x.$$

As V_{DS} increases, ε_x increases but μ_n decreases, b almost remains constant. Therefore I_D remains constant in the pinch off region.

Pinch off voltage (V_{P0} or V_P) is the voltage at which the drain current I_D levels off. When V_{GS} = 0V, if V_{GS} = – 1,V, the voltage at which the current I_D levels off decreases, (This is not pinch off voltage). If V_{GS} is large – 5, V_{DS} at which the current levels off may be zero. So the relationship between V_{DS}, V_{PD} and V_{GS} is

$V_{DS} = V_{P0} + V_{GS}$ V_{DS} is positive n-channel FET.

V_{P0} = P indicates pinch off voltage, zero indicate the voltage when V_{GS} = 0V.

V_{DS} is the voltage at which the current I_D levels off.(See Fig. 2.26).

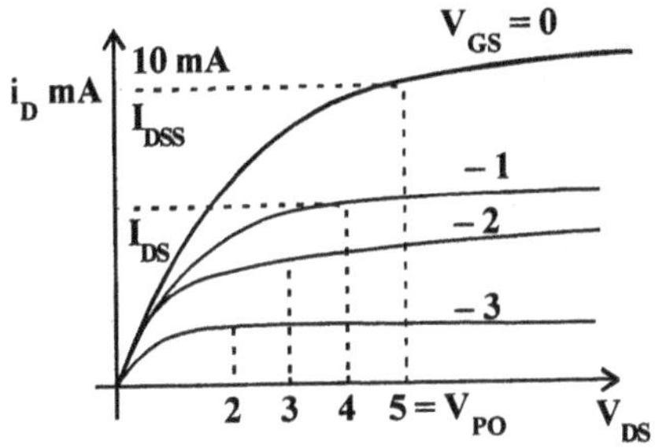

Fig. 2.26 JFET drain characteristics.

2.8 JFET VOLT-AMPERE CHARACTERISTICS

Suppose the applied voltage V_{DS} is small. The resulting small drain current I_D will not have appreciable effect on the channel profile. Therefore, the channel cross-section A can be assumed to be constant throughout, its length.

$\therefore$ $A = 2b.w,$

Where '2b' is the channel width corresponding to negligible drain current and w is channel dimension perpendicular to the 'b' direction.

$$\therefore \quad I_D = Ae \ . \ N_D \ \mu_n \in$$

$$A = 2b \ . \ w$$

$$\in = \frac{V_{DS}}{L}$$

$$= 2bwe \ . \ N_D \ \mu_n \ . \ \frac{V_{DS}}{L} \qquad .....(2.12)$$

Where L is the length of the channel. Eliminating 'b' which is unknown,

$$V_{GS} = \left(1 - \frac{b}{a}\right)^2 \ . \ V_P$$

$$\left(1 - \frac{b}{a}\right) = \left(\frac{V_{GS}}{V_P}\right)^{1/2}$$

$$\frac{b}{a} = 1 - \left(\frac{V_{GS}}{V_P}\right)^{1/2}$$

$$b = a\left[1 - \left(\frac{V_{GS}}{V_P}\right)^{1/2}\right]$$

Substituting, this value in equation (2.12),

$$\boxed{I_D = \frac{2aweN_D\mu_n}{L}\left[1 - \left(\frac{V_{GS}}{V_P}\right)^{1/2}\right] V_{DS}}$$

This is the expression for I_D in terms of V_{GS}, V_P and V_{DS} because the value of b is not directly known.

$$V_P = \frac{e.N_D}{2\varepsilon} \ . \ w^2$$

But, $\quad w = a - b\ (x)$

$b = 0$, at pinch off,

$\therefore \quad w = a =$ The spacing between the two gate dopings.

$$\therefore \quad \boxed{|V_P| = \frac{e.N_D}{2\in}.\, a^2}$$

V_{DS} controls the width of the depletion region (a – b) because for a given V_{GS}, as V_{DS} increases, the reverse potential between drain and gate increase. Therefore, depletion region width increases.

Expression for V_{Gs}

$$V = \frac{e.N_D x^2}{2\in}$$

We can get the expression for V_{GS} if we replace x by (a – b) and V by V_{GS}.

$$\therefore \quad V_{GS} = \frac{e.N_D(a-b)^2}{2\in}$$

$$\text{But} \quad |V_P| = \frac{e.N_D}{2\in}.a^2$$

$$\therefore \quad \frac{e.N_D}{2\in} = \frac{|V_P|}{a^2}$$

Substituting this in expression for V_{Gs},

$$V_{GS} = \frac{V_P}{a^2}(a-b)^2$$

Channel width is controlled by V_{GS}. As V_{GS} increases, channel width reduces.)

Therefore, channel width is controlled by V_{GS} and depletion region width by V_{DS}.

$$\boxed{V_{GS} = \left(1-\frac{b}{a}\right)^2.V_P}$$

The on resistance r_{ds} (ON)

When V_{DS} is small, the FET behaves like an ohmic resistance whose value is determined by V_{GS}.

$$\text{The ratio} \quad \frac{V_{DS}}{I_D} = r_{ds(ON)}$$

$$I_D = 2bwe\, N_D\, \mu_n \cdot \frac{V_{DS}}{L}$$

$$\therefore \quad \frac{V_{DS}}{I_D} = \frac{L}{2a\,we\, N_D\, \mu_n} = r_{ds(ON)}.$$

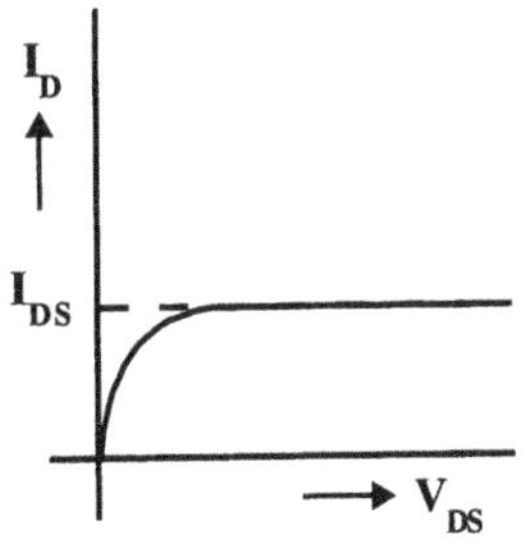

Fig. 2.27 Drain characteristics

2.9 FET AS A VOLTAGE VARIABLE RESISTOR

In general FET is operated in constant current portion of its output characteristics. In the region before pinch. Where 'V'$_{DS}$ is very small, the drain to source resistance can be controlled by 'V'$_{GS}$. Thus FET can be employed as voltage variable resistor.

In JFET, the drain to source conductance $g_d = I_D/V_{DS}$ for small values of V_{DS} can be expressed as

$$g_d = g_{do}\left[1-\left(\frac{V_{GS}}{V_P}\right)^{\frac{1}{2}}\right] \quad .....(2.13)$$

where g_{do} is value of drain conductance when bias voltage V_{GS} is zero.

The variation of r_d with V_{GS} can be given as follows

$$r_d = \frac{r_o}{1 - k\,V_{GS}} \quad .....(2.14)$$

r_o = drain resistance at $V_{GS} = 0$

k = constant depends on FET type

From eq. (2.14) it is clear that $r_d \propto \frac{1}{V_{GS}}$ i.e., as V_{GS} decreases r_d increases and vice versa. Hence FET can be effectively utilized as voltage variable resistor.

2.10 COMPARISON BETWEEN BJT AND FET

The following table lists out the comparison of FET and BJT on different aspects.

BJT	FET
1. In BJT current flow is due to both majority and minority carriers.	1. Current flow is due to only majority carriers.
2. BJT has two junctions (E-B & B-C).	2. It has only one junction.
3. BJT is more noisy.	3. FET is less noisy.
4. BJT has less input impedance.	4. FET offers high input impedance.
5. BJT has loading effect (as its input impedance is less).	5. FET has no loading effect.
6. It has poor thermal stability.	6. FET has good thermal stability.
7. BJT is less immune to radiation.	7. FET is more immune to radiation.
8. Difficult to fabricate as compared to FET.	8. Easy (or) simpler to fabricate as compared to BJT.
9. BJT occupies more space in the IC form. Hence its packing density is less.	9. FET occupies less space in IC form. Hence it offers high packing density.
10. It has more gain-bandwidth product.	10. FET has less gain-bandwidth product.
11. There is no such facility available with BJT.	11. Conductivity of FET can varied by varying gate bias.
12. Reverse saturation current is more in BJT.	12. Reverse saturation current is less in FET and hence it can be operated over wide ranges of temperature.

2.11 MOSFET

The impedence Z of JFET is much higher compared to a junction transistor. But for MOSFETs, the impedence Z is much higher compared to JFETs. So these are very widely used in ICs and are fast replacing JFETs.

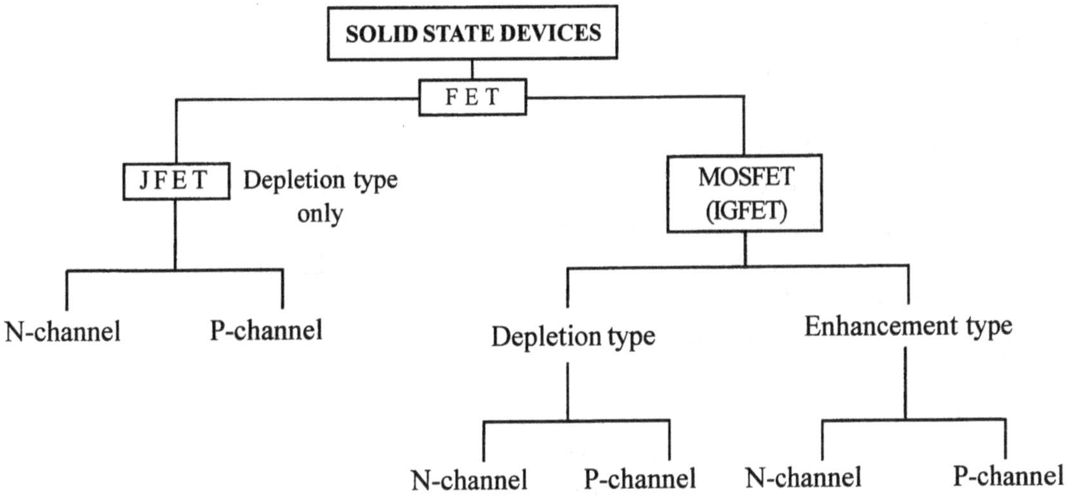

Fig. 2.28 FET Tree.

There are two types of MOSFETs, enhancement mode and depletion mode. These devices, derive their name from Metal Oxide Semiconductor Field Effect Transistor (MOSFET) or MOS transistor (See Fig. 2.28). These are also known as IGFET (Insulated Gate Field Effect Transistor). The gate is a metal and is insulated from the semiconductor (source and drain are semiconductor type) by a thin oxide layer.

In FETs the gate source junctions is a p-n junction. (If the source is n-type, gate will be p-type). But in MOSFETs, there is no p-n junction between gate and channel, but rather a capacitor consisting of metal gate contact, a dielectric of SiO_2 and the semiconductor channel. [two conductors separated by dielectric]. It is this construction which accounts for the very large input resistance of 10^{10} to $10^{15}\Omega$ and is the major difference from the JFET.

The symbols for MOSFETs are,

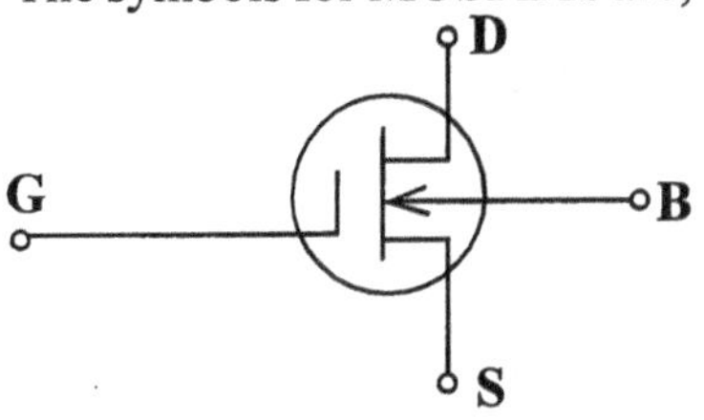

Fig. 2.29 n-channel MOSFET, depletion type

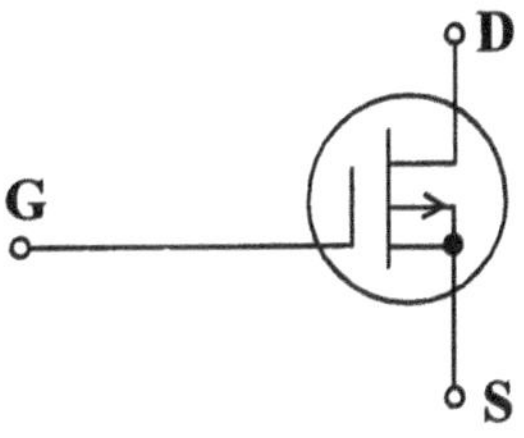

Fig. 2.30 p-channel MOSFET, depletion type

Some manufacturers internally connect the bulk to the source. But in some circuits, these two are to be separated (See Fig. 2.29). The symbol, if they are connected together is, given in Fig.2.30.

for p - channel, the arrow will point outwards.

Bulk is the substrate material taken.

Advantages of MOSFETs (over JFETs and other devices)

1. High package density $\simeq 10^5$ components per square cm.
2. High fabrication yield. p - channel Enhancement mode devices require 1 diffusion and 4 photo masking steps.

   But for bipolar devices, 4 diffusions and 8 – 10 photo masking steps are required.
3. Input impedance is very high $Z_n \simeq 10^{14}\Omega$.
4. Inherent memory storage : charge in gate capacitor can be used to hold enhancement mode devices ON.
5. CMOS or NMOS reduces power dissipation, micropower operation. Hence, CMOS ICs are popular.
6. Can be used as passive or active element.

   Active : As a storage device or as an amplifier etc.

   Passive : As a resistance or voltage variable resistance.
7. Self Isolation : Electrical isolation occurs between MOSFETs in ICs since all p-n junctions are operated under zero or reverse bias.

Disadvantage

1. Slow speed switching.
2. Slower than bipolar devices.
3. Stray and gate capacitance limits speed.

2.11.1 Depletion MOSFET

In enhancement type MOSFET, a channel is not diffused. It is of the same type (p-type or n-type) as the bulk or substrate. But if a channel is diffused between source and drain with the same type of impurity as used for the source and drain diffusion depletion type MOSFET will be formed.

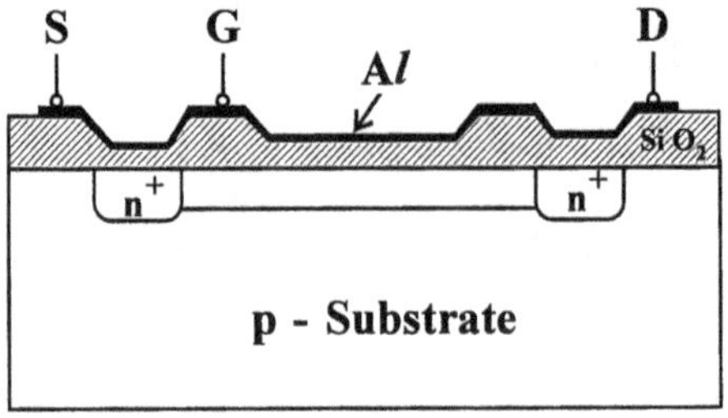

Fig. 2.31 Depletion MOSFET.

The conductivity of the channel in the case of depletion type is much less compared to enhancement type. The characteristics of depletion type MOSFETs are exactly similar to that of JFET.

When V_{DS} is positive and $V_{GS} = 0$, large drain current denoted as I_{DSS} (Drain to source current saturation value) flows. If V_{GS} is made negative, positive charges are induced in the channel through SiO_2 of the gate capacitor. But the current in MOSFET is due to majority carriers. So the induced positive charges in the channel reduces the resultant current I_{DS}. As V_{GS} is made more negative, more positive charges are induced in the channel. Therefore its conductivity further decreases and hence I_D decreases as V_{GS} is made more negative (for n-channel depletion type). The current I_D decreases because, the electrons from the source recombine with the induced positive charges. So the number of electrons reaching the drain reduces and hence I_D decreases. So because of the recombination of the majority carriers, with the induced charges in the channel, the majority carriers will be depleted. Hence, this type of MOSFET is knows as ***depletion type MOSFET***. JFET and depletion MOSFET have identical characteristics.

A MOSFET of depletion type can also be used as enhancement type. In the case of n-channel depletion type (source and drain are n-type), if we apply positive voltage to the gate-source junction, negative charges are induced in the channel. So, the majority carriers (electrons in the source) are more and hence I_D will be very large. Thus, depletion type MOSFET can also be used as enhancement type by applying positive voltage to the gate (for n-channel type).

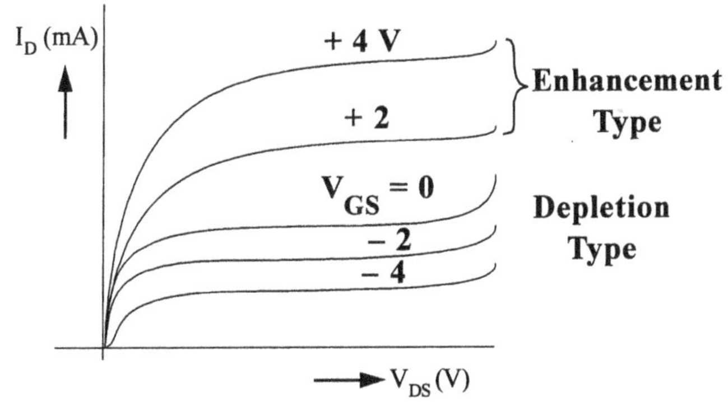

Fig. 2.32 Drain characteristics of DMOSFET.

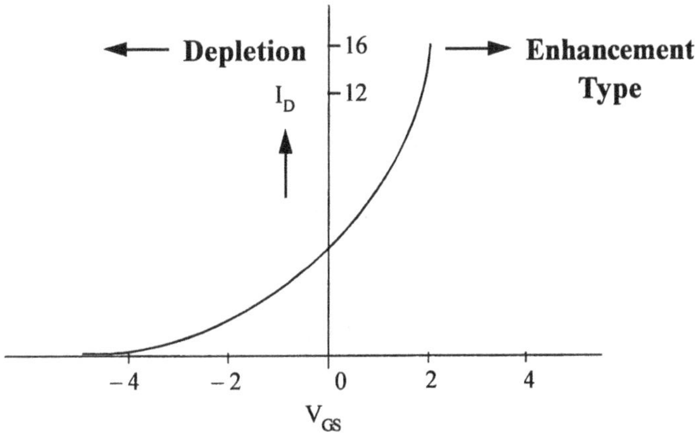

Fig. 2.33 Gate characteristics of depletion type MOSFET.

Problem 2.1

A JFET is to be connected in a circuit with drain load resistor of 4.7 kΩ, and a supply voltage of V_{DD} = 30V. V_D is to be approximately 20V, and is to remain constant within ± 1V. Design a suitable self bias circuit.

Solution

$$V_D = V_{DD} - I_D \cdot R_L$$

$$I_D = \frac{V_{DD} - V_D}{R_L}$$

$$= \frac{30V - 20V}{4.7k\Omega} = 2.1 \text{ mA}$$

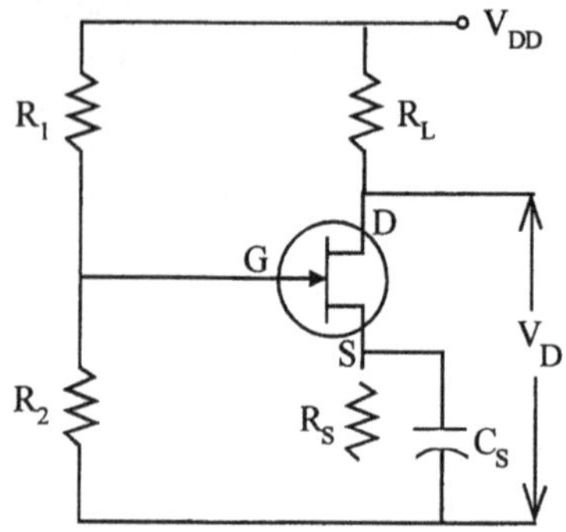

Fig. 2.34 For problem 2.1.

for V_D to be constant, it should be within ± 1V,

$$\Delta I_D = \frac{\pm V}{R_L}$$

$$= \frac{\pm 1V}{4.7k\Omega} \simeq \pm 0.2\text{mA}$$

$$I_D = (2.1 \pm 0.2\text{mA})$$

$$I_D(\min) = (2.1 - 0.2) = 1.9\text{mA}$$

$$I_D(\max) = (2.1 + 0.2) = 2.3\text{mA}$$

Indicate the points A and B on the maximum and minimum transfer characteristics of the FET. Join these two points and extend it till it cuts at point C.

The reciprocal of the slope of the line gives R_S

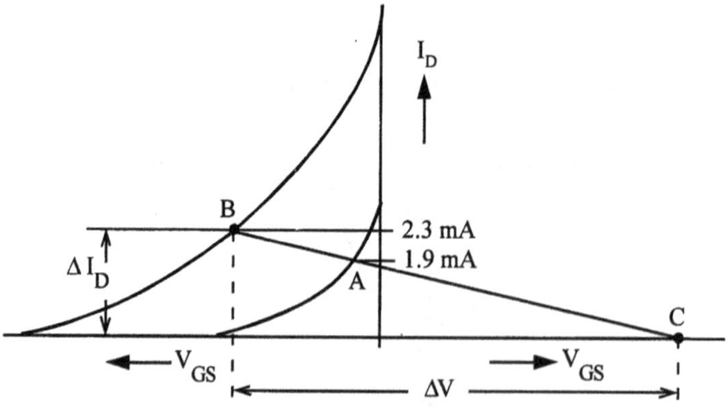

Fig. 2.35 For problem 2.1.

$$R_S = \frac{\Delta V}{\Delta I} = \frac{10V}{2.5mA} = 4k\Omega$$

The bias line intersects the horizontal axis at V_G = 7V

∴ An external bias of 7V is required.

$$V_G = \frac{R_2}{R_1+R_2}.V_{DD}$$

$$\therefore \quad \frac{R_2}{R_1+R_2} = \frac{V_G}{V_{DD}} = \frac{7V}{30v}$$

$$\therefore \quad \frac{R_2}{R_1} = \frac{7}{23}$$

R_2 and R_1 should be large to avoid overloading input signals. If $R_2 = 700\ k\Omega$,

$$R_1 = 2.3\ M\Omega$$

Problem 2.2

(a) For the common source amplifier, calculate the value of the voltage gain, given

$$r_d = 100\ k\Omega,\ g_m = 300\ \mu\ ,\quad R_L = 10k\Omega, \qquad R_o = 9.09k\Omega$$

Solution

$$A_V = \frac{-g_m.r_d.R_L}{rd+R_L} = \frac{-3000\times10^{-6}\times100\times10^3\times10\times10^3}{(100\times10^3)+(10\times10^3)} = -27.3$$

(b) If $C_{DS} = 3pf$, determine the output impedance at a signal frequency of 1MHz.

Solution

$$X_C = \frac{1}{2\pi f\ C_{DS}}$$

$$f = 1\ MHz,$$

$$X_C = \frac{1}{2\pi\times1\times10^6\times3\times10^{-12}} = 53k\Omega$$

$$|Z_0| = \frac{R_0.X_C}{\sqrt{R_0^2+X_C^2}}$$

$$|Z_0| = \frac{9.09k\Omega\times53k\Omega}{\sqrt{(9.09k\Omega)^2+(53k\Omega)^2}} = 8.96k\Omega$$

2.11.2 ENHANCEMENT MOSFET

n-channel is induced, between source and drain. So it is called as n-channel MOSFET.

For the MOSFET shown in Fig. 2.36, source and drain are n-type. Gate is A*l* (metal) in between oxide layer is there. The source and drain are separated by 1 millimeter. Suppose the p substrate is grounded, and a positive voltage is applied at the gate. Because of this an electric field will be directed perpendicularly through the oxide. This field will induce negative charges on the semiconductor side. The negative charge of electrons which are the minority carriers in the p - type substrate form an inversion layer. As the positive voltage on the gate increases, the induced negative charge in the semiconductor increases. The region below the SiO_2 oxide layer has n-type carriers. So the conductivity of the channel between source and drain increases and so current flows from source to drain through the induced channel. Thus, the drain current is enhanced by the positive gate voltage and such a device is called an ***enhancement type MOS***.

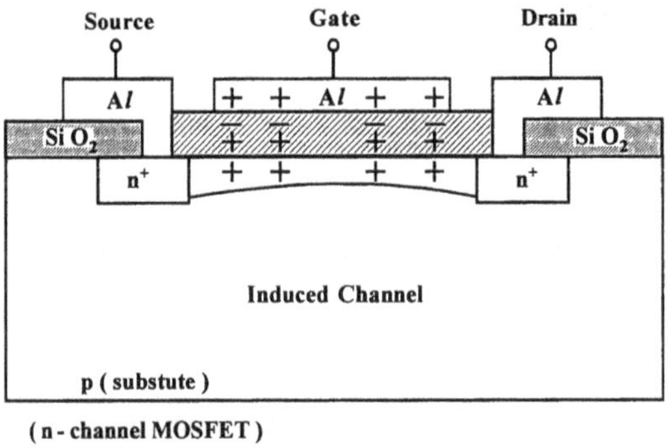

Fig. 2.36 n-channel MOSFET

The volt-ampere drain characteristic of an n-channel enhancement mode MOSFET are as shown in Fig. 2.37.

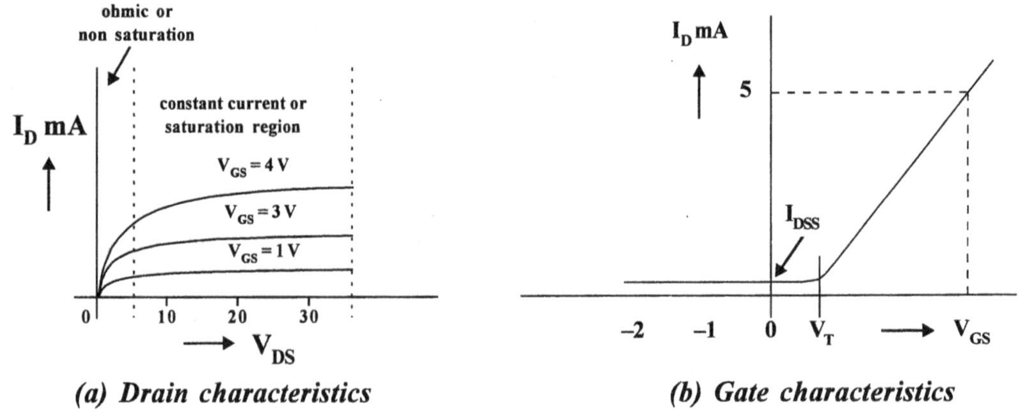

Fig. 2.37

The current I_D for $V_{GS} \leq 0$ is very small being of the order of few nano amperes. As V_{GS} is made positive, the current I_D increases, slowly first and then much more rapidly with an increase in V_{GS} (Fig. 2.37(b)). Sometimes the manufacturers specify gate, source threshold voltage V_{GST} at which I_D reaches some defined small value $\simeq$ 10μA. For a voltage < V_{GST}, (V_{GS} threshold) I_D is very small. $I_{D(ON)}$ of MOSFET is the maximum value of I_D which remains constant for different values of V_{DS}.

In the ohmic region, the drain characteristic is given by

$$I_D = \frac{\mu C_0 . w}{2L} [2(V_{GS} - V_T) V_{DS} - V^2_{DS}]$$

where, μ = Majority carriers mobility.
C_0 = Gate capacitance per unit area.
L = Channel length.
w = Channel width perpendicular to C

The ohmic region in the I_D vs V_{DS} characteristic of FET or MOSFET is also known as ***triode region***, because like in a triode I_D increases with V_{DS}. The constant current region is known as ***pentode region*** because the current remains constant with V_{DS}.

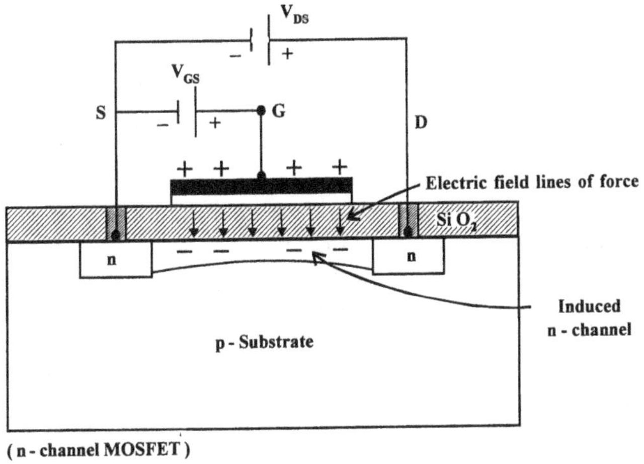

Fig. 2.38 MOSFET structure

2.11.3 MOSFET OPERATION

Consider a MOSFET in which source and drain are of n-type. Suppose a negative potential is applied between gate and source. SiO_2 is an insulator. It is sandwiched between two conducting regions the gate (metal) and the semiconductor p-type substrate. Therefore, equivalent capacitor is formed, with SiO_2 as the dielectric whenever a positive charge is applied to one plate of a capacitor a corresponding negative charge is induced on the opposite plate by the action of the electric field with in the dielectric. A positive potential is applied to the gate. So a negative charge is induced on the opposite plate by the action of electric field within the dielectric, in the p - substrate. This charge results from the minority carriers (electrons) which are attracted towards the area below the gate, in the p - type substrate. The electrons in the p - substrate are attracted towards the lower region of SiO_2 layer because there is positive field acting from the gate through SiO_2 layer, because of the applied positive potential to the gate V_{GS}. As more number of electrons are attracted towards this region, the hole density in the p - substrate (below the SiO_2 layer between n - type source and drain only) decrease. This is true only in the relatively small region of the substrate directly below the gate. An n - type region now extends continuously from source to drain. The n-channel below the gate is said to be induced because it is produced by the process of electric inductor. If the positive gate potential is removed, the induced channel disappears.

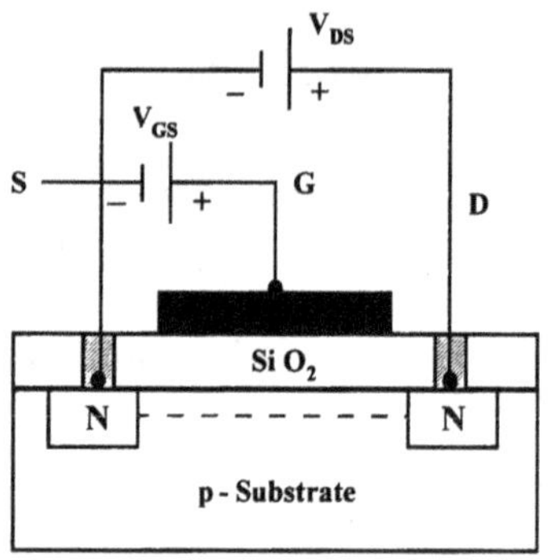

Fig. 2.39 MOSFET biasing.

If the gate voltage is further increased, greater number of negative charge carriers are attracted towards the induced channel. So as the carriers density increases, the effective channel resistance decreases, because there are large number of free carriers (electrons). So the resistance seen by the V_{DS} depends on the voltage applied to the gate. The higher the gate potential, the lower the channel resistance, and the higher drain current I_D. This process is referred to as enhancement because I_D increases, and the resulting MOSFET is called ***enhancement type MOSFET***. The resistance looking into the gate is high since the oxide is an insulator. The resistance will be of the order of $10^{15}\Omega$ and the capacitance value will also be large.

Depletion type MOSFET can also be constructed in the same way. These are also known as DE MOSFETs. Here as the gate voltage increases, the channel is depleted of carriers. So channel resistance increases.

Here the region below the gate is doped n - type. If negative voltage is applied to the gate, negative charge on the gate induces equal positive charge i.e., holes. These holes will recombine with the electrons of the n - channel between sources drain since channel resistivity increases. The channel is depleted of carriers. Therefore I_D decreases as V_{GS} increases. (negative voltage) If we apply positive voltage to V_{GS}, then this becomes enhancement type.

2.11.4 MOSFET CHARACTERISTICS

There are two types of MOS FETs,

1. Enhancement type
2. Depletion type.

Depletion type MOSFET can also be used as enhancement type. But enhancement type cannot be used as depletion type. So to distinguish these two, the name is given as depletion type MOSFET which can also be used in enhancement mode.

As V_{GS} is increased in positive values (0, +1, +2 etc) if I_D increases, then it is enhancement mode of operation because I_D is enhanced or increased.

As V_{GS} is increased in negative values (0, –1, –2, etc) if I_D decreases, then it is depletion mode of operation.

Consider n - channel MOSFET, depletion type. Fig. 2.40.

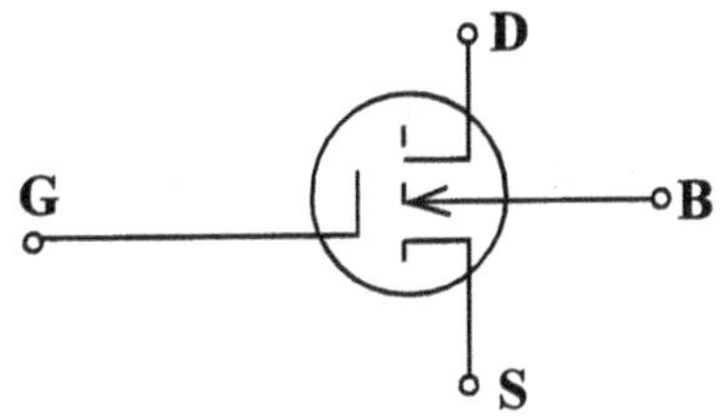

Fig. 2.40 n-channel MOSFET.

Drain characteristic : I_D versus V_{DS}

When $V_{GS} = 0V$, for a given V_{DS}, significant current flows just like in a JFET. When the gate is made negative (i.e., $V_{GS} = -1V$,) it is as if a negative voltage is applied for one plate of plate capacitor. So a positive charge will be induced below the gate in between the n type source and drain. Because it is semiconductor electrons and holes are induced in it below the gate. The channel between the source and drain will be depleted of majority carriers electrons, because these induced holes will recombine with the electrons. Hence, the free electron concentration in the channel between source and drain decreases or channel resistivity increases. Therefore current decreases as V_{GS} is made more negative i.e., –1, –2, etc. This is the depletion mode of operation, because the channel will be depleted of majority carriers as V_{GS} is made negative (for n-channel MOSFET).

In a FET, there is a p-n junction between gate and source. But in MOSFET, there is no such p-n junction. Therefore, positive voltage positive V_{GS} can also be applied between gate and source. Now a negative charge is induced in the channel, thereby increase free electron concentration. So channel conductivity increases and hence I_D increases. Thus, the current is enhanced. So, the device can be used both in enhancement mode and depletion mode.

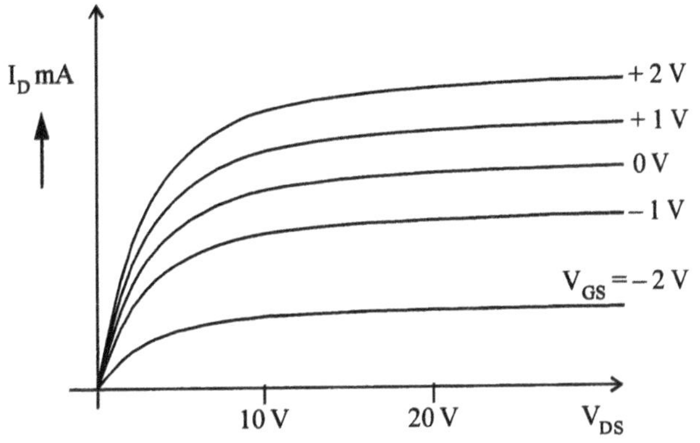

Fig. 2.41 Drain characteristics (N MOS).

In the transfer characteristic as V_{GS} increases, V_D increases. Similarly in the depletion mode as V_{GS} is increases in negative values, I_D decreases

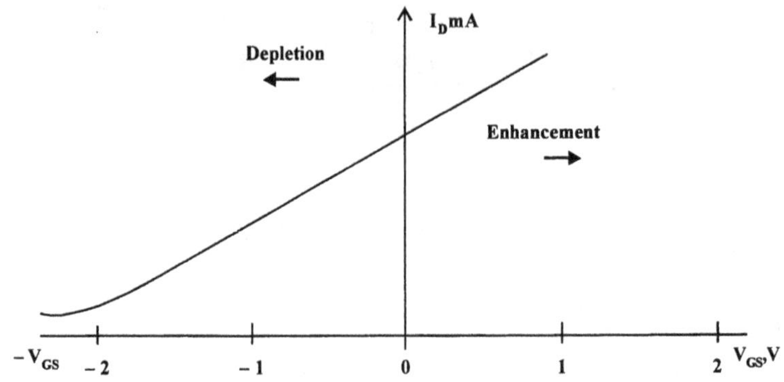

Fig. 2.42 Gate characteristics

A JFET is a depletion type, because as V_{GS} is made positive (+1, +2 etc for n-channel FET) I_D increases. So it can be used in enhancement type. But there is no other type of JFET which is used only in enhancement mode and no depletion mode, hence no distinction is made.

SUMMARY

- The relation between Emitter Efficiency γ, Transportation Factor β^* and Current Gain α is,

$$\alpha = \beta^* \times \gamma$$

- Transistor is an acronym for the words Transfer Resistor. As the input side in forward biased and output side is reverse biased, there is transfer of resistance from a lower value on input side to a higher value on the output side.
- Transitor can be used as an amplifiers, when operated in the Active Region. It is also used as a Switch, when operated in the cut-off and saturation regions.
- The three configurations of Transistor are Common Emitter, Common Base and Common Collector.
- The proper name for this device being referred as transistor is Bipolar Junction Transistor (BJT).
- The three regions of the output characteristics of a transistor are
  1. Active Region
  2. Saturation Region
  3. Cut-off Region
- JFET is UNIPOLAR Device (Unipolar only one type of carriers either holes or electrons)
- JFET device has Higher input resistance compared to BJT and Lower input resistance compared to MOSFET.
- The disadvantage of JFET amplifier circuits is Smaller Gain - Bandwidth product compared to BJT amplifier circuits.
- I_{DSS} = Satuaration value of Drain current when gate is shorted to source.
- For zero drift current in the case of JFET, $0.007|I_D| = 0.0022\ g_m$
- For zero drift current, in the case of JFETs, $|V_P| - |V_{GS}| = 0.63V$
- Expression for g_m for JFET interms of g_{mO} is, $g_m = \left(1 - \frac{V_{GS}}{V_P}\right)$
- The two types of n-channel or p-channel MOSFETs are
  1. Depletion type
  2. Enchancement type
- In JFET terminology, $B\ V_{DGO}$ parameter means the value of breakdown voltage V_{DG} at breakdown when the source is left open i.e., $I_S = 0$.
- The parameters $B\ V_{GSS}$ stands for breakdown voltage V_{GS} when drain is shorted to source.

OBJECTIVE TYPE QUESTIONS

1. For identical construction, ... types of bipolar junction transistors (BJTs) have faster switching times.
2. The arrow mark in the symbols of BJTs indicates ....
3. For NPN Bipolar Junction Transistor Emitter Efficiency γ = ...
4. For PNP Transistor (BJT), the transportation factor β^* = ...
5. Relation between α, β^* and γ for a transistor (BJT). α = ...
6. The expression for α in terms of β for a BJT is ....
7. The expression for β in terms of α for a BJT is ....
8. Expression for I_C in terms of β, I_B and I_{CBO} is ....
9. Expression for I_{CEO} in terms of I_{CBO} and α is ....
10. β' of a BJT is defined as ....
11. β' of a transistor (BJT) is ....
12. Relation between β' and β is ....
13. β' is synonymous to .... and β is same as ....
14. β is .... (for DC Currents).
15. β' is .... (for AC Currents).
16. Base width modulation is ....
17. IGFET is the other name for .... device.
18. In JFET recombination noise is less because it is .... device.
19. The disadvantage of JFET amplifier circuit is ....
20. The D, G, S terminals of JFET are similar to .... terminals of BJT respectively.
21. The voltage V_{DS} at which I_D tends to level off, in JFET is called ....
22. The voltage V_{GS} at which I_D becomes zero in the transfer characteristic of JFET is called ....
23. The range of $r_{DS\,(ON)}$ for JFET is ....
24. For low electric fields E_x of the order of 10^3V/cm, $\mu\ \alpha$ ....
25. Expression for V_{GS} in terms of V_P, ...
26. Expression for I_{DX} in terms of I_{DSS} in I_{DS} ....
27. I_{DSS} in defined as ....
28. JFET can be used as .... resistor.
29. Relation between μ, r_d and g_m for JFET is ....
30. The square law device is ....
31. The MOSFET that can be used in both enhancement mode and depletion mode is ....

ESSAY TYPE QUESTIONS

1. With the help of a neat graph qualitatively explain the Potential distribution through a transistor (BJT).
2. Explain about the different current components in a transistor.
3. Derive the relation between β and β'.
4. Derive the relation between α, β^* and γ.
5. Differentiate between the terms h_{FE} and h_{fe}. Derive the relationship between them.
6. Qualitatively explain the input and output.
7. Explain the V-I characteristics in Common Emitter Configuration.
8. Describe the V-I characteristics of a transistor in Common Collector Configuration and Explain.
9. Compare the input and output characteristics of BJT in the three configurations, critically.
10. Compare BJT, JFET and MOSFET devices in all respects.
11. Obtain the expression for the pinch off voltage V_p in the case of n-channel JFET.
12. Deduce the condition for JFET biasing for zero drift current.
13. Draw the structure and explain the static Drain and Gate characteristics of n-channel JFET. Repeat the same for p-channel JFET.
14. Draw the structure of p-channel MOSFET and Qualitatively explain the static Drain and Gate characteristics of the device.
15. What are the applications of JFET and MOSFET devices?

MULTIPLE CHOICE QUESTIONS

1. As reverse bias voltage is increased, for a diode, the base width at the junction

(a) decreases (b) increases (c) remains same (d) none of these

2. Expression for 'α' of a BJT in terms of I_{pc}, I_E, I_c etc is

(a) $\frac{I_{pc}}{I_E}$ (b) $\frac{I_{pE}}{I_E}$ (c) $\frac{I_{nc}}{I_E}$ (d) $\frac{I_{nE}}{I_C}$

3. Expression for β^* is,

(a) $\frac{I_{pc}}{I_{p_B}}$ (b) $\frac{I_{pc}}{I_{n_B}}$ (c) $\frac{I_{pc}}{I_{p_E}}$ (d) $\frac{I_{p_E}}{I_{pc}}$

4. The relation between I_{CEO}, I_{CBO} and α is, I_{CEO} =

(a) $\frac{I_{CBO}}{\alpha}$ (b) $\frac{I_{CBO}}{(1+\alpha)}$ (c) $\frac{I_{CBO}}{(1-\alpha)^2}$ (d) $\frac{I_{CBO}}{(1-\alpha)}$

5. **For a transistor in C.E configuration to be at cut-off, the condition is,**

(a) $V_{BE} = -V_{BB} + R_{B.} I_{CBO} \leq -0.4$ (b) $V_{BE} = +V_{BB} - R_{B.} I_{CBO} \leq 0.4$

(c) $V_{BE} = -V_{BB} + R_{B.} I_{CBO} \leq -0.1$ (d) $V_{BE} = -V_{BB} - R_{B.} I_{CBO} \leq -0.1$

6. **The condition for test for saturation in a BJT is**

(a) $|I_B| \leq \left|\frac{I_C}{\alpha}\right|$ (b) $|I_B| \leq \left|\frac{I_C}{\beta}\right|$ (c) $|I_B| \geq \left|\frac{I_C}{\alpha}\right|$ (d) $|I_B| \geq \left|\frac{I_C}{\beta}\right|$

7. **Expression for saturation resistance Rcs of a transistor (BJT) is ...**

(a) $Rcs = \frac{V_{CE}}{I_C(sat)}$ (b) $Rcs = \frac{I_C(sat)}{V_{CE}}$

(c) $Rcs = \frac{V_{CE}\ (sat)}{I_C}$ (d) $Rcs = \frac{V_{CE}}{I_C}$

8. **Insulated Gate Field Effect Transistor (IGFET) is a ...**

(a) Normal JFET device (b) n-channel J FET device

(c) p-channel JFET device (d) MOSFET device

9. **Compared to BJT, JFET is**

(a) Less noisy (b) more noisy

(c) Same noisy (d) can't be said

10. **At pinch off voltage in a JFET device,**

(a) Channel width is zero and I_D becomes zero

(b) Channel width is zero, but I_D is not zero

(c) V_{GS} becomes zero

(d) None of these

11. **For high eletric field strengths $\varepsilon > 10^4$ V/cm, mobility of carriers is proportional to**

(a) ε_x (b) $\frac{1}{(\varepsilon_x)^2}$ (c) $\frac{1}{\varepsilon_x}$ (d) None of these

12. **The pinch off voltage Vp in the case of n-channel JFET is proportional to**

(a) $N_D^{\ 2}$ (b) $\frac{1}{N_D}$ (c) $\frac{1}{\sqrt{N_D}}$ (d) N_D

13. **Expression for I_D in the case of JFET is,**

(a) $I_{DSS}\left(1 - \frac{V_{gs}}{V_p}\right)$ (b) $I_{DSS}\left(1 - \frac{V_{gs}}{V_p}\right)^2$

(c) $I_{DSS}\left(1 + \frac{V_{gs}}{V_p}\right)^2$ (d) $I_{DSS}\left(\frac{V_{gs}}{V_p} - 1\right)^2$

14. The condition to be satisfied for zero drift current is,

(a) $0.007\ |I_D| = 0.0022$ gm (b) $0.07\ |I_D| = 0.022$ gm

(c) $0.7\ |I_D| = 0.0022$ gm (d) $0.007\ |I_D| = 0.022$ gm

15. The relation between μ, r_d and g_m for JFET is,

(a) μ = rd/gm (b) rd = μ.²gm (c) μ = rd . gm (d) μ = rd².gm

16. Typical values of μ, r_d and g_m for JFET are,

(a) μ = 5, rd = 100 Ω, gm = 100 m℧ (b) μ = 1000, rd = 10 Ω, gm = 0.1 m℧

(b) μ = 6, rd = 1MΩ, gm = 10 K℧ (d) μ = 7, rd = 1MΩ, gm = 0.5 mA/V

17. The resistor which is connected in series with source resistance Rs to reduce distortion in JFET amplifier circuits is called

(a) Swamping resistor (b) swinging resistor

(c) bias resistor (d) distortion control resistor

3 Biasing and Stabilization

In this Chapter,

- The need for biasing and its significance in amplifier circuits is explained. The Quiescent point, or Operating point or 'Q' point is explained.
- Different types of biasing circuits are given and the expressions for stability factors are derived.
- Variation of 'Q' point with temperature and temperature compensating circuits are given.

3.1 OPERATING POINT

The transistor output characteristics in Common Emitter Configuration are as shown in Fig. 3.1. The point on the characteristics of the transistors indicated by V_{CE}, I_C is called ***Operating Point*** or ***Quiescent Point*** Q. This is shown in Fig. 3.1. The operating point must be in the active region for transistor to act as an amplifier.

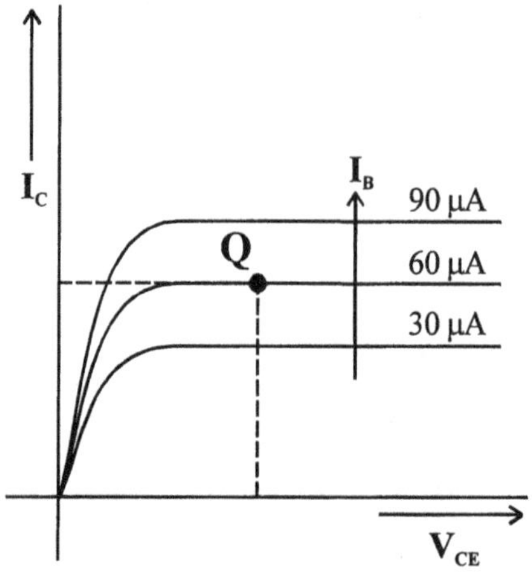

Fig. 3.1 Output characteristics of BJT.

Transistor when being used as an amplifier is operated in the active region. As I_B (in μA) increases, I_C (in mA) increases. So for a small change in I_B, there is corresponding large change in I_C and thus transistor acts as an amplifier. The operating point or the quiescent point has to be fixed when any transistor is being used as an amplifier.

$$A_V = \frac{\alpha' R_L}{r_C} \qquad R_L << r_c.$$

R_L = Load resistance

r_c = Collector resistance

α' = Small signal current gain (less than 1)

A_V = Voltage gain

Therefore, transistor acts as an amplifier.

$$\alpha' = \frac{\Delta I_C}{\Delta I_E}$$

$$\Delta V_C = \alpha' \times R_L \times \Delta I_E$$

$$\Delta V_i = r_C \, \Delta I_E$$

$$\therefore \quad A_V = \frac{\Delta V_C}{\Delta V_i} = \frac{\alpha' R_L \times \Delta I_E}{r_C \times \Delta I_E} = \frac{\alpha' R_L}{r_C}$$

$R_L >> r_c$. So $A_V > 1$

The quiescent point changes with temperature T, because it depends on the parameters β, I_{CO} or V_{BE} etc. β and I_{CO} depend upon temperature. Hence operating point changes with temperature. So by suitably biasing the transistor circuit, the quiescent point may be made stable or not to change with temperature. Compensation techniques are also used in order to make the operating point stable with temperature.

It is very essential to choose the operating point properly so that the transistor is operated in the active region and that the operating point does not change with temperature. Since if the input is sinusoidal wave, the output will also be a true replica of the input or it will also be a perfect sine wave. Otherwise, the output will be distorted, clipped etc., if the quiescent point is not chosen correctly or it is changing with temperature. Hence it is essential to choose the operating point correctly. Operating point is fixed with respect to the output characteristic of a transistor only.

So now the question is how to choose the operating point? Before this let us consider as to how the quiescent point is fixed. Consider the circuit as shown in Fig. 3.2 and 3.3. It is Common Emitter Transistor circuit with a load resistance R_L. PNP transistor is being considered.

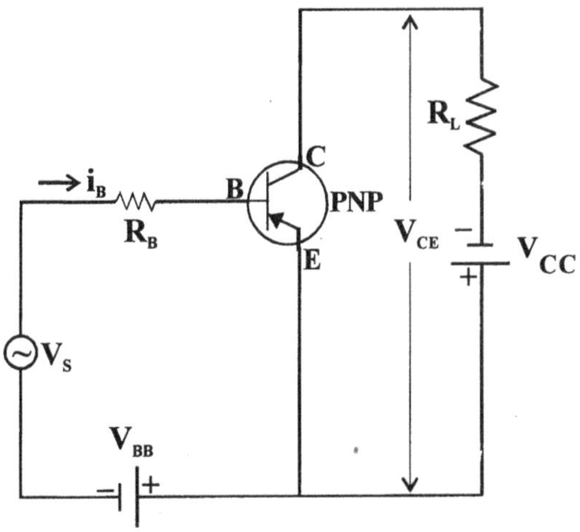

Fig. 3.2 Amplifier circuit.

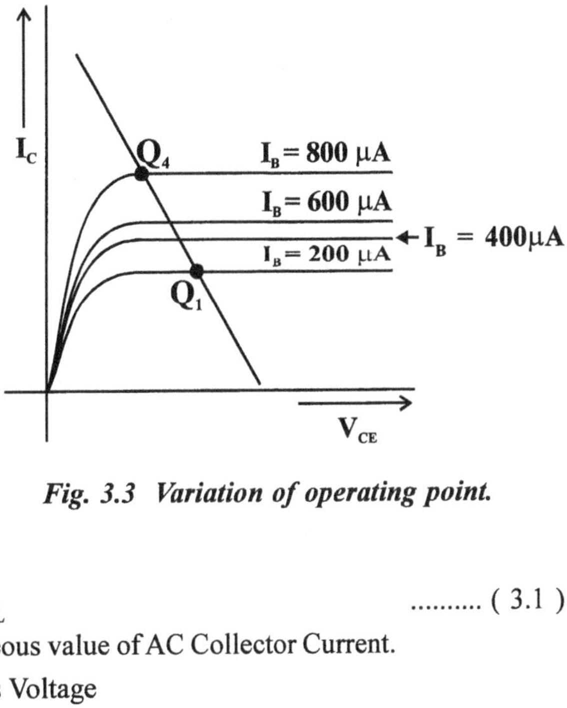

Fig. 3.3 Variation of operating point.

From the above circuit,

$$V_{CE} = V_{CC} - i_C \times R_L \qquad \text{.......... (3.1)}$$

V_{CC} is DC Bias Supply. i_C is instantaneous value of AC Collector Current.

V_{CE} = Instantaneous Voltage

From the Equation (3.1), the load line can be drawn. To draw the load line,

Let $i_C = 0$ then $V_{CE} = V_{CC}$. This is one point of the load line.

Let $V_{CE} = 0$ then $i_C = \frac{V_{CC}}{R_L}$. This another point of the load line.

The value of r_C is very small compared to R_L. Hence it can be neglected.

$$i_C = \left(-\frac{V_{CC}}{R_L + r_C}\right)$$

Neglecting, r_C, $i_C = \left(-\frac{V_{CC}}{R_L}\right)$

So by setting $i_C = 0$, $V_{CE} = 0$, the load line is drawn. It is a straight line independent of the transistor parameters. The slope of the line depends on R_L the load resistance only. V_{CC} is a fixed quantity. For a given value of I_B, I_C is also fixed when the transistor is being operated in the active region. Therefore for $I_B = 200\mu A$, the quiescent point is Q_1. For $I_B = 800\ \mu A$, the operating point is Q_4 and so on. Now, when the input signal applied between base and emitter is a symmetrical signal, varying both on positive and negative sides equally, within the range of the transistor characteristics, then the operating point is chosen to lie in the centre of the output characteristics (i.e., corresponding to $I_B = 300\ \mu A$), so that the output signal is a true replica of the input without distortion (V_{CC} and V_{BB} are bias voltages. Input signal is the a.c signal between B and E). When the input signal is symmetrical, the ***quiescent point is chosen at the centre of the load line***.

The load line as explained above, if drawn, is called as the ***dynamic (AC) load line***. If the output impedance in the circuit is reactive, then the load line will be elliptical. Since the voltage and current are shifted in phase, and the resulting equation will be that of an ellipse. For a resistance the load line is a straight line.

While drawing a load line, if only the collector resistance R_C is considered and load resistance R_L is infinity (∞), then the load line is drawn with the points $\left(\frac{V_{CC}}{R_C} \text{ and } V_{CE}\right)$. It is called as ***Static***.

3.2 DC AND AC LOAD LINES

But if the load line is drawn considering a finite load resistance R_L, and the resistance considered is R_L in parallel with R_C, it is called ***AC Load Line or Dynamic Load Line***, with AC Input.

There are three types of Biasing Circuits

1. Fixed Bias circuit or Base Bias circuit
2. Collector to Base Bias circuit
3. Self Bias or Emitter Bias circuit (Universal bias or Voltage divider bias circuit).

3.3 NEED FOR BIASING/IMPORTANCE OF BIASING

In order that amplification is achieved, the BJT must be operated in the active region. To ensure thus, E-B junction of the transistor (BJT) must be forward biased and collector-Base junction must be reverse biased. The operating point or quiescant point must be in the centre of the active region. To ensure thus, biasing must be done. Similarly to use the BJT as on-off switch, the operation point must change between cut-off and saturation region. So biasing is to be done.

3.4 FIXED BIAS CIRCUIT OR (BASE BIAS CIRCUIT)

Consider the NPN Transistor Circuit as shown in Fig. 3.4. R_C is collector resistance. C_C is coupling capacitor. It blocks DC since capacitor is open circuit for D.C. So the output signal will be pure a.c signal. R_C. limits the current I_C . R_B provides the bias voltage for the base. Since this is NPN transistor, V_{CC} should be positive, because the collector should be reverse biased. The drop across R_b, ($V_{CC} - I_B \times R_B$) is negative and it provides positive bias for the base. Since R_L is connected, dynamic load line has to be considered. Suppose the output characteristics of the transistor are as shown in Fig. 3.5.

Suppose Q_1 is the centre point, and it is chosen as the operating point. But if the input signal is greater than 40 μA on the negative side, with Q_1, the transistor will be cut off since for that swing $I_B = 0$, and the transistor is cut off, so another operating point is to be chosen. Choosing the operating point means, a correct value of I_B, I_C, V_{CE}. (if the transistor is in Common Emitter Configuration) so that we get undistorted output.

Since from the below circuit,

$$I_B = \frac{V_{CC} - V_{BE}}{R_B}$$

$$I_C = \frac{V_{CC} - V_{CE}}{R_C}$$

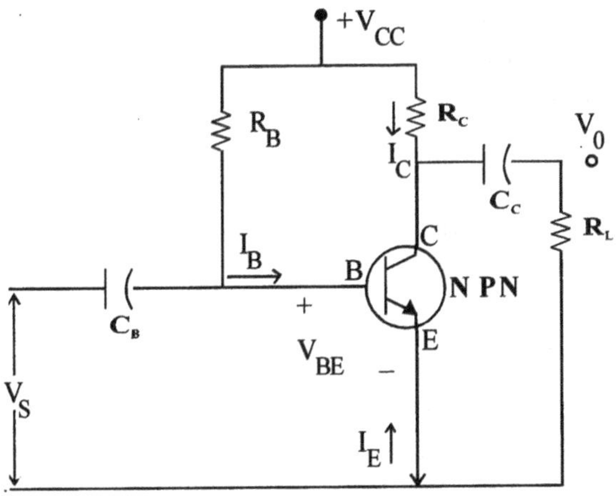

Fig. 3.4 Fixed bias circuit

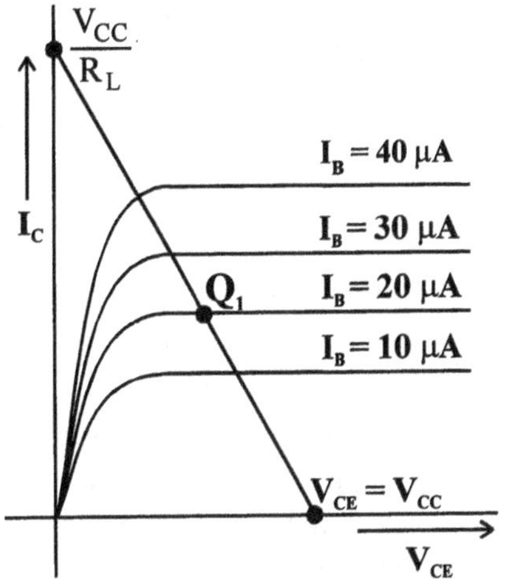

Fig. 3.5 Operating point variation

V_{BE} is the base emitter voltage for a forward biased EB Junction. This will be 0.2V for Germanium and 0.6 for Silicon. In order to get a large value of I_B or the change in the operating point, V_{CC} or R_B has to be changed. Since once V_{CC}, R_B etc., are fixed, I_B is fixed. So the above circuit is called as ***Fixed Bias Circuit***.

I_B is fixed when R_B and V_{CC} are fixed. Because in the expression, for I_B, β is not appearing. So once R_B and V_{CC} are fixed, the biasing point is also fixed. Hence the name Fixed Bias Circuit.

3.5 COLLECTOR TO BASE BIAS CIRCUIT

The circuit shown in Fig. 3.4 is fixed bias circuit, since I_B is fixed, if V_{CC}, R_C and V_{BB} are fixed. But an improvement over this circuit is the circuit shown is Fig. 3.6.

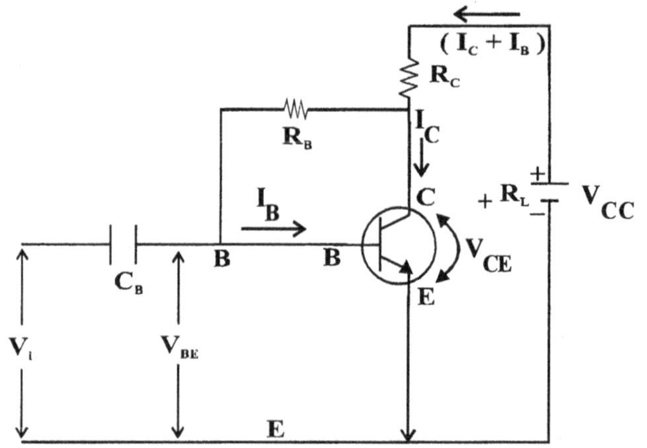

Fig. 3.6 Collector to base bias circuit.

In this circuit, R_B is connected to the collector point C, instead of V_{CC} as in Fig. 3.4. V_i is input, the capacitor will block DC input and allow only AC signal to the base. Since it is NPN transistor, collector junction is reverse biased. R_C is a resistor in the collector circuit, which controls the reverse bias voltage of the collector junction with respect to emitter.

$$\because \quad V_{CE} = V_{CC} - I\,R_C$$
$$I = I_C + I_B$$

This circuit is an improvement over the Fig. 3.4 fixed bias circuit because,

$$V_{CE} = V_{CC} - (I_C + I_B)\,R_C$$
$$I = I_C + I_B$$

With temperature I_C increases. Since V_{CC} is same, I is same and so I_B decreases.

As I_C increases with temperature because of the increases in I_{CO}, $(I_C + I_B)\,R_C$ product increase. So V_{CE} reduces. That means the reverse bias voltage of the collector junction is reduced or number of carriers collected by the collector reduces. Hence I_B also reduces. Therefore, because of the reduction in the bias current I_B, the increase in I_C will not be as much as it would have been considering only increase in I_{CO} with temperature.

We can calculate the stability factor for this circuit. Applying KVL,

$$V_{BE} - V_{CC} + (I_B + I_C)\,R_C + I_B \cdot R_B = 0$$

or

$$V_{CC} - V_{BE} = I_B\,(R_C + R_B) + I_C \cdot R_C$$

$$I_B = \frac{V_{CC} - V_{BE} - I_C R_C}{R_C + R_B}$$

V_{BE} is the cut in voltage and it is 0.6V for Silicon, and 0.2V for Germanium. It is independent of I_C. So differentiate this expression with respect to I_C, we get

$$\frac{\partial I_B}{\partial I_C} = \frac{0 - R_C - 0}{R_C + R_B} = \frac{-R_C}{R_C + R_B}$$

The expression for stability factor S

$$S = \frac{1+\beta}{1-\beta\left(\frac{\partial I_B}{\partial I_C}\right)}$$

But
$$\frac{\partial I_B}{\partial I_C} = \frac{-R_C}{R_C + R_b}$$

$$\therefore \quad \boxed{S = \frac{1+\beta}{1+\beta \times \frac{R_C}{R_C + R_B}}}$$

Therefore, the value of S is less than $(1 + \beta)$.

$(1 + \beta)$ is the maximum value for the fixed bias circuit. Therefore collector bias circuit, is an improvement over the fixed bias circuit. Now let us consider the effect of β on the stability of the circuit, discussed above.

$$I_C = (1+\beta)\, I_{CO} + \beta \times I_B \qquad \text{.......... (3.2)}$$

$$I_B = \frac{V_{CC} - I_C R_C - V_{BE}}{R_C + R_B} \qquad \text{.......... (3.3)}$$

From Equation (3.2), $I_B = \dfrac{I_C - (1+\beta)I_{CO}}{\beta}$ .... (3.4)

Equating (3.3) and (3.4) and transferring I_C to one side

$$\frac{I_C - (1+\beta)I_{CO}}{\beta} = \frac{V_{CC} - I_C R_C - V_{BE}}{R_C + R_B}$$

$$I_C\,(R_C + R_B) - (1 + \beta)(R_C + R_B)\, I_{CO} = \beta[V_{CC} - V_{BE}] - \beta I_C\, R_C$$

$$I_C\,(R_C + R_B) + \beta\, I_C\, R_C = \beta\,([V_{CC} - V_{BE}] + (R_C + R_B)\, I_{CO}\,(1 + \beta))$$

$$\beta >> 1 = 1 + \beta \simeq \beta$$

$$\therefore \quad I_C[R_C + \beta R_C + R_B] = \beta[V_{CC} - V_{BE}] +)(R_C + R_B)\, I_{CO}\, \beta$$

$$I_C\,[(\beta + 1)\, R_C + R_B] \simeq \beta\,[V_{CC} - V_{BE} + (R_C + R_B) I_O]$$

Again $(\beta + 1) \simeq \beta$.

$$\therefore \quad I_C \simeq \frac{\beta[V_{CC} - V_{BE} + (R_C + R_B)I_{CO}]}{(\beta R_C + R_B)} \qquad \text{.......... (3.5)}$$

This expression, in this form is derived to get the condition to make I_C independent of β. Therefore, with the changes in the value of β, with temperature, I_C changes so the operating point also changes.

The above circuit, is an improvement over the fixed bias circuit, I_B is being decreased, with increase in I_C (Because I_{CO} increases with temperature) to keep operating point constant. Since I_C is dependent on the β from the above expression, and since β changes with temperature, I_C also changes and thereby operating point changes. Therefore from the above expression, it is essential to make I_C insensitive to β, so that the modified circuit works satisfactorily to keep the operating point fixed. The assumption to be made in the above eqn (3.5) is,

$$\beta \times R_C >> R_B.$$

Therefore, the above eqn becomes,

$$I_C \simeq \frac{\beta\left[V_{CC} - V_{BE} + (R_C + R_B)I_{CO}\right]}{\beta R_C}$$

$$\therefore \quad \beta R_C >> R_B$$

$$\therefore \quad I_C \simeq \frac{V_{CC} - V_{BE} + (R_C + R_B)I_{CO}}{R_C}$$

So if we choose R_B very small compared to βR_C, the circuit will work satisfactorily. But in all cases, this may not work out. If R_B value is ***not so small***, then the above assumption is not valid and hence the circuit won't work satisfactorily.

For a given circuit, to determine the operating point, draw the load line on the output characteristics of the given transistor. Load line is drawn as given below :

$$V_{CC} = I_C R_L + V_{CE}$$

If $I_C = 0$, then, $\quad V_{CC} = V_{CE}$

If $V_{CE} = 0$, then $\quad I_C = \dfrac{V_{CC}}{R_L}$

With these points, a straight. line is drawn on the output characteristics of the transistor. Corresponding to a given value of I_B, where this line cuts the output characteristic, it is called as the operating point.

In the collector to base bias circuit we said that the stability is good since, I_B is decreased with increase in I_C to keep operating point fixed. In that circuit, there is feedback from the output to the input through the resistor R_B. If the signal voltage causes an increase in I_B, I_C increases. But V_{CE} decreases. So because of this, the component of base current coming through R_B decreases. Hence the increase in I_B is less than what it would have been without feedback. In feedback circuits the amplification factor will be less than what it would have been without feedback. But the advantage is stability is improved. Such an amplifier circuit with feedback is called as ***Feedback Amplifier Circuit***. This type of circuits are discussed in subsequent chapters.

3.6 SELF BIAS OR EMITTER BIAS CIRCUIT

In the fixed bias circuit, since I_C increases with temperature, and I_B is fixed, operating point changes. In collector to base bias circuit, though I_C changes with temperature, I_B also changes to keep operating point fixed. But since β changes with temperature and I_C depends on β, the assumption has to be made that $\beta R_C >> R_B$ to make I_C independent of β. But if in one circuit, R_C is small, the above assumption will not hold good. [In transformer coupled circuits R_C will be small].

So collector to base circuit is as bad as fixed bias circuit. A circuit which can be used even if there is zero DC resistance in series with the collector terminal, is the ***Self Bias on Emitter Bias Circuit***. The circuit is as shown. This circuit is also known as ***Voltage Divider Bias or Universal Bias Circuit.***

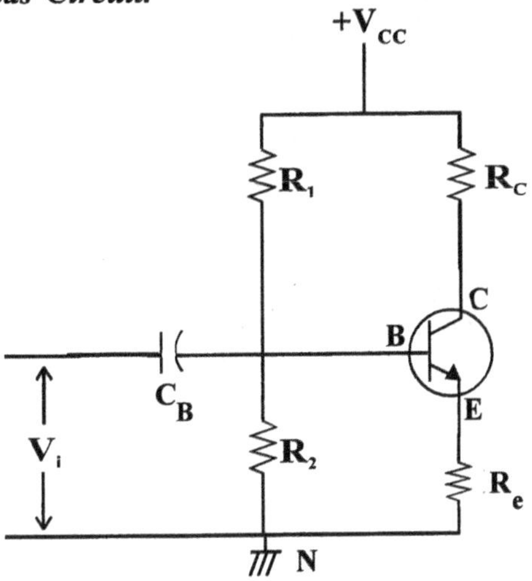

Fig. 3.7 Self Bias circuit.

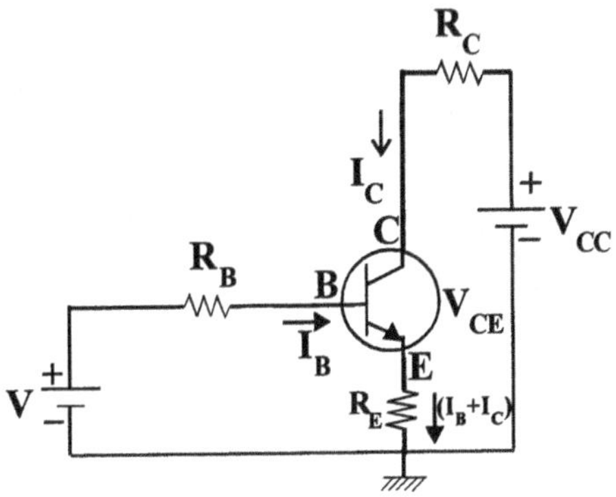

Fig. 3.8 Thevenin's equivalent.

The current in the emitter lead causes a voltage drop across R_e in such a direction, that it forward biases the emitter-base junction. It is NPN transistor emitter is *n-type*. Negative polarity should be at the emitter. If I_C increases due to increase in I_{C0} with temperature the current in R_C also increases.

$$I = I_C + I_B;\ V_{BE} = V_{BN} - W\ V_{EN}$$

As I_B increases, I_C also increases.

So current I or emitter current increases. Therefore IR_e drop increases. Since the polarity of this voltage is to reverse bias the E – B junction, I_B decrease. Therefore I_C will increase less than it would have been, had there been no self biasing resistor.

[The voltage drop across R_C provides the self bias for the emitter]. Hence stability is good.

With reference to Fig. 3.8, it is NPN transistor. So the conventional current is flowing out of the transistor, from the emitter. That is why the symbol is with arrow mark pointing outwards. So emitter point is at positive with respect to N. Hence the drop $(I_B + I_C)\ R_C$ has the polarity such as to reverse bias or to oppose the forward bias voltage V_{BE} junction. R_B can be regarded as parallel combination of R_1 and R_2.

3.7 BIAS STABILITY

The circuit shown in Fig. 3.4 is called as a ***Fixed Bias Circuit***, because I_B remains constant for given values of V_{CC} and V_{BB} and R_B. ($I_B \simeq V_{BB}/R_B$). So the operating point must also remain fixed. Suppose the transistor in the circuit is AC128, and we replace this transistor by another AC 128 transistor. The characteristics of these two transistors will not be exactly the same. There will be slight difference. It is because of the limitations in the fabrication technology.

The doping concentration, diffusion length etc., in fabrication cannot be controlled precisely. If we say impurity concentration is 1 in $10^6/cm^3$, it need not be the same. The diffusion length will also be not the same since the temperature inside the furnace during drive in diffusion may vary. The furnace may not have constant temperature profile along the length of the furnace. Because of these reasons, transistors of the same type may not have exactly the same characteristics. Hence in the above circuit, if one AC 128 transistor is replaced by another transistor, the operating point will change, since the characteristics of the transistor will be slightly different.

Hence for the same circuit when one transistor is replaced by the other, operating point changes. In some cases, because of the change in the operating point, the transistor may be cut off. In the above circuit, the operating point is not fixed, since I_B is fixed, and β changes. On the other hand I_B should be changed to account for the change in β so that the operating point is fixed or I_C and V_{CE} are fixed.

3.8 STABILIZATION AGAINST VARIATIONS IN I_{CO} AND V_{BE} AND β

S is calculated assuming that no AC signal is impressed and only DC voltages are present. In the Fig. 3.8, the voltage V is the drop across R_2. Combination acts as a potential divider. So the drop across R_2 which is equal to V, is given by the expression,

$$V = \frac{V_{CC}.R_2}{R_1 + R_2}$$

R_B is the effective resistance looking back from the base terminal

$$\therefore \quad R_B = \frac{R_1 R_2}{R_1 + R_2}$$

Applying Kirchhoff's Voltage Law around the base circuit,

$$V = I_B R_B + R_e (I_B + I_C) + V_{BE}$$

Assuming V_{BE} to be independent of I_C, differentiating I_B with respect to I_C, to get 'S'.

$$I_B, R_B + I_B R_e = +V - V_{BE} - I_C \times R_e$$

$$I_B (R_B + R_e) = V - V_{BE} - I_C \times R_e$$

$$\therefore \quad I_B = \frac{(V - V_{BE} - I_C R_e)}{(R_B + R_e)}$$

$$\frac{\partial I_B}{\partial I_C} = -\frac{R_e}{R_e + R_B}$$

The expression for stability factor $S = \dfrac{1+\beta}{1-\beta\left(\dfrac{\partial I_B}{\partial I_C}\right)}$

$$\therefore \quad S = \frac{1+\beta}{1+\beta\left(\dfrac{R_e}{R_e + R_B}\right)}$$

If $\dfrac{R_B}{R_e}$ is very small, $S \simeq 1$.

The fixed bias circuit, collector to base bias circuit and Emitter bias circuits provide stabilisation of I_C against variations in I_{CO}. But I_C also varies with V_{BE} and β. ***V_{BE} decreases at the rate of 2.5 mV/°C for both Germanium and Silicon transistors.*** Because, with the increase in T, the potential barrier is reduced and more number of carriers can move from one side (p) to the other side (n). β also increases side with temperature. Hence I_C changes.

3.9 BIAS COMPENSATION

The collector to base bias circuit and self bias circuit are used for stability of I_C with variation in I_{CO}, V_{BE} and β. But we said that there is feed back from the output to the input. Hence the amplification will be reduced, even though the stability is improved.

Problem 3.1

An NPN transistor with β = 50, is used in Common Emitter Circuit with V_{CE} = 10V and R_C = 2KΩ. The bias is obtained by connecting a 100 kΩ resistance from collector to base. Assume V_{BE} = 0. Find

(a) *The Quiescent Point*

(b) *The Stability Factor*

Solution

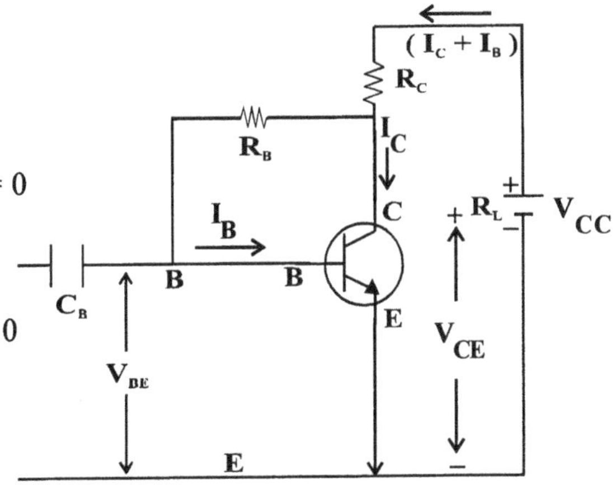

Fig. 3.9 For problem 3.1.

$$V_{CC} - (I_C + I_B)\,R_C - I_B\,R_B = 0$$

Neglecting I_{CO}, $I_C = \beta \times I_B$

Then $(\beta + 1)\,I_B R_C + I_B R_B = V_{CC}$

$$(50 + 1) \times 2 \times 10^3 + 100 \times 10^3 \times I_B = 10$$

$$\therefore\ I_B = \frac{10}{202 \times 10^3} = 49.5\ \mu A$$

$$I_C = 50 \times 49.5 \times 10^{-6} = 2.475\ mA$$

$$V_{CE} = I_B \times R_B = 4.95$$

$$S = \frac{\beta+1}{1+\dfrac{\beta.R_C}{R_C+R_B}} = \frac{51}{1+\dfrac{50\times 2}{102}} = 25.6$$

Problem 3.2

A transistor with β = 100 is to be used in Common Emitter Configuration with collector to base bias. R_C = 1KΩ V_{CC} = 10V. Assume V_{BE} = 0. Choose R_B so that quiescent collector to emitter voltage is 4V. Find Stability Factor.

Solution

$$V_{CC} - (\beta + 1)\,I_B \times R_C - V_{CE} = 0$$

$$S = \frac{101}{1+\dfrac{100\times 1}{68.3}} = 41$$

$$10 - 101 \times 10^3\,I_B - 4 = 0$$

$$I_B = \frac{6}{101\times 10^3} = 59.4\ \mu A$$

$$V_{CE} = I_B R_B = 4V$$

$$\therefore \quad R_B = \frac{4}{59.4 \times 10^{-6}} = 67.3K$$

Problem 3.3

One NPN transistor is used in the self biasing arrangement. The circuit components values are $V_{CC} = 4.5V$, $R_C = 1.5K$, $R_C = 0.27\ K\Omega$, $R_2 = 2.7\ K\Omega$ and $R_1 = 27K$. If $\beta = 44$. Find the Stability Factor. Also determine the Quiescent point $Q\ (V_{CE}, I_C)$.

Solution

The Quiescent Point

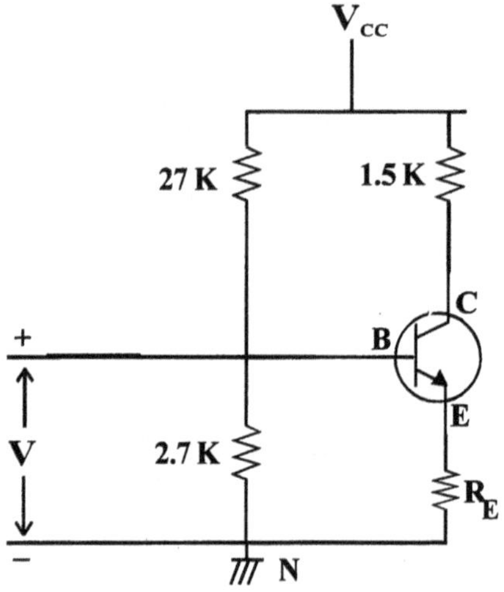

Fig. 3.10 For problem 3.3.

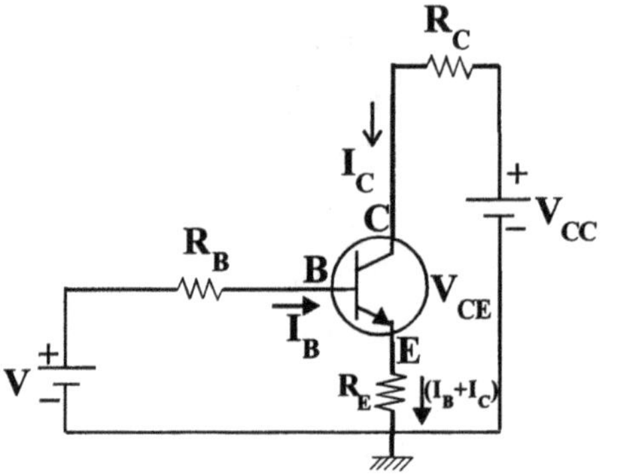

Fig. 3.11 For problem 3.3.

$$V = \frac{R_2 . V_{CC}}{R_1 + R_2} = \frac{2.2 \times 4.5}{27 + 2.7} = 0.409V$$

$$R_B = \frac{R_1 R_2}{R_1 + R_2} = \frac{27 \times 2.7}{27 + 2.7} = 2.46k\Omega$$

$$S = \frac{(1+\beta)1 + \frac{R_B}{R_e}}{1 + \beta + \frac{R_B}{R_e}} = 8.4$$

Applying KVL over the loop.

$$+V = I_B \times R_B + V_{BE} + (1 + \beta)\, R_C I_B$$

$$+0.409 = I_B \times (2.46 + 45 \times 0.27) + 0.2$$

or $$I_B = \frac{0.409 + 0.2}{14.6} = 1.1\text{mA}$$

$$I_C = \beta \times I_B = 44 \times 1.1 = 48.4 \text{ mA}$$

$$V_{CE} = V_{CC} - I_C \times R_C - (I_B + I_C) R_C = 2.4\text{V}$$

If the reduction in gain is not tolerable then compensation techniques are to be used. Thus both stabilization and compensation technique are used depending upon the requirements.

3.9.1 Diode Compensation For V_{BE}

The circuit shown in Fig. 3.12 employs self biasing technique for stabilization. In addition a diode is connected in the emitter circuit and it is forward biased by the source V_{DD}. Resistance R_d limits the current through the diode. The diode should be of the same material as the transistor. The negative voltage V_0 across the diode will have the same temperature coefficient as V_{BE}. Since the diode is connected as shown, if V_{BE} decreases with increase in temperature (V_B is cut in voltage), V_0 increases with temperature. Since with temperature, the mobility of carriers increases, so current through the diode increases and V_0 increases. Therefore, as V_{BE} decreases, V_0 increases, so that I_C will be insensitive to variations in V_{BE}.

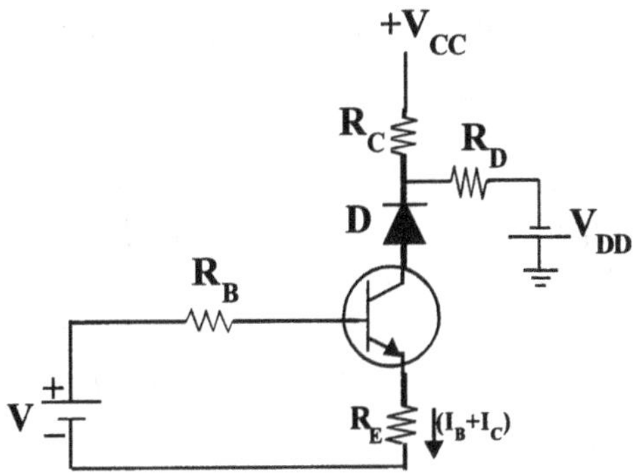

Fig. 3.12 Diode compensation for V_{BE}.

The decrease in V_{BE} is nullified by corresponding increase in V_D. So the net voltage more or less remains constant. Thus the diode circuit compensates for the changes in the values of V_{BE}.

3.10 THERMAL RUNAWAY

The max power which a transistor can dissipate (power that can be drawn from the transistor) will be from mW to a few hundred Watts. The maximum power is limited by the temperature also that the collector to base junction can withstand. For Silicon transistor the max temperature range is 150 °C to 205 °C and for Germanium it is 60 °C to 100 °C. For Germanium it is low since the conductivity is more for Germanium compared to Silicon. The junction temperature will increase either because of temperature increase or because of self heating.

As the junction tempetature rises, collector current increases. So this results in increased power dissipation. So temperature increases and I_C still further increases. This is known as ***Thermal Runaway***. This process can become cumulative to damage the transistor. Now

$$\Delta T = T_J - T_A = \theta \times P_D$$

where T_J and T_A are the temperature of the junction and ambient. P_D is the power dissipated at the junction in Watts. θ is the so called ***Thermal Resistance*** of the transistor varying from 0.2° C/W for a high power device to 1000 °C/W for a low power device. The condition for thermal stability is that the rate at which heat is released at the collector junction must not exceed the rate at which heat can be dissipated.

$$\frac{\partial P_C}{\partial T_J} < \frac{\partial P_D}{\partial T_J}$$

or
$$\frac{\partial P_C}{\partial T_J} < \frac{1}{\theta}$$

$$\because \quad \theta = \frac{\partial T_J}{\partial P_D}$$

So, heat Sink limit the temperature rise. Heat sink is a good conducting plate which absorbs heat from the transistor and will have large area for dissipation.

Problem 3.4

Calculate the value of thermal resistance for a transistor having P_C (max) = 125 mW at free air temperature of 25°C and T_J (max) = 150^{0}C. Also find the junction temperature if the collector dessipation is 75 mW.

(i) $T_J - T_A = \theta \times P_D$ or $150 - 25 = \theta \times 125$ or $\theta = 1$°C/mW

(ii) $T_J - 25 = 1 \times 75$; $T_J = 100$°C

Solution

It has been found that, if the operating point is so chosen that

$$V_{CE} = \frac{V_{CC}}{2}$$

then the effect of thermal runaway is not cumulative. But if

$$V_{CE} > \frac{V_{CC}}{2}$$

then temperature increases with a increase in I_C. Again with increase in I_C, T increases and hence the effect is cumulative. As I_C increases, P_C increases. But it follows a certain law. If I_C is so chosen that

$$V_{CE} = \frac{V_{CC}}{2}$$

rate of increase of P_C with I_C is not large and Thermal Runaway problem will not be there.

Problem 3.5

A silicon type 2N780 transistor with $\beta = 50$, $V_{CC} = 10$V and $R_C = 250\Omega$ is connected in collector to base bias circuit. It is desired that the quiescent point be apporoximately at the middle of the load line, so I_B and I_C are chosen as 0.4 mA and 21mA respectively. Find R_B and Calculate 'S', the stability factor.

Solution

$$V_{CE} = V_{CC} - (I_C + I_B)\, R_C$$

I_B is very small compared to I_C. So it can be neglected.

$$\therefore \quad V_{CE} = V_{CC} - I_C \cdot R_C$$
$$= 10V - (21 + 0.4)\,250$$
$$V_{CE} \simeq 4.6\text{ V}$$

Since it is Silicon transistor,

$$V_{BE} = V_\gamma = 0.6V$$
$$\therefore \quad R_B = \frac{V_{CE} - V_{BE}}{I_B}$$
$$= \frac{4.6 - 0.6}{0.4} = 10\text{ k}\Omega$$
$$S = \frac{\beta + 1}{1 + \beta . \frac{R_c}{R_c + R_b}}$$
$$S = \frac{(50+1)}{1 + 50 \times \frac{250}{(250 + 10 \times 10^3)}} = 23$$
$$\therefore \quad S = 23$$

Problem 3.6

A transistor with $\beta = 100$ is to be used in Common Emitter Configuration with collector to base bias. The collector circuit resistance is $R_C = 1k\Omega$ and $V_{CC} = 10V$. Assume $V_{BE} = 0$.

(a) *Choose R_B so that the quiescent collector to emitter voltage is 4V.*

(b) *Find the stability factors.*

Solution

$$\beta = 100\,; \quad R_C = 1k\Omega \quad V_{CC} = 10V \quad V_{BE} = 0$$
$$R_b = ? \quad S = ?$$

V_{CE} should be 4V

Applyining KVL around the loop,

$$V_{CC} - (I_C + I_B)\,R_C - V_{CE} = 0.$$

V_{CC} is voltage these negative to positive. V_{CE} is voltage drop positive to negative. But I_C value is not given.

$$\frac{I_C}{I_B} = \beta$$
$$\therefore \quad I_C = \beta \times I_B$$
$$\therefore \quad V_{CC} - (\beta + 1)I_B\,R_C - V_{CE} = 0.$$
$$V_{CC} = 10V \quad V_{CE} = 4V \quad R_C = 1k\Omega \quad \beta = 100 \quad I_B = ?$$

$$10 - (101)\, I_B \times 1 \times 10^3 - 4 = 0$$

or $$I_B = \frac{6}{101 \times 10^3} = 59.4\ \mu A$$

But $$V_{CE} = I_B \cdot R_B$$

$\therefore$ $$4\ V = 59.4\ \mu A \times R_B$$

$\therefore$ $$R_B = \frac{4}{59.4 \times 10^{-6}} = 67.3 k\Omega$$

Therefore, Stability Factor,

$$S = \frac{\beta + 1}{1 + \beta . \dfrac{R_C}{R_C + R_B}}$$

$$S = \frac{101}{1 + 100 \times \dfrac{1 \times 10^3}{10^3 (1 + 67.3)}} = 41$$

$\therefore$ $$S = 41$$

Problem 3.7

Two identical Silicon transistor with $\beta = 48$, $V_{BE} = 0.6V$ at $T = 25^0$ C, $V_{CC} = 20.6V$, $R_1 = 10K$ and $R_C = 5K$ are connected as shown. (a) Find the currents I_{B1}, I_{B2}, I_{C1} and I_{C2} at $T = 25^0C$

(b) Find I_{C2} at $T = 175^0$ C when $\beta = 98$ and V_{BE} 0.22V

Solution

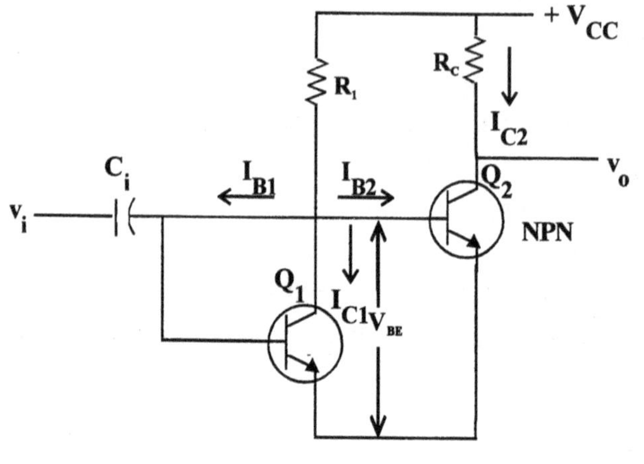

Fig. 3.13 For problem 3.7.

(a) Since the two transistors are identical, we can assume that $I_{B1} = I_{B2}$.

$\therefore$ $$I = \frac{V_{CC} - V_{BE}}{R_1}$$

$$I = \frac{20.6 - 0.6}{10k} = 2 \text{ mA}$$

$\because \quad V_{BE} = 0.6V \text{ at } 25\,^0C$

Since $I_{B1} = I_{B2} = I_B$, then $I = 2I_B + I_{C1} = 2I_B + \beta I_B$ or $2mA = (2 + 48)I_B$

$$\therefore \quad I_B = \frac{2}{50} \text{ mA} = 40 \text{ mA}$$

$$\therefore \quad I_{C1} = I_{C2} = \beta \times I_B = 48 \times 40 \times 10^{-3} = 1.92 \text{ mA}$$

β is same for the two transistors. I_B is same. Hence I_C should be the same.

(b) At 175°C, V_{BE} decrease with temperature. Hence V_{BE} at 175 °C = 0.22 V and $\beta = 98$

$$\therefore \quad I = \frac{V_{CC} - V_{BE}}{10k} = 2.038 \text{ mA} = (2 + \beta)I_B$$

or
$$I_B = \frac{2.038}{2+98} \text{ mA} = 20.38\ \mu\text{A}.$$

$$I_{C1} = I_{C2} = \beta \times I_B = 98 \times 20.38 \times 10^{-6} = 2 \text{ mA}$$

Problem 3.8

Determine the quiescent currents and the collector to emitter voltage for a Ge transistor with $\beta = 50$ in the self biasing arrangements. The circuit component values are $V_{CC} = 20V$, $R_C = 2K$, $R_e = 0.1k$, $R_1 = 100k$ and $R_2 = 5k\Omega$. Find the stability factor S.

Solution

$$V = \frac{R_2.V_{CC}}{R_1 + R_2} = \frac{5 \times 20}{100 + 5} = 0.952 \text{ V}$$

$$R_B = \frac{R_1 R_2}{R_1 + R_2} = \frac{5 \times 100}{100 + 5} = 4.76 \text{ K}\Omega$$

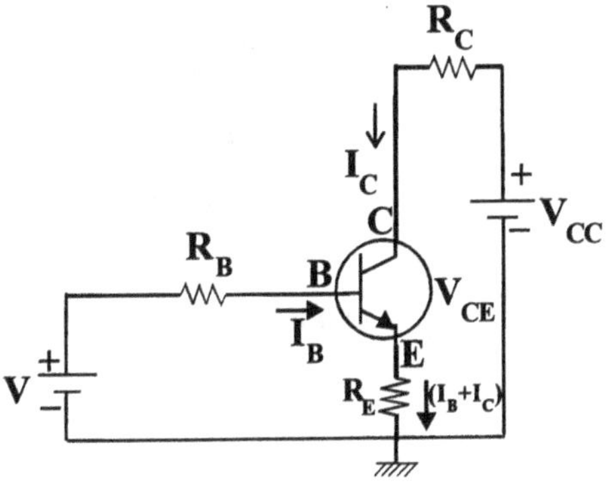

Fig. 3.14 For problem 3.8.

For Germanium transistor, $V_{BE} = 0.2V$

$$\therefore \quad V = I_B \times R_B + V_{BE} + (I_B + I_C)R_e$$

$$V = I_B \times R_B + V_{BE} + (\beta + 1)\, I_B\, R_e$$

$$= I_B\, [R_B + R_e(1+\beta)\,] + V_{BE}$$

$$\therefore \quad 0.952 = I_B[4.76 \times 10^3 + 51 \times 0.1] + 0.2$$

$$\therefore \quad I_B = \frac{0.952 - 0.2}{9.86 \times 10^3} = 76.2 \text{ mA}$$

$$I_C = \beta \times I_B = 3.81 \text{ mA}$$

$$I_E = (I_B + I_C) = 3.81 \text{ mA}$$

$$V_{CE} = V_{CC} - I_C R_C - I_E\, R_e$$

$$= 20 - 3.81 \times 2 - 3.89 \times 0.1 = 12V$$

$$S = (1 + \beta) \times \left(\frac{1 + \frac{R_B}{R_e}}{1 + \beta + \frac{R_B}{R_e}} \right) = 51 \times \frac{1 + \frac{4.76}{0.1}}{51 + \frac{4.76}{0.1}} = 25$$

Problem 3.9

Design a circuit with Ge transistor in the self biasing arrangement with $V_{CC} = 16V$ and $R_C = 1.5K$. The quiescent point is chosen to be $V_{CE} = 8V$ and $I_C = 4mA$. Stability factor $S = 12$ is desired $\beta = 50$.

Solution

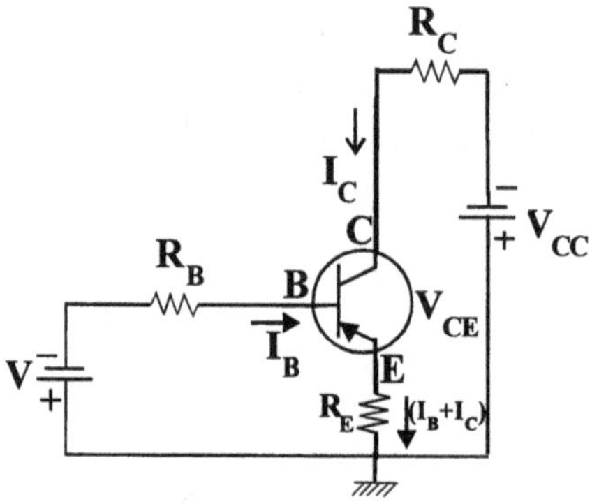

Fig. 3.15 For problem 3.9.

$$I_B = \frac{I_C}{\beta} = \frac{4}{50} = 80\mu A$$

$$R_E = \frac{V_{CC} - V_{CE} - I_C R_C}{(I_B + I_C)}$$

$$= \frac{16 - 8 - 4 \times 1.5}{4.08} = 0.49\text{k}$$

$$S = 12 = \frac{(\beta+1)1 + \frac{R_B}{R_E}}{1 + \beta + \frac{R_B}{R_E}}$$

or $$\frac{R_B}{R_E} = 14.4$$

$\therefore$ $R_B = 14.4 \times R_E = 7.05\text{k}.$

V_{BN} = Base to ground voltage

$= V_{BE} + I_E R_E$

Therefore, I_E is negative.

$V_{BN} = 0.2 + 4.08 \times 0.49 = 2.2\text{V}$

$V = V_{BN} + I_B R_b$

$= 2.2 + 0.08 \times 7.05 = 2.76\text{V}$

$$V = \frac{R_2 . V_{CC}}{R_1 + R_2}; R_B = \frac{R_1 R_2}{R_1 + R_2}$$

$\therefore$ $$\frac{V}{R_B} = \frac{V_{CC}}{R_1} = \frac{16}{R_1}$$

$\therefore$ $$R_1 = \frac{16 R_B}{V}$$

$$= \frac{16 \times 7.05}{2.76} = 41\text{k}$$

$V = 2.76$ V

$$= \frac{R_2 \times V_{CC}}{R_1 + R_2} = \frac{R_2 \times 16}{41\text{k} + R_2}$$

$\therefore$ $R_2 = 8.56\text{k}$

3.11 CONDITION FOR THERMAL STABILITY IN CE CONFIGURATION

As to avoid thermal runaway, first of all we have to study about thermal resistance which is found practically that the steady state temperature rise at the collector junction is proportional to the power dissipated at the junction, or

$$\Delta T = T_j - T_A = \textcircled{H}\ P_O \qquad .....(3.6)$$

where T_j and T_A are the junction and ambient temperatures, respectively, in degrees centigrade, and P_o power in watts dissipated at the collector junction and here constant of proportionality $\textcircled{H}$ is called thermal resistance.

Here the required condition to prevent thermal runaway is that the rate at which heat is released at the collector junction must not exceed the rate at which the heat can be dissipated, that is,

$$\frac{\partial P_C}{\partial T_j} < \frac{\partial P_D}{\partial T_j} \qquad ......(3.7)$$

Difference eq.(3.6) with respect to T_j and substitute in eq. (3.7), we obtain

$$\frac{\partial P_C}{\partial T_j} < \frac{1}{\textcircled{H}} \qquad .....(3.8)$$

This condition must be satisfied to prevent thermal runaway

3.11.1 Derivation for Condition for Thermal Stability for CE Configuration

In Fig. 3.16 is the self bias common emitter cofiguration. Here the power generated at the collector junction with no signal is

$$P_C = I_C V_{CB} \simeq I_C V_{CE} \qquad .....(3.9)$$

Let us consider emitter currents and quiesent collector currents are equal eq.(3.9) becomes,

$$P_C = I_C V_{CC} - I_C^2 (R_e + R_c) \qquad .....(3.10)$$

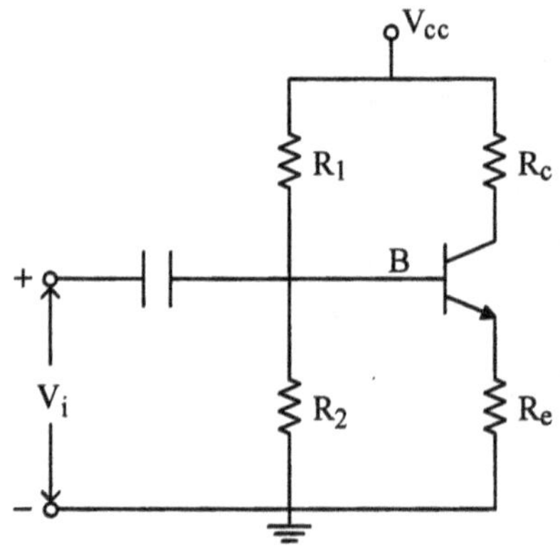

Fig. 3.16

In eq. (3.8) is the condition to avoid thermal runaway, can be rewritten as follows:

$$\frac{\partial P_C}{\partial I_C}\frac{\partial I_C}{\partial T_j} < \frac{1}{Ⓗ} \quad .....(3.11)$$

Here $\frac{\partial P_C}{\partial I_C}$ can be obtained from eq. (3.10),

$$\frac{\partial P_C}{\partial I_C} = V_{CC} - 2I_C(R_e + R_c) \quad .....(3.12)$$

and the $\frac{\partial I_C}{\partial T_j}$ is given as,

$$\frac{\partial I_C}{\partial T_j} = S\frac{\partial I_{CO}}{\partial T_j} + S'\frac{\partial V_{BE}}{\partial T_j} + S''\frac{\partial B}{\partial T_j}\lim_{x\to\infty} \quad .....(3.13)$$

Here for any transistor derivatives are known, then it is required to satisfy eq. (3.13) by the proper section of S, S′, S″ and Ⓗ

From eq. (3.9) and eq. (3.13),

$$\frac{\partial P_C}{\partial I_C}\left(S\frac{\partial I_{CO}}{\partial T_j}\right) < \frac{1}{Ⓗ} \quad .....(3.14)$$

It is noted that the reverse saturation current for either Si or Ge increases about 7 percent/°C

or

$$\frac{\partial I_{CO}}{\partial T_j} = 0.07I_{CO} \quad .....(3.15)$$

sub eq. (3.12) and eq. (3.15) in eq. (3.14) results in,

$$[V_{CC} - 2I_C(R_e + R_c)](S)(0.07\,I_{CO}) < \frac{1}{Ⓗ} \quad .....(3.16)$$

The eq. (3.16) is always satisfied provided that the quantity in the brackets is negative, or provided that

$$I_C > \frac{V_{CC}}{2(R_e + R_c)} \quad ....(3.17)$$

This is the condition of thermal stability.

3.12 BIASING OF FET

One of the problem in using FET is that each device type does not have a single transfer characteristic. This is because I_{DSS} and V_P cannot be specified accurately. For the same FET, BFW 10, I_{DSS} and V_P will not be constant. They vary, because, the voltage V_{GS} for which pinch off occurs varies. This cannot be exactly defined. Recombination of carriers can cause variation in depletion region thickness. So, V_P and I_{DSS} will not be constant. Hence the manufacturers specify maximum and minimum values for I_{DSS} and V_P. Accordingly, two transfer characteristics are drawn for FET, Maximum transfer characteristics which is a plot between maximum I_{DSS} and maximum V_P and minimum transfer characteristic which is a plot between minimum V_P and minimum I_{DSS}.

The DC load line for a FET circuit is drawn upon the device characteristic in exactly the same way as was done with bipolar transistor circuits.

$$V_{DS} = V_{DD} - I_D \cdot R_L \qquad .....(3.18)$$

R_L is known, So by choosing any convenient value of I_D in eq (3.18), V_{DS} can be calculated. V_{DS} is fixed, (known).

when $I_D = 0$, $V_{DS} = V_{DD}$. So we get point A as the characteristic X-axis.

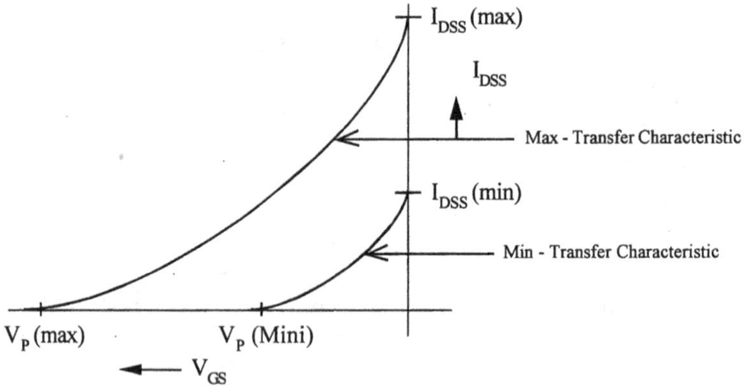

Fig. 3.17 Transfer characteristics.

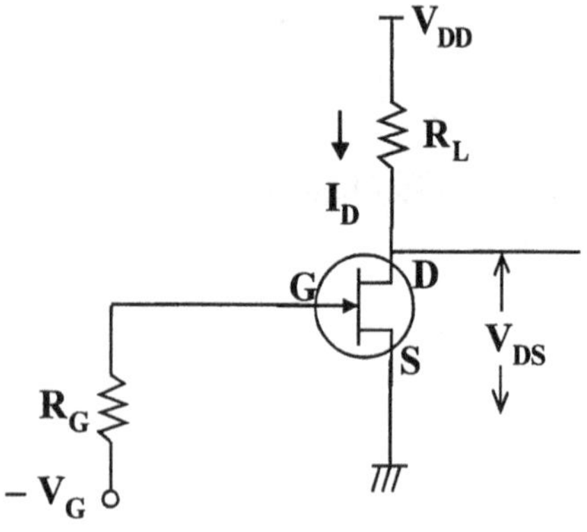

Fig. 3.18 FET biasing circuit.

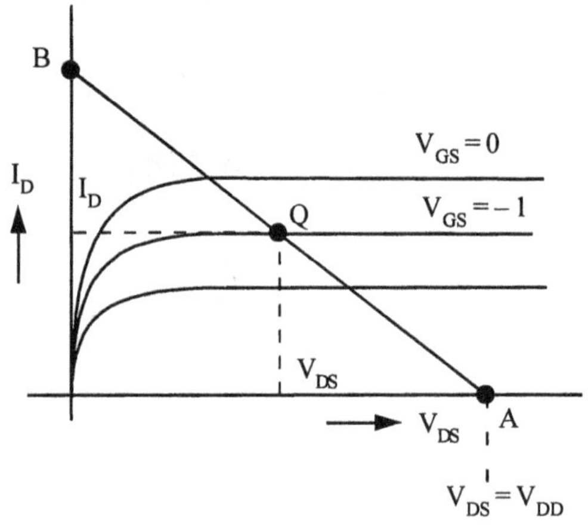

Fig. 3.19 DC loadline.

when $V_{DS} = 0, I_D = \frac{V_{DD}}{R_L}$. So we get point B on y-axis. Join points A and B. This will be the load line. Choose any value of V_{GS} say –1. Then the point at which the load line intersects the corresponding drain characteristics is the operating point Q. For a FET, V_p and I_{DSS} are not fixed. The variation of maximum and minimum values can be as much as $\pm$ 50% or more over the typical values.

3.12.1 Fixed Bias Circuit

The circuit is shown in Fig.3.20. This is an example of fixed bias circuit. The gate is biased via resistance R_G to a negative voltage V_G. So the maximum and minimum levels of I_D for a given bias voltage can be best determined by a graphical first plot maximum and minimum transfer characteristics as shown in Fig. 3.21.

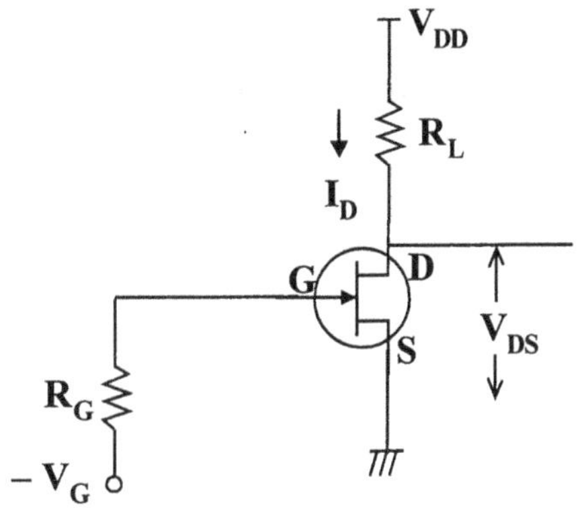

Fig. 3.20 JFET Fixed bias circuit.

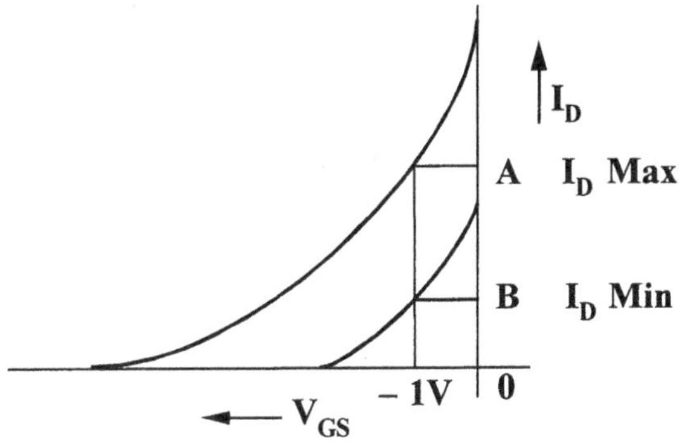

Fig. 3.21 Transfer characteristics.

Let V_{GS} is chosen as – 1V. Then draw a base line (vertical line) through the –1V V_{GS} value. It cuts the maximum and minimum transfer characteristics at two points A and B. A gives maximum I_D, B gives minimum I_D. Corresponding to these values, the maximum and minimum values of V_{DS} can be calculated using the expression,

$$V_{DS} = V_{DD} - I_D \cdot R_L$$
$$V_{DS(max)} = V_{DD} - I_{D(min)} \cdot R_L$$
$$V_{DS(min)} = V_{DD} - I_{D(max)} \cdot R_L$$

As it is evident, the fixed bias circuit is not useful. The Q point (V_{DS} and I_D values) vary over a wide range. So, the operating point is not fixed in a fixed bias circuit.

3.12.2 Self Bias Circuit

In self bias circuit, a resistance in series with source terminal provides the gate bias voltage. R_S will stabilize the drain current. R_S will also tend to stabilize I_D against signals applied to the gate i.e, R_S will reduce the AC voltage gain of the circuit. C_S is the bypass capacitors (C_E in BJT circuits). The total DC load is $(R_L + R_S)$. Therefore total AC load with R_S bypassed is R_L.

Therefore AC load line must be drawn to describe the AC performance of the circuit.

(Because AC current is not passing through R_S but only through C_s).

(V_{RS}) voltage drop across $R_S = I_D \times R_s$.

Gate is grounded via R_G. Therefore gate is negative with respect to source.

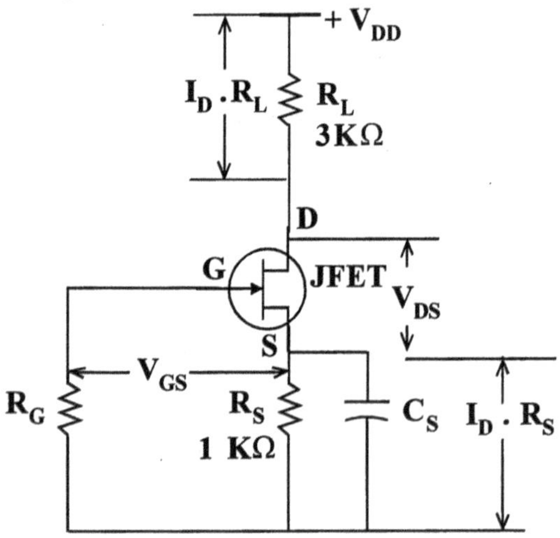

Fig. 3.22 JFET self bias circuit.

Let $I_D = 1mA; R_S = 1k\Omega$

$\therefore$ $V_{RS} = 1mA \times 1k\Omega = 1V$

$\therefore$ Source is 1V positive with respect to ground.

$\because$ Gate is grounded through R_G. Gate is 1V negative with respect to source. Hence Gate-Source junction is forward biased.

$$V_{GS} = -I_D \times R_S$$
$$V_{DD} = I_D \cdot R_L + V_{DS} + I_D \cdot R_S$$

From the maximum and minimum transfer characteristics, if V_{GS} is fixed, maximum and minimum values of I_D are determined. Then, maximum and minimum values of V_{DS} can be calculated.

Self bias technique minimises variations in I_D and V_{DS}, compared to fixed bias circuit. I_D variations will be less if larger value of R_S is chosen.

3.12.3 Self Bias with External Voltage

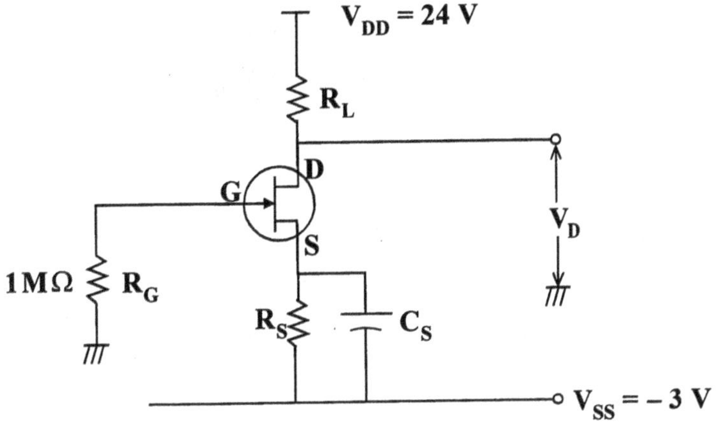

Fig. 3.23 JFET self bias circuit.

$$V_{SS} = I_D \cdot R_S + V_{GS} \qquad \because \text{ G is grounded.}$$

$$\therefore \quad V_{GS} = V_{SS} - I_D \cdot R_S$$

Another self bias circuit is

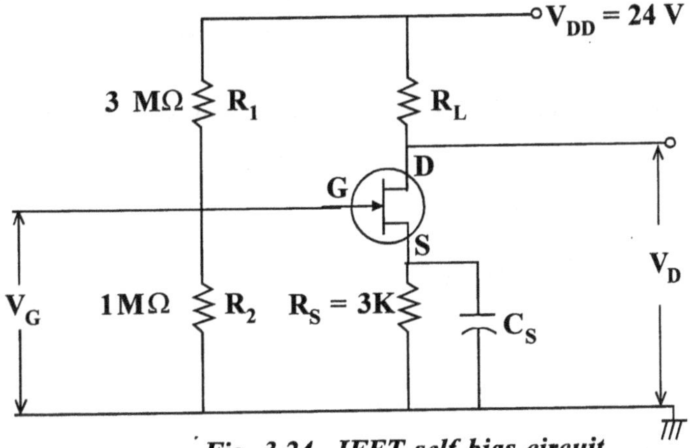

Fig. 3.24 JFET self bias circuit.

A potential divider R_1 and R_2 is used to derive a positive bias voltage from V_{DD}. (V_G should be positive because G-S junction should be reverse biased. Gate is p - type).

$$V_G = V_{GS} + I_D R_S$$

$$V_G = \frac{V_{DD} \cdot R_2}{R_1 + R_2}$$

$$\therefore \quad V_{GS} = V_G - I_D R_S = \frac{V_{DD} \cdot R_2}{R_1 + R_2} - I_D R_S.$$

Bias circuit which have an external voltage source V_{SS} and source resistances maintain I_D within closer limits i.e., Q point will be more stable.

3.12.4 Source Self Bias

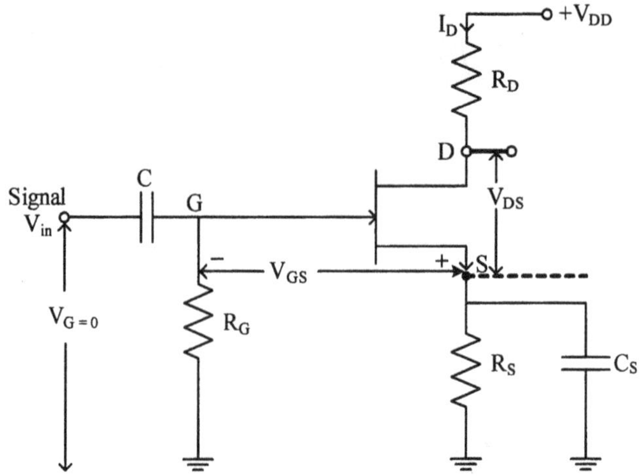

Fig. 3.25 Source self-bias circuit

The circuit for source self-bias is as shown in Fig. 3.25.

- We know that no gate current flows through the reverse-biased gate-source, the gate current $I_G = 0$ and therefore $V_G = i_G R_G = 0$
- It is implemented in bias junction FET devices or depletion-mode MOS transistors.
- For specified drain current I_D, the corresponding voltage V_{GS} are achieved using Eq. (3.19).
- The equation for I_{DS} is given as follows,

$$I_{DS} = I_{DSS}\left(1 - \frac{V_{GS}}{V_P}\right)^2 \qquad .....(3.19)$$

3.13 BIASING OF ZERO CURRENT DRIFT

To discuss about biasing of FET when drift current in zero we have to observe transfer characteristics on temperature.

- Here in FET the main parameters are V_{GS} and I_D.
- In Fig 3.26 the value V_{GS} exists when $I_D = I_4$ does not changes with temperature T.
- So, it is possible to bias a FET when drain current is zero.
- There are two factors affect the variation of I_D with T.
- The initial factor when there is a decrease in majority carrier mobility with temperature T.
- When T increases, the crystal lattice ions vibrate more vigoursly so carrier cannot move freely for a given field strength leads to decrease in velocity which reduces the current.
- Practically it has been found that the reduction in I_D is 0.7 percent/ °C.

- The next factor is width of the gate-to-channel barrier is decrease and with increasing temperature and hence this pass the current I_D to increase and has been found that the increase in I_D is equivalent to a change of 2.2 mV/°C in V_{GS}.
- when a current of $g_m \Delta V_{GS}$, for zero drift the condition is,

$$0.07|I_D| = 0.0022\, g_m \quad .....(3.20)$$

$$\Rightarrow \frac{|I_D|}{g_m} = 0.314 \quad .....(3.21)$$

$$\rightarrow g_m = g_{m0}\left(1 - \frac{V_{GS}}{V_p}\right) \quad .....(3.22)$$

where $$g_{m0} = -2I_{DS}|V_p \quad .....(3.23)$$

$$I_{DS} = I_{DSS}\left(1 - \frac{V_{GS}}{V_p}\right)^2 \quad .....(3.24)$$

If we substitute eq.(3.24), (3.25), (3.26) in (3.20), we get,

$$|V_p| - |V_{GS}| = 0.63\,V \quad .....(3.25)$$

Eq. (3.25) gives the value of V_{GS} for zero drift if V_P is known

→ If $V_P = 0.63$ V, $V_{GS} = 0$ & ID = I_{DSS},

$$I_D = I_{DSS}\left(\frac{0.63}{V_p}\right)^2 \quad .....(3.26)$$

$$g_m = g_{md}\frac{0.63}{|V_p|} \quad .....(3.27)$$

Eq. (3.26) & (3.27) are used to specify the drain current and transcoductance for zero drift of I_D with T.

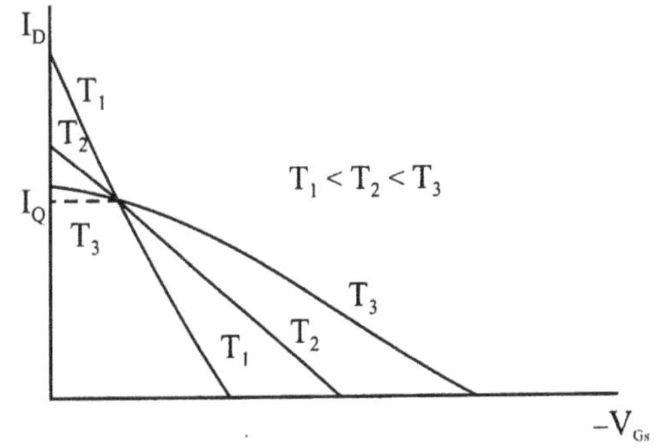

Fig. 3.26 Transfer characteristics of an n-channel FET function of temperature.

3.14 BIASING AGAINST DEVICE VARIATION

There are FET specifications supplied by FET manufactures which also gives information about maximum and minimum values of current I_{DSS} and voltage V_P at room temperature.

- The values of temperature device variations are obtained from the transfer characteristics of n-channel FET which are indicated at the top and bottom of the curve.
- We know that, it is necessary to bias the device at a drain current which will not drift outside of $I_D = I_A$ and $I_D = I_B$.
- Here bias line V_{GS} in given as $V_{GS} = - I_D R_S$ must intersect the transfer characteristics between the points A and B as shown in Fig 3.27a.
- Here slope is the bias line is determined by the source resistance R_S.

  i.e., Bias line $V_{GS} = - I_D R_S$ compare with line equation

  $$y = mx$$

  Since V_{GS} is proportional to I_D here slope is nothing but R_S.
- The transfer characteristics of any FET when two extremes indicated, the I_Q must lie between I_A and I_B i.e., $I_A < I_Q < I_B$ should be satisfied for desirable operation.

  Let us consider the physical situation indicated in Fig 3.27b, where ever a line drawn passes between A and B not through origin the bias line is $V_{GS} = V_{GG} - I_D R_S$.

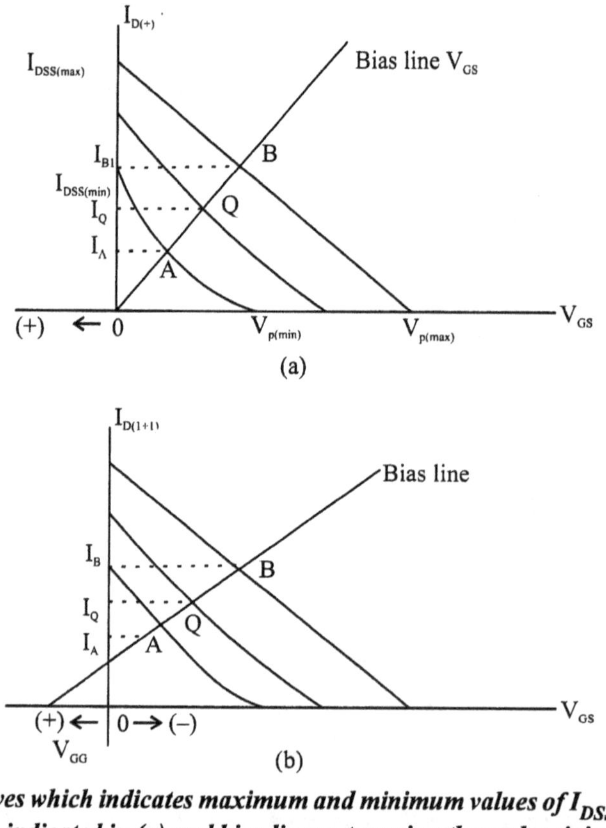

Fig. 3.27 Transfer curves which indicates maximum and minimum values of I_{DSS} and V_P. The bias line passing through origin indicated in (a) and bias line not passing through origin is indicated in (b).

- This bias relationship can be obtained by adding a fixed bias to the gate in addition to source self bias
- when one power supply is used, we rearrage Fig 3.28(a) into Fig 3.28(b). For this circuit,

$$V_{GG} = \frac{R_2 V_{DD}}{R_1 + R_2}; \; R_g = \frac{R_1 R_2}{R_1 + R_2}$$

The Gate current is zero (say). This condition is possible V_{GG} to fall in reverse bias region so line Fig 3.28(b) intersects the axis at abscissa to the right of the origin under these conditions we must use two power supplies.

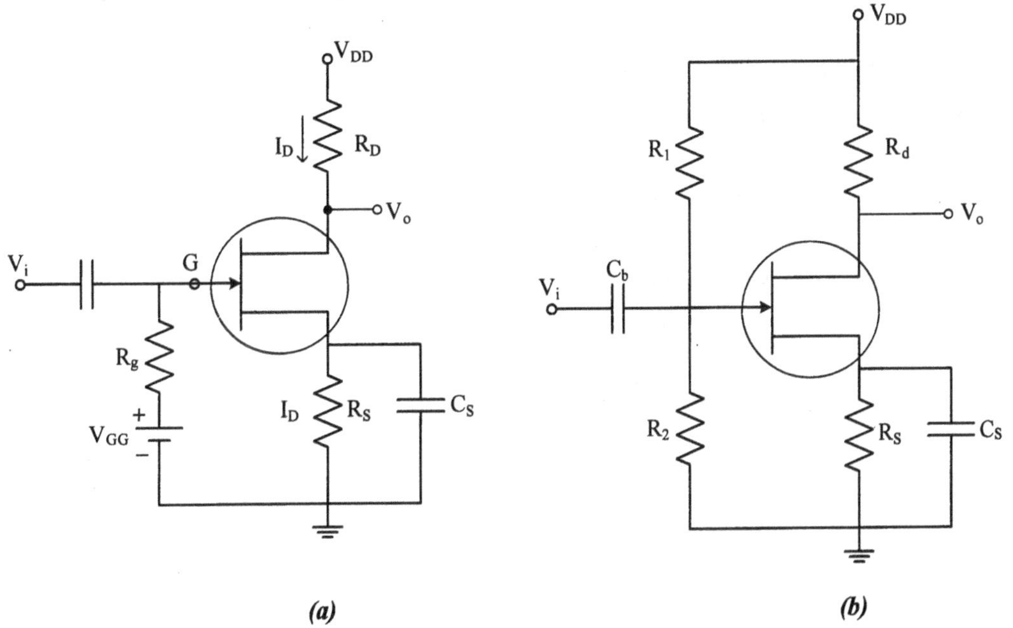

Fig 3.28 (a) Biasing a FET with a fixed bias V_{GG} combined with self-bias through R_S

(b) When single power-supply used which is equivalent to the circuit in (a)

SUMMARY

- In order that for a transistor, Emitter-Base junction is forward biased and collector-Base junction is reverse biased, biasing circuit must be employed. The three types of biasing circuits are
  1. Fixed bias or Base bias circuit.
  2. Collector to base bias
  3. Self bias or Emitter bias
- Self bias circuit is commonly used. It ensures that the operating point is in the centre of the active region. The operating point is defined by I_C, V_{CE} and I_B.

Usually V_{CE} is chosen to be equal to $\frac{V_{CC}}{2}$.

- Stability factor is a function of I_{Co}, β, V_{BE}

$$S = \left.\frac{\partial I_C}{\partial I_{co}}\right|_{\beta \text{ and } V_{BE} = \text{Constant}}$$

General Expression for,

$$S = \frac{1+\beta}{1-\beta\left(\frac{\partial I_B}{\partial I_C}\right)}$$

For a Self bias, circuit,

$$S = (1+\beta)\left[\frac{1+\frac{R_B}{R_e}}{1+\frac{R_B}{R_e}+\beta}\right]$$

- Quiescent point Q, (or operating point or biasing point) changes with temperature. So to nullify for the effect of temperature, compensating techniques must be used. The circuits are :
  1. Diode Compensation for V_{BE}
  2. Diode Compensation for I_{CO}
  3. Thermistor and sensistor compensation.

OBJECTIVE TYPE QUESTIONS

1. Operating point is also known as ....
2. Operating point depends upon the following BJT parameters ....
3. For a transistor to function as an amplifier, the operating point must be located at ....
4. For normal amplifier circuits, the operating point is chosen such that V_{CE} = ...V_{CC}.
5. Fixed bias circuit is also known as ....
6. Self bias circuit is also known as ....
7. Stability factor S is defined as S = ...
8. S' is defined as S' = ...
9. Stability factor S'' is defined as S'' = ...
10. General expression for S = ...
11. Expression for S for self bias circuit is.................. .
12. For both Silicon and Germanium, V_{BE} or V_r cut in voltage decreases at the rate of ....
13. Semiconductor materials having positive temperature coefficient of R (PTCR) are called.................. .
14. Materials used for Thermistors are ....
15. For most configurations, the D.C. analysis of the circuit begins with ...... current of the BJT.
16. A BJT circuit is most stable, and least sensitive to temperature changes if the circuit has .... stability factor.
17. 'β' of a BJT is very sensitive to .... and V_{BE} .... at the rate of 2.5 mv for each 1° increase in temperature.
18. If a transistor (BJT) is in active region, V_{CE} will be about .... % of V_{CC}.

ESSAY TYPE QUESTIONS

1. Explain about the need for biasing in electronic circuits, what are the factors affecting the stability factor.
2. Draw the Fixed bias circuit and derive the expression for the stability factor S. What are the limitations of this circuit ?
3. Draw the Collector to Base bias circuit and derive the expression for the stability factors What are the limitations of this circuit ?
4. Draw the Self bias circuit and obtain the expression for the stability factor S. What are the advantages of this circuit ?
5. Compare all the three biasing circuits.
6. Explain the terms: Thermal Runaway and Thermal Resistance.

7. Explain the terms Bias Stabilization and Bias Compensation.
8. Draw the circuits and explain the principles of working of Diode Compensation for V_{BE} and I_{CO}.

MULTIPLE CHOICE QUESTIONS

1. When the input is symmetrical, to operate the BJT in active region, the quiescent point is chosen ....

(a) at the top edge of the load line
(b) at the bottom edge of the load line
(c) at the centre of the load line
(d) can be chosen any where on the load line

2. AC load line is also known as ....

(a) dynamic load line (b) variable load line
(c) quiescent load line (d) active load line

3. For better stability of the amplifier circuit, value of stability factor 'S' must be

(a) large (b) as small as possible (c) 1 (d) 0

4. For fixed bias circuit, the expression for stability factor 'S' is

(a) $1-\beta$ (b) $1+\beta^2$ (c) $\dfrac{1+\beta}{1-\beta\frac{(\partial I_C)}{(\partial I_B)}}$ (d) $1+\beta$

5. Voltage divider bias or universal bias circuit is also known as

(a) self bias circuit (b) collector bias circuit
(c) collector to base bias circuit (d) Fixed bias circuit

6. By definition, expression for stability factor S" is

(a) $S'' = \left.\dfrac{\partial I_C}{\partial \beta}\right|_{\substack{V_{BE}=K \\ I_{CO}=K}}$ (b) $S'' = \left.\dfrac{\partial \beta}{\partial I_C}\right| V_{BE}, I_{CO}=K$

(c) $S'' = \dfrac{\partial I_\beta}{\partial I_C}$ (d) $S'' = \dfrac{\partial I_\beta}{\partial \beta}$

7. The units of stability factor S are

(a) ohms (b) constant (c) mhos (d) volt-amperes

8. Compared to silicon BJT, thermal stability for germanium transistors is ...

(a) good (b) poor (c) same (d) can't be said

CONCEPTUAL QUESTIONS (Interview Questions)

1. Why transistor biasing is to be done?

Ans : In order that amplification is achieved, the BJT must be operated in the active region. To ensure thus, E-B junction of the transistor (BJT) must be forward biased and Collector - Base junction must be reverse biased. The operating point or quiescant point must be in the centre of the active region. To ensure thus, biasing must be done. Similarly to use the BJT as On - Off switch, the operating point must change between cut-off and saturation regions. So biasing is to be done.

4 Small Signal Analysis of FET and BJT Amplifiers

In this Chapter,

- A Transistor (BJT) can be represented as a group of elements consisting of Current Source, Voltage Source, Resistor, Conductor, etc. Using this equivalent circuit, it is easy to analyse Transistor Amplifier Circuits.
- The Transistor is represented in terms of h - Parameters or Hybrid Parameters.
- The values of these parameters vary with the configuration of the transistor namely Common Emitter (CE), Common Base (CB) and Common Collector (CC).
- The expressions for Voltage Gain A_V, Current Gain A_I etc., are derived using these h - Parameters.

4.1 BJT MODELING USING h-PARAMETERS

In drawing the equivalent circuit of a transistor, we assume that the Transistor Parameters are constant. With the variation in operating points, transistor parameters will also vary. But we assume that the Quiescent or operating point variation is small.

The advantages of considering h-parameters for transistor are

1. *They are easy to measure.*
2. *Can be obtained from Transistor characteristics.*
3. *They are real numbers at Audio frequencies.*
4. *Convenient to use in circuit design and analysis.*

Manufactures specify a set of ***h-parameters*** for transistors.

4.2 DETERMINATION OF h-PARAMETERS FROM TRANSISTOR CHARACTERISTICS

The output characteristics of a Transistors in Common Emitter Configuration are as shown in Fig. 4.1.

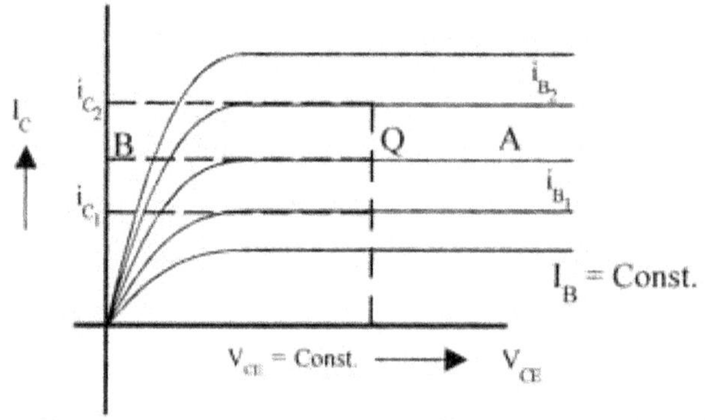

Fig 4.1 Transistor Output characteristics (To determine h_{fe}, h_{oe} graphically).

$$h_{fe} = \left.\frac{\partial I_C}{\partial I_B}\right|_{V_C=K} = \left.\frac{\Delta I_C}{\Delta I_B}\right|_{V_C=0} \qquad V_{CE} = \text{Constant} = K$$

h_{fe}, h_{oe} can be determined from Output characteristics. h_{ie}, h_{re} can be determined from input characteristics.

Suppose Q is the Quiescent Point around the Operating point h_{fe} is to be determined. For h_{fe}, we have to take the ratio of incremental change in I_B keeping V_{CE} constant. From the output characteristics two values i_{c1}, and i_{c2} are taken corresponding to i_{b_1} and i_{b_2}. Then,

$$\mathbf{h_{fe}} = (i_{c2} - i_{c1})/(i_{b2} - i_{b1})$$

h_{fe} is also known as ***Small-Signal Beeta*** (β) of the transistors. Since we are determining the incremental values of Current Ratio of I_C and I_B. h_{fe} is represented as β.

β is represented as h_{fe} and is called large signal β. $\beta = I_C / I_B$, since here no incremental values are being considered.

$$h_{oe} = \left.\frac{\partial I_C}{\partial v_C}\right|_{I_B = K}$$

Output characteristics of a transistor are plotted between i_c and v_c. Therefore, h_{oe} is the slope of the output characteristics, corresponding to a given value of I_b at the Operating point. For Common Emitter Configuration, in exactly the same way we have drawn for amplifier system, the h-parameter equivalent circuit can be drawn. The Transistor in Common Emitter Model is shown in Fig. 4.2. The h-parameter equivalent is shown in Fig. 4.3.

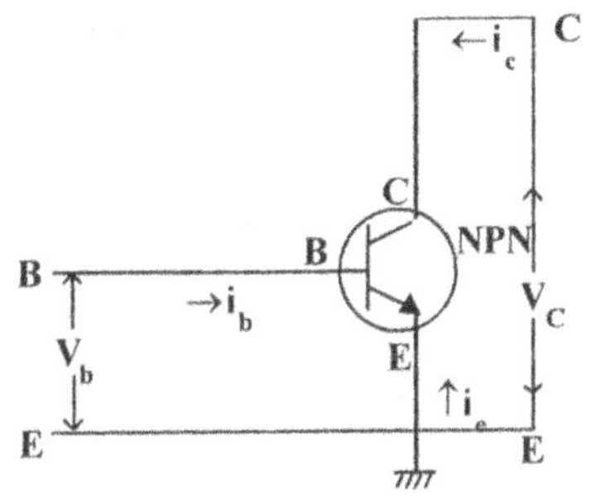

Fig 4.2 Common Emitter configuration.

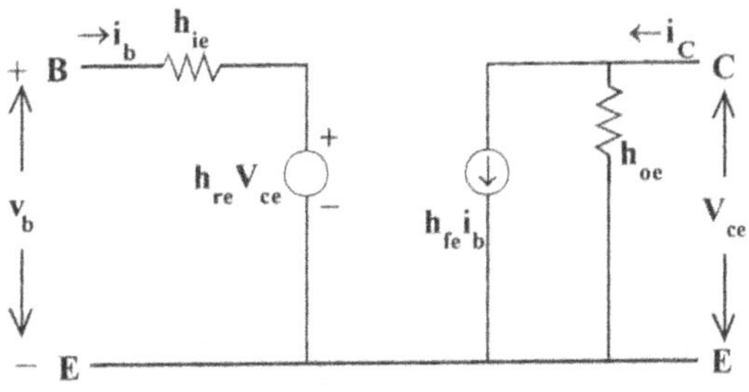

Fig 4.3 Equivalent circuit in C.E. configuration.

Equations are
$$v_b = h_{ie} i_b + h_{re} v_C \qquad \text{.......... (4.1)}$$
$$i_C = h_{fe} i_b + h_{oe} v_C \qquad \text{.......... (4.2)}$$

4.3 MEASUREMENT OF h-PARAMETERS

In the circuit shown in Fig. 4.4 to determine the h-parameter, PNP transistors are taken into account. The Operating point is chosen by suitably adjusting R_2, V_{EE} and V_{CC}, there by adjusting the values of I_C and V_{CB}. The h-parameters depend upon the frequency of operation. So the signal generator is adjusted to be 1KHz, at which frequency normally all the transistor h parameters are specified by the manufactures. 1 KHz is in the audio frequency range. ***So the h-parameters are assumed to be real at audio frequency***. The tank Circuit impedance will be very high at this frequency, of the order of 500kΩ. This is very high compared to the input resistance of the transistor. The tank circuit is tuned to the same frequency of the signal generator (1KHz). This will prevent any stray pick up. The stray signal will pass through the LC tank circuit, since it offers least impedance. So spurious signal will not affect the net voltages and currents of the transistor, while determining the h-parameters.

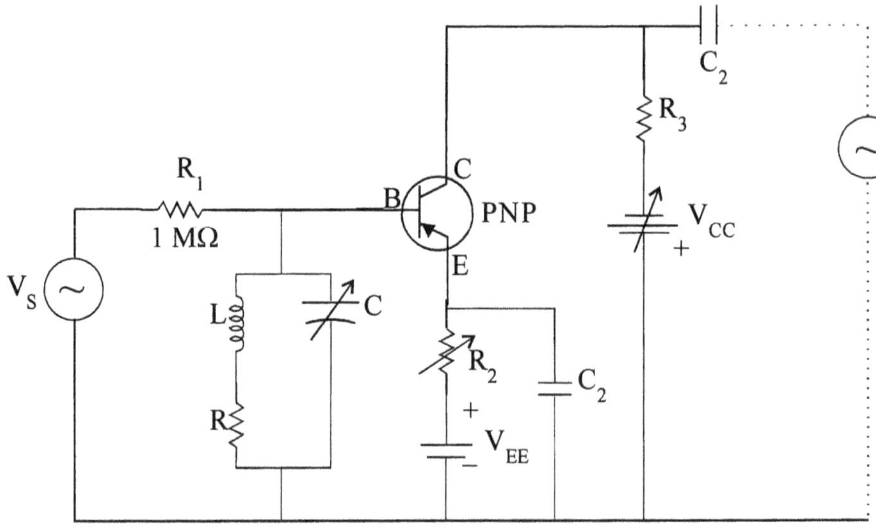

Fig. 4.4 Circuit for determining h - parameters.

The name, tank circuit, is used for LC Parallel circuit, because energy is stored in these elements, as water is stored in a tank.

$$h_{ie} = \left.\frac{V_b}{I_b}\right|_{V_c} ; I_b = \frac{V_S}{R_1}$$

Since the input resistance of the transistor is small compared to the Tank Circuit Resistance, all the current I_b flows into the transistors alone. R_S is negligible compared to R_1 (1MΩ).

$$\therefore \qquad I_b = \frac{V_S}{R_1}$$

Now h_{ie} and h_{fe} are determined when $V_c = 0$ or the collector is shorted to ground. But if the collector is directly connected to the ground, I_c will be different and V_{CE} will be different. The operating point will change h parameters. Depending on the Operating point h - parameters differ. Hence AC short circuit is to be done . C_2 will block D.C. Hence the D.C values of V_{CE}, I_c will not change at all. But the A.C. Voltage of V_c is = 0 or the collector is at ground potential for A.C. voltages . So this is called AC short circuit.

$$\text{Now} \qquad h_{ie} = \left.\frac{V_b}{I_b}\right|_{V_c=0} \qquad \because I_b = \frac{V_S}{R_1}$$

$$h_{fe} = \left.\frac{I_c}{I_b}\right|_{V_c=0} \qquad I_b = \frac{V_S}{R_1}$$

$$\therefore \qquad h_{fe} = \frac{I_C R_1}{V_S}$$

where V_o is the voltage across R_3.

h_{re} and h_{oe} are defined when $I_b = 0$ or input is open circuited. Therefore, the tank circuit impedance is very large. The input side can be regarded as open circuit. The signal generator is connected on the output side.(as shown by the dotted lines).

$$h_{re} = \left.\frac{V_b}{V_c}\right|_{I_b=0} = \frac{V_b}{V_c}$$

$$h_{oe} = \left.\frac{I_c}{V_c}\right|_{I_b=0} = \frac{I_c}{V_c}{}' = \frac{V_O}{R_L V_C} \qquad \because I_C = \frac{V_O}{R_L}$$

4.4 ANALYSIS OF CE, CB AND CC CONFIGURATIONS USING h-PARAMETERS

To form transistor amplifier configuration, we connect a load impedance Z_L and a signal source as shown in Fig. 4.5.

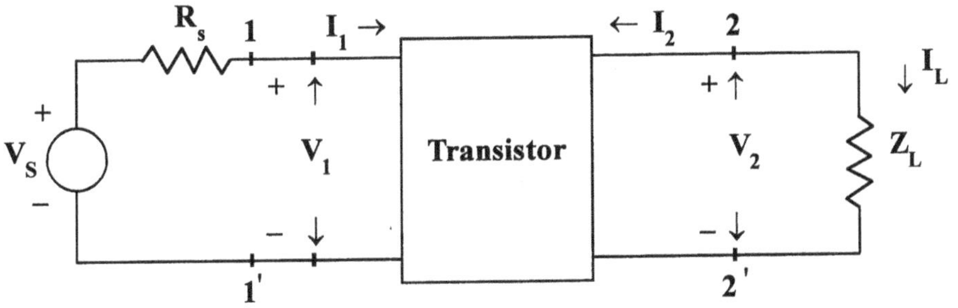

Fig. 4.5 Amplifier circuit.

V_s is the signal source, R_s is the resistance of the signal source, and Z_L is the load impedance. Transistor can be connected in C.E, C.B. and C.C. Configuration. To analyse these circuits i.e., to determine the current gain A_I, Voltage gain A_v, input impedance, output impedance etc, we can use the h parameters. So the equivalent circuit for the above transistor amplifier circuit in general form without indicating C.E, C.B, or C.C. Configuration, can be denoted as in the Fig. 4.6.

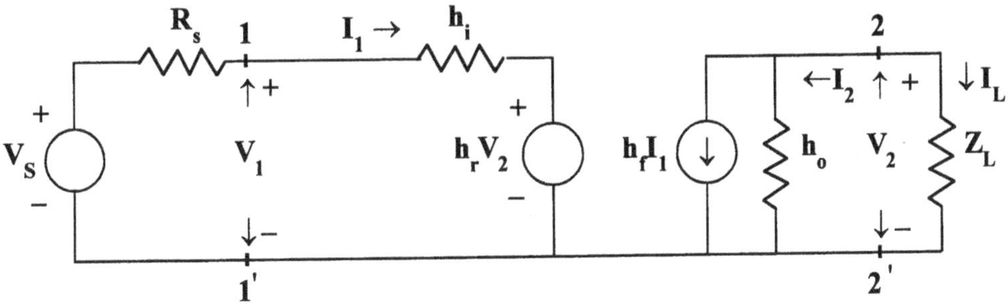

Fig. 4.6 h-parameter equipvalent circuit (general repersentation).

Current Amplification,

$$A_I = \frac{I_L}{I_1};$$

But $I_L = -I_2$

$$\therefore \quad A_I = -\frac{I_2}{I_1};$$

Negative sign is because I_2 is always represented as flowing into the current source. For a transistor PNP or NPN type, if it is flowing out of the transistor, I_2 is represented as $-I_2$. Now voltage across Z_2 is taken with 2 as +ve and 2^1 as –ve The directions of I_L are as shown and $I_L = -I_2$.

From the above circuit we have,

$$I_2 = h_f I_1 + h_o V_2$$

But $V_2 = I_L Z_L = -I_2 . Z_L$ (Since $I_L = -I_2$)

$$\therefore \quad I_2 = h_f I_1 - h_o . Z_L . I_2$$

or $I_2 + h_o Z_L I_2 = h_f I_1$.

$$I_2 (1 + h_o Z_L) = h_f I_1$$

$$\therefore \quad A_I = -\frac{I_2}{I_1} = \frac{-h_f}{1 + h_o Z_L} \quad \text{.......... (4.3)}$$

Negative sign indicates that I_1 and I_2 are out of phase by 180^0.

4.4.1 Input Impedance (Z_i)

Input impedance of any circuit is the impedance we measure looking back into the amplifier circuit. Now amplifier terminals are 1and 1^1. R_s is the resistance of the signal source. So to determine the output z of the amplification alone, it need not be considered.

$$Z_i = \frac{V_1}{I_1}$$

But $V_1 = h_i I_1 + h_r V_2$

Hence

$$Z_i = \frac{V_1}{I_1} = h_i + h_r \frac{V_2}{I_1} \quad \text{.......... (4.4)}$$

But

$$V_2 = -I_2 . Z_L, \quad \therefore \frac{V_2}{I_1} = \left(\frac{-I_2}{I_1}\right) Z_L = A_I Z_L \quad \text{.......... (4.5)}$$

Substituting the values of V_2 in Equation (4.4),

$$Z_i = h_i + h_r . A_I . Z_L$$

But

$$A_I = \frac{-h_f}{1 + h_o z_L}$$

$$\therefore \quad Z_i = h_i - \frac{h_f h_r z_L}{1 + h_o z_L}$$

$$= h_i - \frac{h_f . h_r}{\frac{1}{Z_L} + h_o}$$

$$Z_i = h_i - \frac{h_f . h_r}{Y_L + h_o} \qquad \text{.......... (4.6)}$$

Y_L is load admittance. Therefore, Input impedance is a function of load impedance. If $Z_L = 0$, $Y_L = \infty$,

$\therefore \quad Z_I = h_i.$

4.4.2 Voltage Gain (A_V)

$$A_v = \frac{V_2}{V_1}$$

But $\quad V_2 = + I_L . Z_L = - I_2 . Z_L \qquad$ Since, $I_2 = - I_L$

But $\quad I_2 = - A_I . I_1$

$\therefore \quad V_2 = + A_I . I_1 Z_L$

$$\frac{V_2}{V_1} = \frac{A_I . I_1 Z_L}{V_1}$$

But $\quad \frac{V_1}{I_1} = Z_i$

$$Z_i = h_i - \frac{h_f . h_r}{\frac{1}{R_L} + h_o}$$

$$\therefore \quad A_V = \frac{A_I . Z_L}{Z_i}$$

$$A_V = - \frac{-h_f . z_L}{h_i + z_L (h_i h_o - h_f h_r)} \qquad \text{.......... (4.7)}$$

4.4.3 Output Admittance (Y_O)

The output impedance can be determined by using two assumptions $Z_L = \infty$, and $V_s = 0$

Y_o is defined as $\frac{I_2}{V_2}$ with $Z_L = \infty$

But $\quad I_2 = h_f I_1 + h_o V_2$

Dividing by V_2,

$$\frac{I_2}{V_2} = Y_O = \frac{h_f . I_1}{V_2} + h_o. \qquad \text{.......... (4.8)}$$

From the equivalent circuit with $V_s = 0$,

$$R_S . I_1 + h_i I_1 + h_r V_2 = 0.$$

Dividing by V_2 Through out, we get

$$\frac{R_S.I_1}{V_2} + \frac{h_i I_1}{V_2} + h_r = 0$$

or $$\frac{I_1}{V_2} = \frac{-h_r}{h_i + R_S}$$ .... (4.9)

Substitute this value in the equation for $Y_{o.}$

$$Y_o = h_f \left(\frac{-h_r}{h_i + R_s} \right) + h_o$$

$$Y_o = h_o - \frac{h_f . h_r}{h_i + R_s}$$

Therefore, $Z_o = 1/Y_o$

Therefore, Output admittance is a function of Source Resistance R_s where as Z_i is a function of Y_L. If $R_s = 0$, $Y_o \cong h_o$. Since ($h_f h_r / h_i$) is very small.

4.4.4 VOLTAGE GAIN (A_{VS}) CONSIDERING SOURCE RESISTANCE R_S

$$\therefore \quad A_{vs} = \frac{V_2}{V_S} = \frac{V_2}{V_1} \times \frac{V_1}{V_S} = A_v . \frac{V_1}{V_S}.$$

This is the equivalent circuit of the input side of the amplifier circuit.

$$\therefore \quad V_1 = \frac{V_S . Z_i}{Z_i + R_S}$$

$$\therefore \quad A_{vs} = \frac{A_V . Z_i}{Z_i + R_S}$$

$$A_v = \frac{V_2}{V_1}$$

$$\therefore \quad A_V Z_i = \frac{V_2}{V_1} . Z_i$$

But $$V_2 = I_L . Z_L ; V_1 = I_1 Z_i$$

$$\therefore \quad A_v . Z_i = \frac{I_L . Z_L . Z_i}{I_1 . Z_L} = A_I . Z_L$$

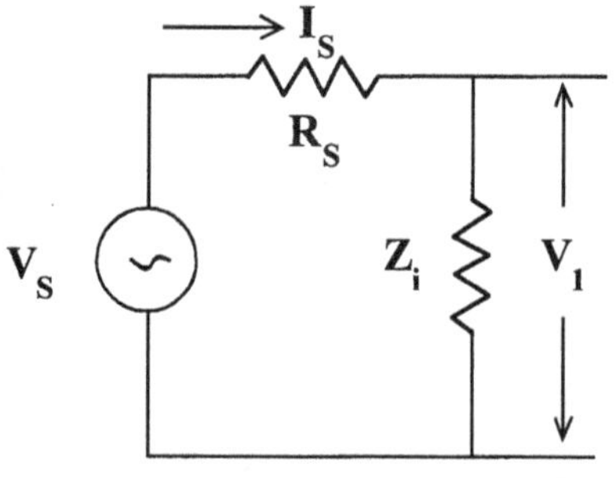

Fig 4.7 Voltage source.

$$\therefore \quad A_{vs} = \frac{A_I . Z_L}{R_S + Z_i}$$ If $R_s = 0$, $A_{vs} = A_v$ .... (4.10)

Hence A_V is the voltage gain of an ideal voltage source with zero terminal resistance.

4.4.5 CURRENT GAIN (A_{IS}) CONSIDERING SOURCE RESISTANCE

The input source can also be represented as a current source I_S in parallel with resistance R_s or Voltage source V_s in series with resistance R_s. Now let us consider the input source as a current source in parallel with R_s. The equivalent circuit is

$$A_{IS} = \frac{I_L}{I_S};\ I_L = -I_2$$

$$\therefore \quad A_{IS} = \frac{I_L}{I_S} = -\frac{I_2}{I_S}$$

$$A_{IS} = -\frac{I_2}{I_S} = -\frac{I_2}{I_1}\cdot\frac{I_1}{I_S}$$

$$= A_I \cdot \frac{I_1}{I_S} \qquad \text{.......... (4.11)}$$

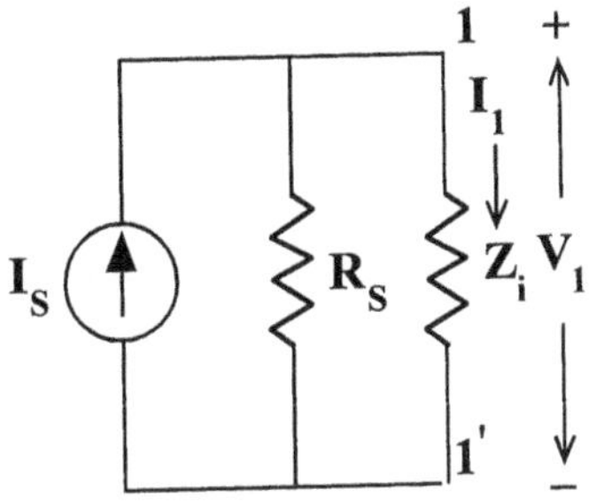

Fig 4.8 Current source.

$$\therefore \quad A_I = \frac{-I_2}{I_1}$$

From the Fig 4.8, $I_1 = \dfrac{I_S . R_S}{R_S + Z_i}$

$$\therefore \quad A_{IS} = \frac{A_I \times R_S}{(R_S + Z_i)} = \frac{A_I . R_S}{(R_S + Z_i)}$$

If $R_S = \infty$, $A_I = A_{IS}$. Therefore, A_I is the current gain for an ideal source

Now $$A_{VS} = \frac{A_I . Z_L}{Z_i + R_s} \qquad \text{.......... (4.12)}$$

$$A_{VS} = \frac{A_I . R_S}{Z_i + R_S} \qquad \text{.......... (4.13)}$$

Dividing 1 and 2, □ $\dfrac{A_{vs}}{A_{IS}} = \dfrac{Z_L}{R_S}$

or $$A_{vs} = A_{IS} \cdot \frac{Z_L}{R_S} \qquad \text{.......... (4.14)}$$

This equation is independent of the transistor parameters. This is valid if the equivalent Current and Voltage sources have the same resistance.

4.4.6 POWER GAIN (A_P)

$$A_P = \frac{P_2}{P_1} = \frac{V_2 I_L}{V_1 I_1}$$

$$= A_v . A_I$$

$$\because \quad A_v = \frac{A_I . Z_L}{Z_i},$$

$$\therefore \quad A_P = \frac{A_I^2 Z_L}{Z_i} \qquad \text{........ (4.15)}$$

4.5 COMPARISON OF CE, CB, CC CONFIGURATIONS USING h-PARAMETERS

C.E : Of the three, it is the most versatile. Its voltage and current gain are > 1. Input and output resistance vary least with R_s and R_L. R_i and R_o values lie between maximum and minimum for all the three configurations. Phase shift of 180^0 between V_i and V_o. Power Gain is Maximum.

C.B : $A_I < 1$ $A_V > 1$. R_i is the lowest and R_o is the highest . It has few applications. Sometimes it is used to match low impedance source V_i and V_o. No phase shift.

C.C : $A_V < 1$. A_I is very high. It has very ***high input Z and low output Z. So it is used as a buffer*** between high Z source and low impedance load. It is also called as ***Emitter follower***.

To analyse circuit consisting of a number of Transistors, each Transistor should be replaced by its equivalent circuit in h-parameters. The emitter base and collector points are indicated and other circuit elements are connected without altering the circuit Configuration. This way the circuit analysis becomes easy.

Problem 4.1

Find the Common Emitter Hybrid parameters in terms of the Common Collector Hybrid parameters for a given transistor.

Solution

We have to find h_{ie}, h_{re}, h_{fe} and h_{oe} in terms of h_{ic}, h_{rc}, h_{fc} and h_{oc}.

The transistor circuit in Common Collector Configuration is shown in Fig. 4.9. The h-parameter equivalent circuit is shown in Fig. 4.10.

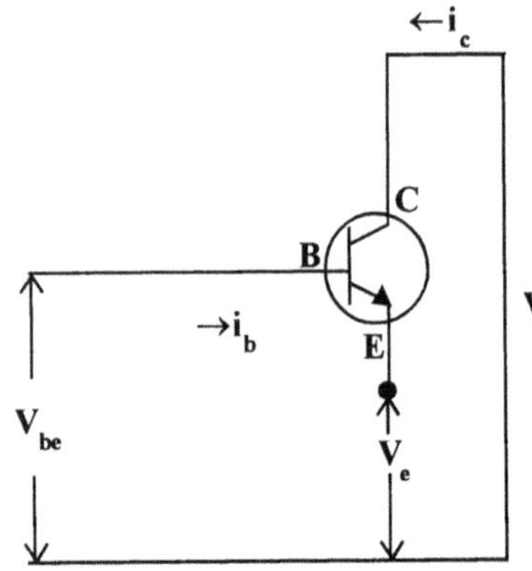

Fig. 4.9 For problem 4.1.

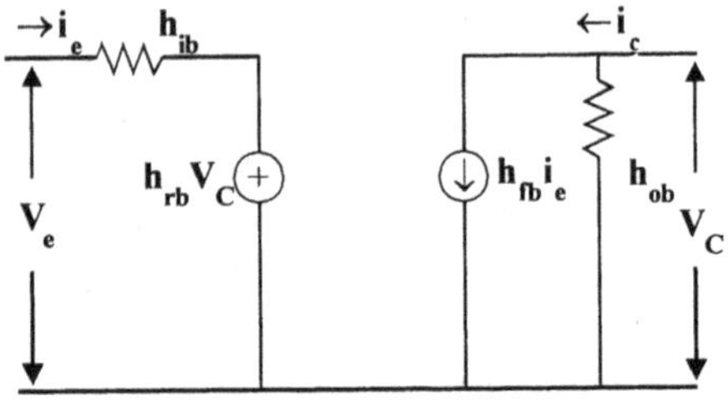

Fig. 4.10 Equivalent circuit.

C.C :

$$v_b = h_{ic} \cdot i_b + h_{rc} v_e$$
$$i_e = h_{fc} i_b + h_{oc} v_e$$
$$i_e = h_{fc} i_b - i_1$$
$$= (h_{fe} i_b - V_{ce} h_{oc})$$

$$V_{be} = i_b \cdot h_{ic} - h_{rc} v_{ce} + v_{ce} = i_b \cdot h_{ic}$$ Since $v_{ce} = 0$

But $h_{ie} = \left.\dfrac{V_{be}}{I_b}\right|_{V_{ce}=0}$

$\therefore$ $h_{ie} = h_{ic}$ .... (4.16)

$$h_{fe} = \left.\frac{i_c}{i_b}\right| V_{ce} = 0$$

$$i_c = -i_b - i_e = -i_b - h_{fc} \cdot i_b . + h_{oc} v_{ce}$$

But $v_{ce} = 0$

$$= -i_b - (h_{fc} i_b - h_{oc} v_{ce})$$

$$= -i_b - h_{fc} i_b + h_{oc} v_{ce} \qquad v_{ce} = 0$$

$$h_{fe} = \frac{-i_b - h_{fc} i_b}{i_b} = -(1 + h_{fc})$$

$$h_{fe} = -(1 + h_{fc})$$ .... (4.17)

$$h_{re} = \left.\frac{V_{be}}{V_{ce}}\right|_{i_{b=0}}$$

$$V_{be} = i_b h_{ie} - h_{rc} V_{ce} + V_{ce}$$

$$i_b = 0$$

$\therefore$ $h_{re} = -\dfrac{h_{rc} V_{ce} + V_{ce}}{V_{ce}} = 1 - h_{rc}$

$$h_{re} = 1 - h_{rc}$$ .... (4.18)

$$h_{oe} = \left.\frac{i_c}{V_{ce}}\right| i_b = 0$$

$$i_c = -i_e - i_b$$

But $i_b = 0.$

$$i_c| = i_e$$

$$i_e = V_{ce}\,hoc$$

$\therefore$ $\dfrac{i_e}{V_{ce}} = \dfrac{i_e}{V_{ce}} = h_{oc}$

$\therefore$ $i_c = -i_e = -h_{fc} i_b + V_{ce} h_{oc}$

$\therefore \quad h_{oe} = h_{oc}$ .... (4.19)

Since, $\quad i_b = 0$

$\therefore \quad i_c = V_{ce} \times h_{oc}$

$\therefore \quad h_{oe} = h_{oc}$ is admittance in mhos (℧).

4.6 SIMPLIFIED TRANSISTOR HYBRID MODEL

The analysis of transistors can be done using h-parameters which includes detailed calculation of all the four parameters related to transistor device such as voltage amplification, current amplification, input and output impedences. However in practice appropriate approximations can be allowed provided that if approximations are within tolerable limits of devices (considering all the parameters of the devices such as aging, temperature dependence of characteristics etc).

Out of available four parameters of transistors, two parameters i.e., h_{ie} and h_{fe} (CE configuration) are sufficient for simplified analysis of transistor for low frequency circuits provided that load resistance is small enough to satisfy the following condition.

$$h_{oe} R_L < 0.1$$

The simplified hybrid model for different configurations of transistors CB, CE and CC are shown in Fig. 4.12, Fig. 4.11, and Fig. 4.13 respectively.

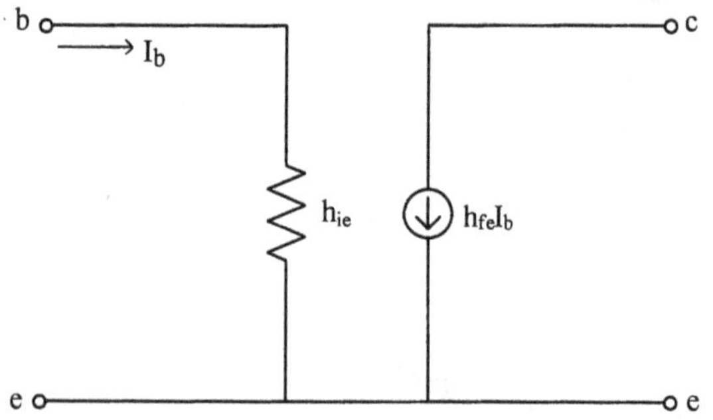

Fig. 4.11 Simplified hybrid model of CE configuration.

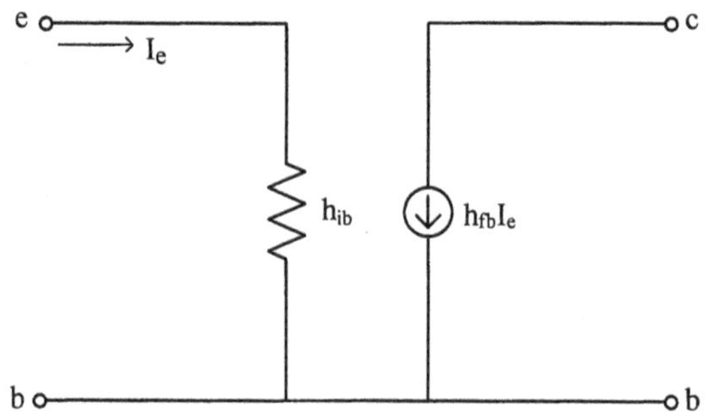

Fig. 4.12 Simplified hybrid model of CB configuration.

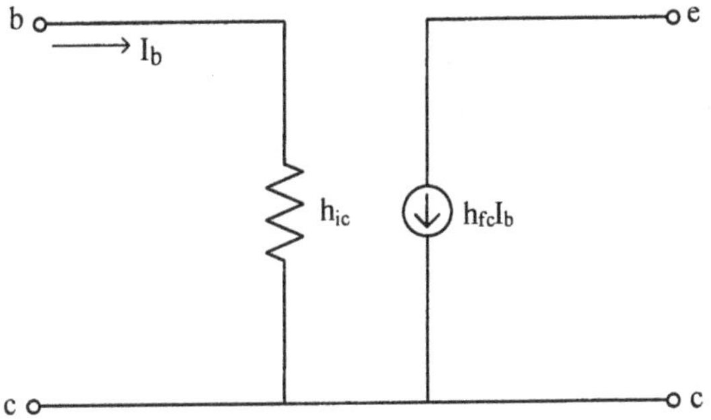

Fig. 4.13 Simplified hybrid model of CC configuration.

Simplified CE Configuration

The Fig. 4.11 shows simplified hybrid model of CE transistor.

Current Gain: The current gain of CE transistor is given by

$$A_I = \frac{I_2}{I_1} = \frac{-h_{fe}}{1 + h_{oe}Z_L} \qquad .....(4.20)$$

By applying condition $h_{oe}R_L < 0.1$, the denominator term in eq.(4.20) becomes $\simeq 1$

Hence $\qquad A_I = -h_{fe} \qquad .....(4.21)$

Input impedance: The input impedance of CE transistor is given by

$$Z_i = h_i + h_r A_I Z_L \qquad .....(4.22)$$

$$= h_{ie}\left(1 - \frac{h_{re}h_{fe}|A_I|}{h_{ie}h_{oe}h_{fe}} h_{oe}R_L\right) \qquad .....(4.23)$$

For any transistor $h_{re}h_{fe}/h_{ie}h_{oe} = 0.5$, $|A_I| \Box h_{fe}$ (from eq. (4.20)) and by applying the condition $h_{oe}R_L < 0.1$, the second term in brackets of eq. (4.23) becomes negligible. And hence

$$Z_i \Box h_{ie} \qquad .....(4.24)$$

Voltage gain: The voltage gain (A_V) of CE configuration transistor is given by

$$A_V = A_I(R_2/R_i) \qquad .....(4.25)$$

Substituting values of A_I from eq. (4.21) and R_i from eq. (4.24) we get

$$A_V = -h_{fe}R_L/h_{ie} \qquad ......(4.26)$$

Output impedance: To calculate output impedance the input signal $(V_s$ assume) is to be made zero and load resistance (assume 'R_L' is made infinite) if $V_s = 0$, output impedence R_L is infinite. There is one dependent ideal current source i.e., $h_{fe}I_b$, in the output circuit and so output impedence will be infinite and hence output impedance will be equal to load resistance i.e., $R_o = R_L$.

$$\text{Output impedance } (Z_o) = \text{Load Resistance } R_L \qquad .....(4.27)$$

By employing corresponding convertions of h-parameters the difference parameters like A_I, Z_i, A_V, Z_o etc., for different transistor configurations (CB, CE, CC) can be easily obtained.

4.7 MILLER'S THEOREM

Consider an arbitrary circuit shown in Fig 4.14(a). Here they are two nodes which are isolated named as 1 and 2, between which an impedance z is connected.

Nodes 1 and 2 are connected to other parts of network, as signified by the broken lines emanating from two nodes.

Furthermore here it is assumed that voltage at node 2 equals to voltage at node 1 is given by,

$$V_2 = kV_1 \quad .....(4.28)$$

Here 'k' is referred as gain factor that may be positive or negative whose magnitude usually greater than unity (i.e., $k > 1$).

MILLER'S THEOREM STATEMENT

The single impedance Z can be replaced by two impedances: First impedance Z_1 connected between node 1 and ground and second impedance Z_2 connected between node 2 and ground, where

$$Z_1 = \frac{Z}{(1-k)} \quad .....(4.29)$$

and

$$Z_2 = \frac{Z}{\left(1-\frac{1}{k}\right)} \quad .....(4.30)$$

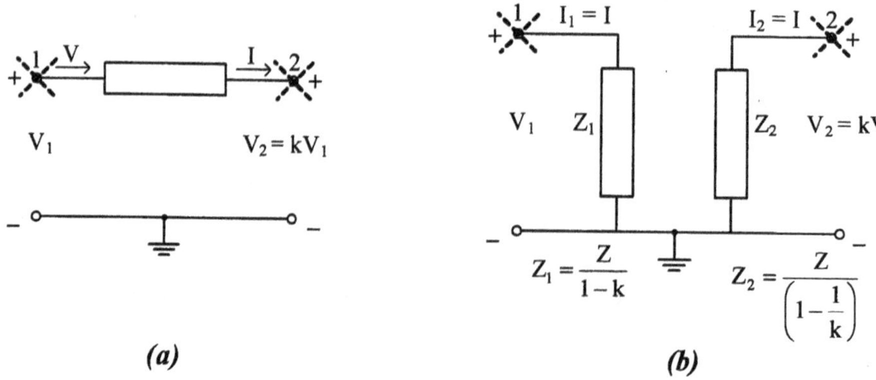

Fig. 4.14 The Miller's equivalent circuit

and equivalent circuit with two impedances as shown in Fig. 4.14(b)

Proof:

Miller's theorem can be proved by deriving eqs. (4.29) and (4.30) as shown below.

In the circuit in Fig 4.14(a), the only way that nodes "assumes the existence" of impedance Z is through the current I that Z draws away from node 1.

Therefore, in order to keep this current I unchanged in the equivalent circuit, the value of Z_1 must be chosen so that it should draw an equal current.

$$I_1 = \frac{V_1}{Z_1} = I = \frac{V_1 - kV_1}{Z}$$

$$\frac{V_1}{Z_1} = \frac{V_1 - kV_1}{Z}$$

$$Z_1 = \frac{V_1}{V_1 - kV_1}.Z$$

$$Z_1 = \frac{Z}{1-k} \text{ as eq. (4.29)}$$

similarly to keep the current into node 2 should remain same we must select the value of Z_2 so that

$$I_2 = \frac{0 - V_2}{Z_2} = \frac{0 - kV_1}{Z_2} = I = \frac{V_1 - kV_1}{Z}$$

or $$Z_2 = \frac{-kV_1}{V_1 - kV_1}$$

$$\frac{Z}{1 - \frac{1}{k}} \text{ as eq. (4.30)}$$

The Miller's equivalent circuit derived valid only as long as the rest of the circuit remains unchanged, otherwise the ratio V_2 to V_1 might change.

4.7.1 Dual of Miller's Theorem

Consider an arbitrary network Fig. 4.15 assumes active or passive elements between nodes 1, 2 and 3 and with an impedance Z' between node 3 and ground N.

Here two meshes are joined by a common element Z'. Here from the circuit current ratio as $A_I = -I_2/I_1$.

Dual of Miller's theorem is nothing but the element Z' is connected in parallel which is opposite as Miller's theorem.

Here single shunt impedance Z' is replaced by an impedance Z_1 is placed in mesh 1 and an impedance Z_2 is added to mesh 2.

Here in Fig. 4.19 the current Z_1I_1 should equals to $(I_1 + I_2)\,Z'$ to satisfy $Z_1I_1 = Z'(I_1+I_2)$ then $Z_1 = Z'(1 - A_I)$.

Hence the voltage between node 1' and N is same in both circuits for equal currents I_1 and I_2.

Similarly, for the mesh 2, the currents $Z_2I_2 = (I_1+I_2)\,Z'$, then

$$Z_2 = \left[\frac{(A_I - 1)}{A_I}\right] Z'$$

Here two networks are identical, when the voltage drop between node 1' and ground and the voltage drop between node 2' and ground and the same current I_1 and I_2 flow in both meshes.

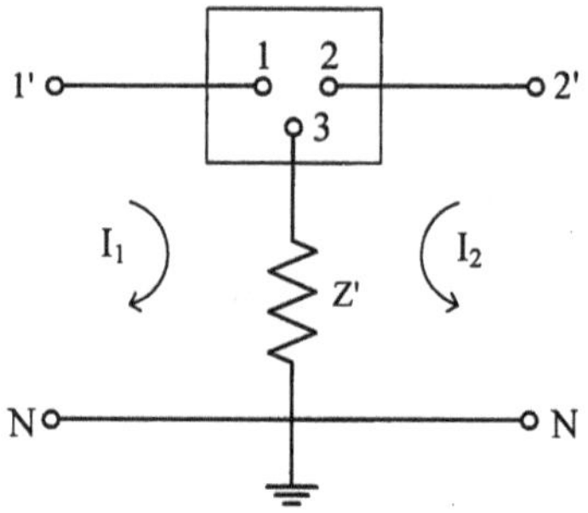

Fig. 4.15 Dual of Miller's theorem

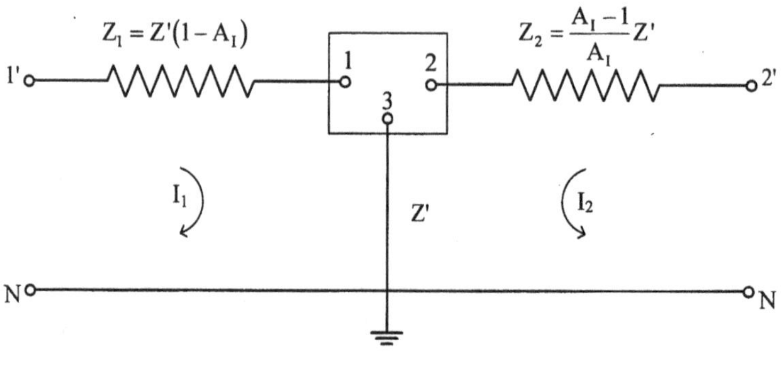

Fig. 4.16 Equivalent circiut

4.8 SMALL SIGNAL MODEL OF JFET

In a FET, instantaneous drain current i_D is a function of the instantaneous gate-source voltage V_{GS} and instantaneous drain-source voltage V_{DS} and is therefore, expressed as

$$i_D = f(V_{GS}, V_{DS}) \quad .....(4.31)$$

If both the gate and drain voltages are varied, the change in drain current is given approximately by the first two terms in the Taylor's series expansion of eq. (4.31) and are

$$\Delta i_D = \left.\frac{\delta i_D}{\delta V_{GS}}\right|_{V_{DS}} \Delta V_{GS} = \left.\frac{\delta i_D}{\delta V_{Ds}}\right|_{V_{GS}} . \Delta V_{DS} \quad .....(4.32)$$

Using the conventional small signal notations, Δi_D, ΔV_{GS} and ΔV_{DS} may be replaced respectively by time varying components i_d, V_{gs} and V_{ds}. Now eq. (4.32) becomes

$$i_d = g_m v_{gs} + \frac{1}{r_d} v_{ds} \quad .....(4.33)$$

where
$$g_m = \left.\frac{\delta i_D}{\delta V_{GS}}\right|_{V_{DS}} \simeq \left.\frac{\Delta i_D}{\Delta V_{GS}}\right|_{V_{DS}} = \left.\frac{i_d}{V_{gs}}\right|_{V_{DS}} \quad .....(4.34)$$

Parameter g_m is the mutual conductance or transconductance

and
$$\frac{1}{r_d} = \left.\frac{\delta i_D}{\delta V_{DS}}\right|_{V_{GS}} \simeq \left.\frac{\Delta i_D}{\Delta V_{GS}}\right|_{V_{GS}} = \left.\frac{i_d}{V_{ds}}\right|_{V_{GS}} \quad .....(4.35)$$

The reciprocal of r_d is the drain conductance g_d.

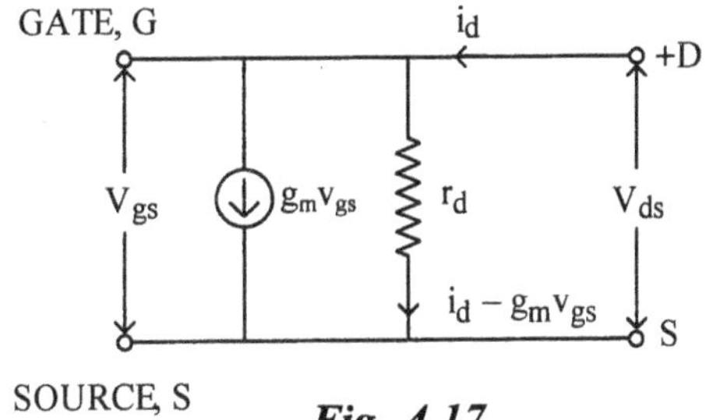

Fig. 4.17

Circuit shown in Fig. 4.17 is made satisfying eq. (4.33) giving the incremental drain current i_d in form of g_m, r_d, v_{gs} and v_{ds}. This circuit forms the low frequency small signal model for FET. This model consists of one dependent current generator whose current $g_m v_{gs}$ is proportional to the time varying gate source voltage v_{gs} and proportionality constant is the transconductance g_m. The arrow points from drain to source to indicate a phase change of 180° between output and input voltages as will occur in actual operation. In low-frequency model, as the gate is reversed biased, the gate current becomes zero.

So, in low-frequency model, gate to source junction is represented by an open-circuit and no current is drawn by the input terminal of the FET.

- The reason is that input resistance, is very large.
- The output resistance is represented by the resistance r_d.

4.9 SMALL SIGNAL MODEL OF MOSFET AMPLIFIERS

In small-signal model for a MOSFET includes the transconductance of the device.

From the transfer characteristics, transconductance of the enhancement-type MOSFET is determined whose definition is

$$g_m = \left.\frac{\Delta I_D}{\Delta V_{GS}}\right|_{V_{DS}=\text{constant}}$$

Here g_m gives the slope of a line drawn tangent to the characteristics at the operating Q-point.

Here value of g_m changes as Q-point changes.

Hence signal variation around the Q-point confined to a limited range over which there is negligible change in g_m is required for small-signal analysis.

It is all sure that device should be operated in active region and the following inequivality should be satisfied for the variations which is given by,

$$|V_{DS}| > |V_{GS} - V_T|$$

The transconductance of an enhancement MOSFET under small signal conditions can be determined from,

$$g_m = \beta\,(V_{GS} - V_T) \text{ siemens.}$$

Fig. 4.18 shows MOSFET amplifier and its equivalent and here using voltage divider method the MOSFET is biased,

here, $r_i = R_1 \parallel R_2$

$$\frac{V_L}{V_S} = \left(\frac{R_1 \parallel R_2}{r_s + R_1 \parallel R_2}\right)(-g_m)(r_d \parallel R_D \parallel R_L)$$

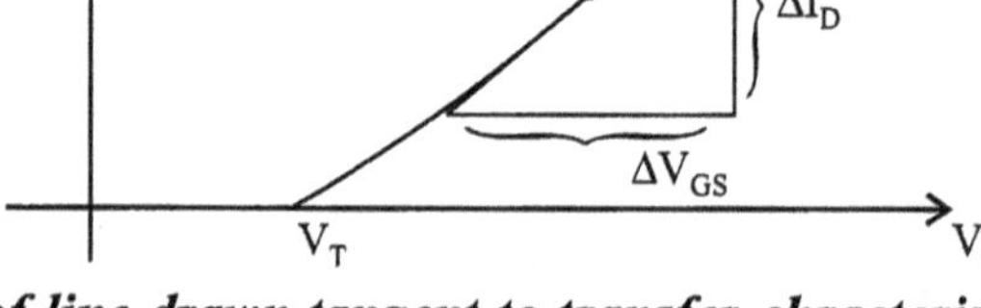

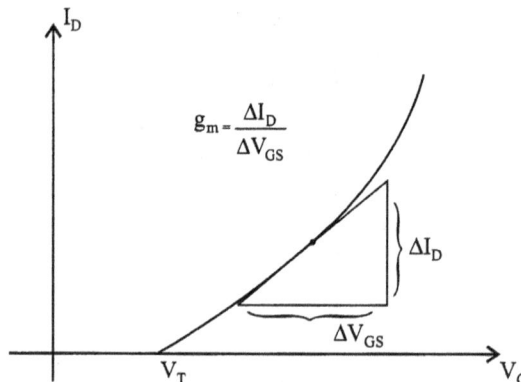

Fig 4.18 The slope of line drawn tangent to transfer chracteristic at the Q-point gives transconductance

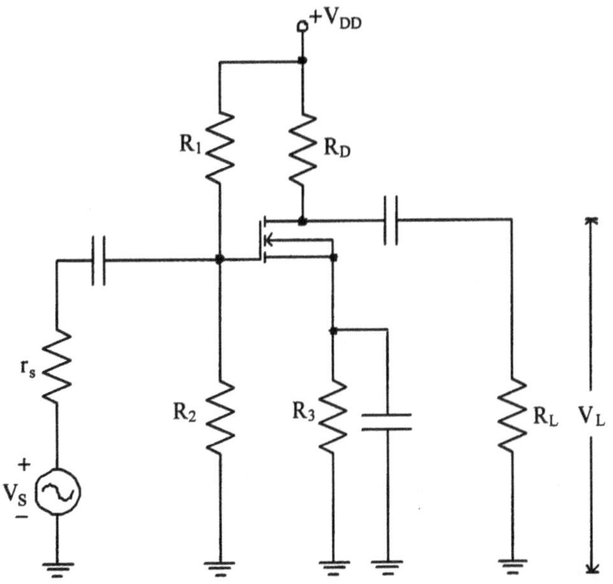

(a) MOSFET amplifier

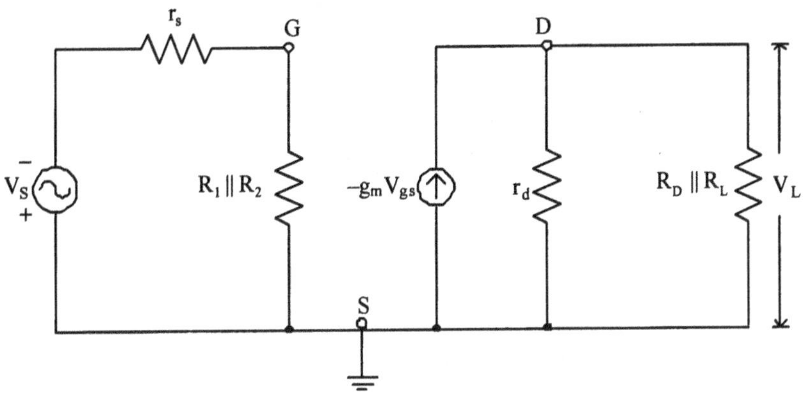

(b) The small signal equivalent circuit

Fig 4.19

If the small bulk resistances of the source and drain are negelected, the small-signal equivalent circuit of the MOSFET between terminals $G(G_1)$ S, and D is identical with that for the JFET and is given in Fig. 4.20.

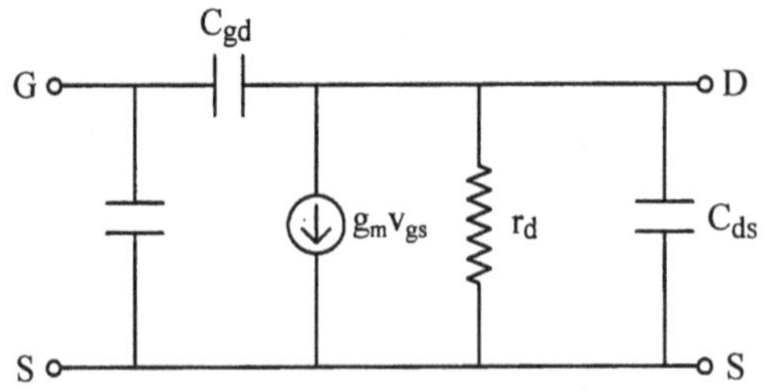

Fig. 4.20

The transconductance and the interelectrode capacitances have comparable values for the two devices, as obvious from the figures given in Table 4.1.

If the substrate terminal G_2 is not conncected to the source, the model of Fig. 4.14 must be generalized as follows.

Between node G_2 and S, a diode D_1 is added to represent the P-N junction between the substrate and the source. Similarly, a second diode D_2 is included between G_2 and D to account for the PN junction formed by the substrate and drain.

Table 4.1 Range of parameter values for JFET and MOSFET.

Parameter	JFET	MOSFET
g_m	0.1-10 ms	0.1-20 ms or more
r_d	0.1-1 MΩ	1.0-50 KΩ
C_{ds}	0.1-1 PF	0.1-1 PF
C_{gs}, C_{gd}	1.0-10 PF	1.0-10 PF
r_{gs}	$> 10^8\ \Omega$	$> 10^{10}\ \Omega$
r_{gd}	$> 10^8\ \Omega$	$> 10^{14}\ \Omega$

4.10 SMALL SIGNAL ANALYSIS OF COMMON SOURCE JFET AMPLIFIER

The common source configuration is the mostly widely employed configuration.

- The main characteristics of this configuration are high input impedance, good voltage gain and moderate output impedance.
- The circuit of a common source of N-channel JFET amplifier using self bias is shown in Fig. 4.21.
- The signal source V_{in} in connected to JFET gate through coupling capacitor C_1 and external load R_L is connected to the drain terminal D via coupling capacitor C_2.
- The purpose of R_G is used to provide leakage path to the gate currrent.
- The purpose of R_S for providing gate bias and C_S is used to provide ac ground to the input signal.

- When we analyze for ac considerations, by pass capacitor C_S effectively links the FET source terminal to the ground level.
- Hence input voltage V_{in} is developed across the gate source terminals.
- As input voltage increases or decreases then gate source voltage V_{gs} changes and corresponding change in V_{gs} makes the drain current I_D to increase or decrease correspondingly.
- The phase relationship between the circuit input and output voltages can be determined by taking the postive going input signal.
- As input voltage V_{in} increases in positive direction. Hence it reduces the negative gate-source voltage $-V_{gs}$.
- The reduction in $-V_{gs}$ leads to increase the level of drain current I_D and consequently increases the voltage drop across drain resistance R_D.

$$\because \qquad V_D = V_{DD} - I_D R_D$$

  Here I_D increases results in a drop in drain voltage V_D.
- Here drain voltage nothing but output voltage.
- Thus as V_{in} increases in a positive direction, V_{out} goes in a negative direction.
- In the same way, when V_{in} goes negative, the resultant increase in $-V_{gs}$ reduces the drain current I_D which in turn reduces the voltage drop across R_D hence develops a positive going output voltage V_{out}.
- Thus, the amplifier output voltage is 180° out of phase with the input voltage.

***Analysis*:** The first step in small signal analysis is to draw its equivalent in small signal.

This is done by replacing all the capacitors with short-circuits and reducing dc supply voltages to zero. Small signal equivalent circuit for common source amplifier is g_1 shown as in Fig. 4.21.

***Input resistance*:** In an ideal FET R_{gs} is infinite but practically R_{gs} is not infinite but extremely high (100 MΩ or so) as compared to R_G. Thus input resistance

$$R_{in} = R_G \parallel R_{gs} \parallel R_G$$

***Output resistance*:** Looking into source and drain terminals, the large drain resistance r_d is

$$Zd \simeq rd$$

Z_d is the device output impedance, the circuit output impedance in R_D in parallel with Z_d, so

$$Z_{out} = RD \parallel Z_d = r_d$$

since $r_d >> R_D$, the circuit output impedance is given as R_D.

***Voltage gain*:** Output voltage,

$$V_{out} = I_D (r_d \parallel R_D \parallel R_L)$$

$$= -g_m V_{in} (r_d \parallel R_D \parallel R_L) \qquad \because I_D = -g_m V_{in}$$

and voltage gain,

$$A_v = \frac{V_{out}}{V_{in}} = -g_m (r_d \parallel R_D \parallel R_L)$$

usually $r_d >> R_D \parallel R_L$

$$\because \quad A_v = -g_m (R_D \parallel R_L)$$

Here minus sign indicates V_{out} is 180° with input voltage V_{in}.

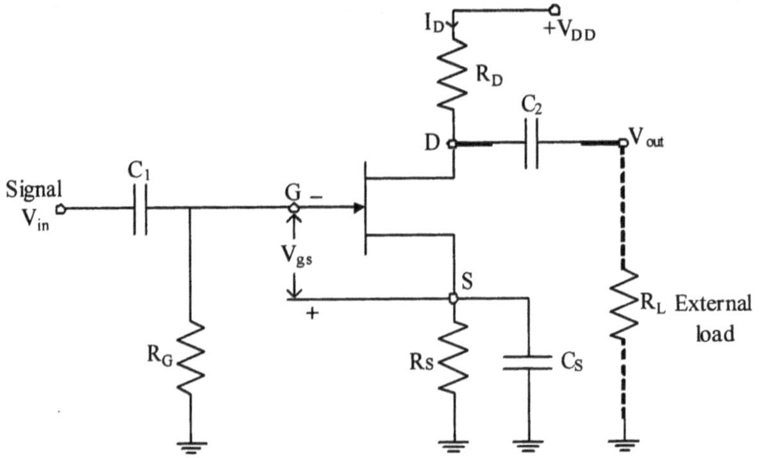

Fig 4.21 Common source amplifier circuit

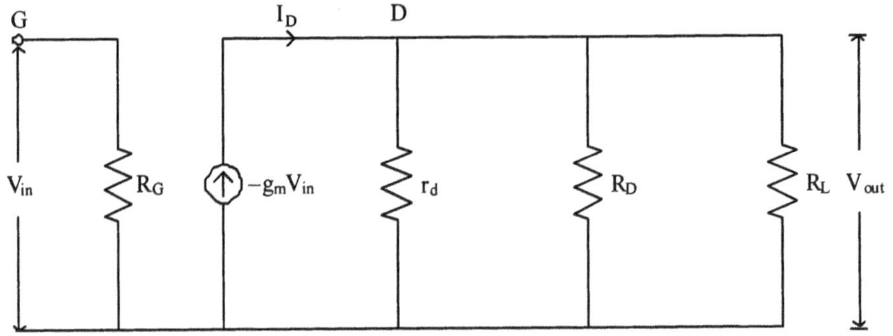

Fig 4.22 Small signal equivalent for Common Source FET amplifier

4.11 SMALL SIGNAL ANALYSIS OF COMMON DRAIN FET AMPLIFIER

In CD amplifier (source follower), the output voltage is taken across source resistor R_S.

External load resistor R_L is connected to FET source terminal S by means of coupling capacitor C_2, and gate bias voltage V_G is derived from V_{DD} by using potential divider R_1 and R_2. Here input signal is applied to the gate through coupling capacitor C_1.

***Operation of Common Drain Circuit*:** Here initially, note that gate voltage V_G is a constant quantity. The source voltage, $V_s = V_g + V_{gs}$.

When a signal is applied to the FET gate through the coupling capacitor C_1, gate voltage increases when input signal goes positive and gate voltage decreases when input signal goes negative.

Here V_{gs} is always constant because $V_s = V_g + V_{gs}$ here V_s increases or decreases according with increase or decrease with V_g.

Since the FET source is the output terminal, it is observe that output voltage from a common drain circuit is equal to as input voltage.

Thus, the common drain circuit can be said to have approximately unity gain. Because output voltage of FET source terminal follows the signal which is applied to gate so it also called Source follower.

Analysis: The current generator is $g_m V_{gs}$ where $V_{gs} = V_{in} - V_{out}$.

Moreover $R_G = R_1 \parallel R_2$

$$V_{out} = I_D(r_d \parallel R_S \parallel R_L) \text{ where } I_D = g_m V_{gs}$$

$$V_{out} = g_m V_{gs} (r_d \parallel R_S \parallel R_L)$$

$$\because \quad V_{in} = V_{gs} + V_{out}$$

$$V_{in} = V_{gs} + V_{gs} g_m (r_d \parallel R_S \parallel R_L)$$

$$\because \quad V_{gs} = V_{in} - V_{out}$$

$$\text{Voltage gain } A_V = \frac{V_{out}}{V_{in}} = \frac{g_m \not{V}_{gs} (r_d \parallel R_S \parallel R_L)}{\not{V}_{gs} 1 + g_m (r_d \parallel R_S \parallel R_L)}$$

$$\therefore \quad A_V = \frac{g_m (r_d \parallel R_S \parallel R_L)}{1 + g_m (r_d \parallel R_S \parallel R_L)}$$

Normally $r_d >> R_S \parallel R_L$

$$\therefore \quad A_V = \frac{g_m (R_S \parallel R_L)}{1 + g_m (R_S \parallel R_L)}$$

If $g_m (R_S \parallel R_L) >> 1$ then $A_V = 1$

So value of voltage gain becomes unity.

Input Impedance: The circuit input impedance is $R_G = R_1 \parallel R_2$.

Output Impedance: The voltage V_{gs} is given as,

$$V_{gs} = V_{out} \times \frac{R_{gs}}{(R_S \parallel R_G) + R_{gs}}$$

Normally $R_{gs} >> R_S \parallel R_G$, so $V_{gs} = V_{out}$ and $I_D = g_m V_{gs}$

Also $\quad Z_s = V_{out}/I_D$

$$= \frac{V_{gs}}{g_m V_{gs}} = \frac{1}{g_m}$$

Actually, $r_d \,||\, \frac{1}{g_m}$ $\because$ $r_d >> 1/g_m$ & R_S is the output impedance of the device. The circuit output impedance.

$$Z_{out} = R_S \,||\, 1/g_m$$

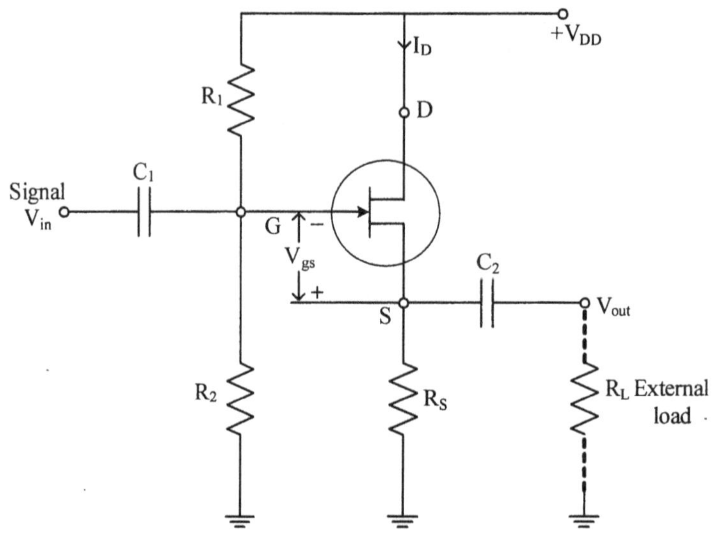

Fig 4.23(a) Common drain amplifier circuit

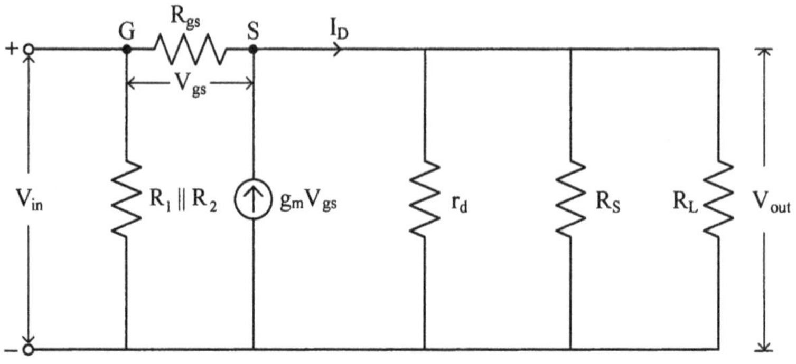

Fig 4.23(b) Small signal equivalent circuit for Common drain FET amplifier.

SUMMARY

- Hybrid Parameters are h - parameters, so called because the units of the four parameters are different. These are used to represent the equivalent circuit of a transistor.
- The h - parameters and their typical values, in Common Emitter Configuration are,

  $h_{ie} = 500\ \Omega;\quad h_{re} = 1.5 \times 10^{-4};\quad h_{oe} = 6\ \mu\Omega;\quad h_{fe} = 250$
- The second subscript denotes the configuration. similarly b and c represent Common Base and Common Collector Configurations.
- The Transistor Amplifier Circuit Analysis can be done, using h-parameters, replacing the transistor by h-parameter equivalent.
- Expressions for Voltage Gain A_V, Current Gain A_I, Input Resistance R_i, Output Resistance R_o etc., can be derived in terms of h-parameters.
- $r_{bb'}$, is called the base spread resistance. It is the resistance between the fiction base terminal B′ and the outside base terminal B.
- C.E. Amplifier circuit characteristics are :

  (a) Low to moderate R_i (300 Ω – 5 kΩ)

  (b) Moderately high R_0 (100 kΩ – 800 kΩ)

  (c) Large current amplification

  (d) Large voltage amplification

  (e) Large power gain

  (f) 180° phase shift between V_i and V_0.

OBJECTIVE TYPE QUESTIONS

1. The units of h-parameters are ....
2. h-parameters are named as hybrid parameters because ....
3. The general equations governing h-parameters are

   V_1 = ...

   I_2 = ...
4. The parameter h_{re} is defined as h_{re} = ...
5. h-parameters are valid in the .... frequency range.
6. Typical values of h-parameters in Common Emitter Configuration are ....
7. The units of the parameter h_{rc} are ....

8. Conversion Efficiency of an amplifier circuit is ....
9. Expression for current gain A_I in terms of h_{fe} and h_{re} are A_I = ...
10. In Common Collector Configuration, the values of $h_{rc} \simeq$ ....
11. In the case of transistor in Common Emitter Configuration, as R_L increases, R_i ....
12. Current Gain A_I of BJT in Common Emitter Configuration is high when R_L is ....
13. Power Gain of Common Emitter Transistor amplifier is ....
14. Current Gain A_I in Common Base Configuration is ....
15. Among the three transistor amplifier configurations, large output resistance is in .... configuration.
16. Highest current gain, under identical conditions is obtained in .... transistor amplifier configuraiton.
17. C.C Configuration is also known as .... circuit.

ESSAY TYPE QUESTIONS

1. Write the general equations in terms of h-parameters for a BJT in Common Base Amplifiers configuration and define the h-parameters.
2. Convert the h-parameters in Common Base Configuration to Common Emitter Configuration, deriving the necessary equations.
3. Compare the transistor (BJT) amplifiers circuits in the three configurations with the help of h-parameters values.
4. Draw the h-parameter equivalent circuits for Transistor amplifiers in the three configurations.
5. With the help of necessary equations, discuss the variations of A_V, A_I, R_i R_O, A_P with R_S and R_L in Common Emitter Configuration.
6. Discuss the Transistor Amplifier characteristics in Common Base Configuration and their variation with R_S and R_L with the help of equations.
7. Compare the characteristics of Transistor Amplifiers in the three configurations.

MULTIPLE CHOICE QUESTIONS

1. The h-parameters have

(a) Same units for all parameters (b) different units for all parameters
(c) do not have units (d) dimension less

2. The expression defining the h-parameter h_{22} is,

(a) $h_{22} = \left.\frac{I_1}{I_2}\right| I_1 = 0$ (b) $h_{22} = \left.\frac{I_2}{I_1}\right| I_1 = 0$

(c) $h_{22} = I2.I1$ (d) $h_{22} = \left.\frac{I_2}{V_2}\right| I_1 = 0$

3. h - Parameters are valid over a ...... frequency range

(a) R.F. (b) For DC only
(c) Audio frequency range (d) upto 1 M Hz

4. By definition the expression for h_{oe} is,

(a) $\left.\frac{\partial I_C}{\partial I_B}\right|_{I_E=K}$ (b) $\left.\frac{\partial I_C}{\partial V_C}\right|_{I_B=k}$ (c) $\left.\frac{\partial V_C}{\partial I_C}\right|_{I_B=k}$ (d) $\left.\frac{\partial I_B}{\partial V_B}\right|_{I_C=k}$

5. Typical value of h_{fb} is

(a) –0.98 (b) –101 (c) + 2.3 (d) 26.6 Ω

6. The expression for h_{ie} interms of h_{fb} and h_{ib} is, $h_{ie} \simeq$

(a) $h_{ie} \simeq \frac{h_{fb}}{1-h_{fb}}$ (b) $\frac{h_{fb}}{1+h_{fb}}$ (c) $\frac{h_{ib}}{1+h_{fb}}$ (d) $\frac{h_{ib}}{1-h_{fb}}$

7. The ratio of AC signal power delivered to the load to the DC input power to the active device as a percentage is called

(a) conversion (b) Rectification η (c) power η (d) utilisation factor

8. The general expression for A_I interms of h_f h_o and Z_L is ... A_I =

(a) $-\frac{h_o}{1+h_f Z_L}$ (b) $-\frac{h_f}{1+h_o Z_L}$ (c) $\frac{h_o}{1-h_f Z_L}$ (d) $\frac{h_f}{1-h_o Z_L}$

9. Expression for Avs in terms of A_s, R_s, Z_L, Z_i, Z_o is .....

(a) $\frac{A_I}{R_s + Z_i}$ (b) $\frac{A_I \, . Z_L}{R_s - Z_i}$ (c) $\frac{Z_L}{R_s + Z_i}$ (d) $\frac{A_I \, Z_L}{R_s + Z_i}$

10. In the case of BJT, the resistance between fictious base terminal B' and outside base terminal B is, called

(a) Base resistance (b) Base drive resistance
(c) Base spread resistance (d) Base fictious resistance

11. In large signal analysis of amplifiers ....

(a) The swing of the input signal is over a wide range around the operating point.
(b) Operating point swimgs over large range
(c) stability factor is large
(d) power dissipation is large

12. The units of the parameter h_{oe} ...

(a) ℧ (b) Ω (c) constant (d) V-A

13. Typical value of h_{re} is

(a) 10^{-4} V/A (b) 10^{-4} (c) 10^{-6} A/V (d) 10^{4}

14. CASCODE transistor amplifier configuration consists of

(a) CE-CE amplifier stages (b) CE-CC stages
(c) CE-CB stages (d) CB-CC stages

15. For CASCODE amplifier, on the input side, the amplifier stage is in ....

(a) C.B configuration (b) C.C. configuration
(c) emitor follower (d) C.E. configuration

16. CASCODE amplifier characteristics are

(a) Low voltage gain, large current are
(b) Low voltage gain, low current gain
(c) large voltage gain, large current gain, high output resistance
(d) Large current gain, Large voltage gain Low putput resistance

17. Characteristics of Darlington circuit are typical values

(a) Ri = MΩ, A_I = 10 k, R_o = MΩ, Av = 100
(b) Ri = MΩ, A_I = 10 k, R_o = 10Ω, Av < 1
(c) Ri = 10Ω, A_I = 1, R_o = 1.0Ω, Av > 1
(d) Ri = MΩ, A_I = 1, Ro = 10 Ω, Av < 1

CONCEPTUAL QUESTIONS (Interview Questions)

1. Why low frequency equivalent circuit is different from High frequency equivalent circuit?

Ans : At low frequencies, capacitive reactance X_c is open circuit. Because, $X_C = \frac{1}{2\pi fC}$, when f is low, X_C is high or like open circuit. So junction capacitances of a transistor can be neglected. At high frequencies, when f is large, X_c is finite and can not be neglected. So the junction capacitances are to be considered. As such high frequency equivalent circuit contains capacitances, where as these are neglected (open circuited) in low frequency equivalent circuits.

Specifications of Amplifiers

Parameters		*Typical Values*
1. Type of Amplifier	:	Audio / Typical Values / Power / RF / Video
2. Frequency Range	:	15 Hzs - 100 KHzs
3. Output Power	:	200 mW
4. Voltage gain	:	20 db
5. Current gain	:	100
6. Power gain	:	9
7. Input impedance	:	10 kΩ 5 pf
8. Output impedance	:	500Ω 1 pf
9. Band width	:	100 KHzs

5 Special Purpose Electronic Devices

In this Chapter,

- The basic characteristics of Semicondutor Diodes – Tunnel diode, Varactor Diode and Schottky Barrier Diode and their operations are given in detail.
- After Semiconductor Diodes, we shall study DIAC and TRIAC and Unijunction Transistor.
- We shall study also Semiconductor Photo Diodes, LDR, LED and their operations.

5.1 THE TUNNEL DIODE

In an ordinary p-n junction diode the doping concentration of impurity atoms is 1 in 10^8. With this doping, the depletion layer width, which constitutes barrier potential is 5μV. If the concentration of the impurity atoms is increased to say 1 in 10^3(This corresponds to impurity density of ≈ $10^{19}/m^3$), the characteristics of the diode will completely change. Such a diode is called ***Tunnel Diode***. This was found by ***Esaki*** in ***1958***.

5.1.1 PRINCIPLE OF OPERATION

Barrier Potential V_B :

$$V_B = \frac{e.N_A}{2\epsilon} \cdot W^2$$

or

$$W = \sqrt{\frac{2\epsilon.V_B}{e.N_A}}$$

where W = Width of Depletion Region in μ

N_A = Impurity Concentration N_O/m^3.

So, the width of the junction barrier varies inversely as the square root of impurity concentration. Therefore as N_A increases W decreases.

Therefore, in tunnel diodes, by increasing N_A, W can be reduced from 5μ to 0.01μ. According to classical mechanics, a particle must possess the potential which is at least equal to, or greater than, the barrier potential, to move from one side to the other. When the barrier width is so thin as 0.01μ, according to Schrodinge equation, there is much probability that an electron will penetrate when a forward bias is applied to the diode, so that potential barrier decreases below E_0. The *n-side* levels must shift upward with respect to those on the *p-side*. So there are occupied states in the conduction band of the n material, which are at the same energy level as allowed empty states in the valence band of the *p-side*. Hence electrons will tunnel from the *n-side* to the *p-side* giving rise to forward current. As the forward bias is increased further, the number of electrons on *n-side* which occupy the same energy level as that ***vacant energy*** state existing on *p-side*, also increases. So more number of electrons tunnel through the barrier to empty states on the left side giving rise to peak current I_p. If still more forward bias is applied, the energy level of the electrons on the *n-side* increases, but the empty states existing on the *p-side*, reduces. So the tunneling current decreases. In addition to the ***Quantum Mechanical Tunneling Current***, there is regular *p-n junction* injection current also. The magnitude of this current is considerable only beyond a certain value of forward bias voltage.

Therefore, the current again starts beyond V_V. The graph is shown in Fig. 5.1.

I_p = Peak current

I_V = Valley current

V_F = Peak forward voltage

V_p ≈ 50 mV

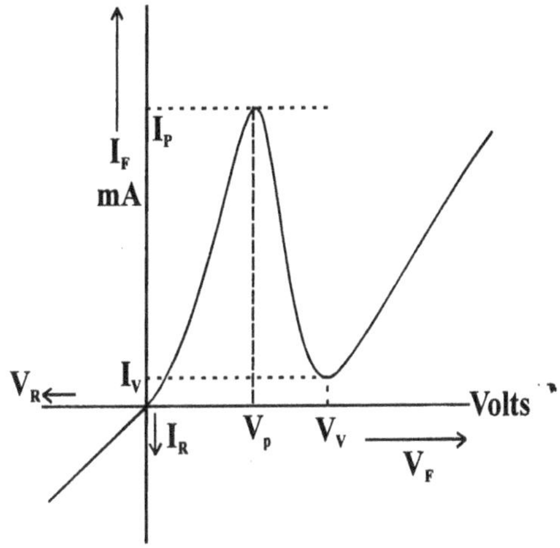

Fig. 5.1 V - I Characteristics of a tunnel diode.

5.1.2 Characteristics of a Tunnel Diode

For small forward voltage (ie., $V_p \approx$ 50mV for Ge), forward resistance is small $\approx$ 5Ω and so current is large. At the ***Peak Current*** I_p, corresponding to voltage V_p, $\frac{dI}{dV}$ is zero. If V is beyond V_p, the current decreases. So the diode exhibits ***Negative Resistance Characteristics*** between I_p and I_V, called the ***Valley Current***. The voltage at which the forward current again equals I_V is called as *peak* forward voltage V_V. Beyond this voltage, the current increases rapidly.

The symbol for tunnel diode is shown in Fig. 5.2(a). Typical values of a tunnel diode are :

$V_p = 50$ mV, $\quad V_V = 350$ mV,

$V_F = 0.5$V, $\quad \frac{I_P}{I_V} = 8$

$I_P = 10$ mA, $\quad$ Negative resistance $R_n = -30\Omega$.

Series Ohmic resistance $R_s = 1\Omega$. The series inductance L_s depends upon the lead length and the geometry of the diode package. $L_s \approx$ 5 nH. Junction capacitance C= 20 pF. Its circuit equivalent is shown in Fig. 5.2(b).

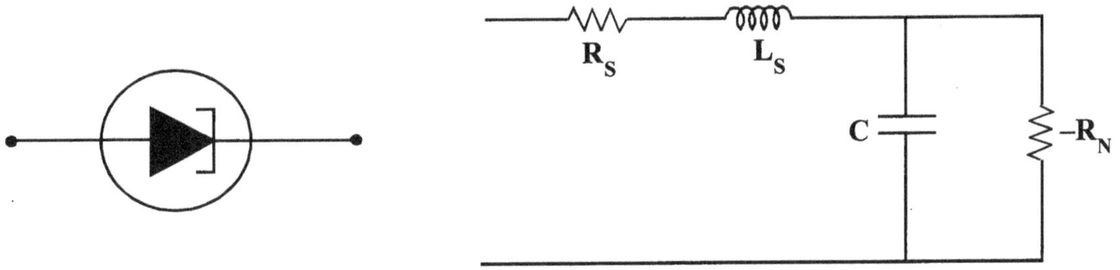

(a) Symbol of tunnel Diode *(b) Equivalent circuit*

Fig. 5.2

Advantages :

1. Low Cost
2. Low Noise
3. Simplicity
4. High Speed
5. Low Input Power
6. Environmental Immunity.

Disadvantages :

1. Low output voltage swings, even for small voltage, while the current goes to large values. So the swing is limited.
2. Circuit design difficulty, since it is a two terminal device and there is no isolation between input and output.

Applications :

1. As a high frequency oscillator (GHz)
2. As a fast switching device. Switching time is in nano-seconds. Since tunneling is a quantum mechanical phenomenon, there is no time lag between the application of voltage and consequent current variation. So it can be used for high frequencies.

If the barrier width is $\approx$ 3A°, the probability that electrons will tunnel through the barrier is large. The barrier width will be $\approx$ 3A°, when impurity doping concentration is $\geq 10^{19}/cm^3$. If width $\approx 0.01\mu$ electrons will tunnel without the application of field. But if width $\approx$ 3A°, very small applied voltage is sufficient for the electron to tunnel.

The two conditions to be satisfied for tunneling phenomenon to take place are :

Necessary Condition :

1. The effective depletion region width near the junction must be small, of the order of 3A° by heavy doping.

Sufficient Condition :

2. There must be equivalent empty energy states on the *p-side* corresponding to energy levels of electrons on the *n-side*, for these electrons to tunnel from *n-side* to *p-side*.

When the tunnel diode is reverse biased, there will be some empty states on *n-side* corresponding to filled states on the *p-side*. So electrons will tunnel through the barrier from the p side to n side (since when the diode is reverse biased the energy levels on n side will decrease). As bias voltage is increased the number of electrons tunneling will increase, so forward current increases. When the diode is forward biased, electrons from *n-side* will increase, and tunnel through the barrier to *p-side*. This is called quantum mechanical current. Apart from this, normal p-n diode current will also be there. Beyond valley voltage V_V, it is normal diode forward current.

5.2 VARACTOR DIODE

Barrier of transition capacitance C_T varies with the value of reverse bias voltage. The larger the reverse voltage, the larger the W.

$$C_T = \frac{\in \times A}{W}$$

So C_T of a *p-n junction* diode varies with the applied reverse bias voltage.

Diodes made especially for that particular property of variable capacitance with bias are called ***Varactors***, ***Varicaps*** or ***Voltacaps***. These are used in LC Oscillator Circuits.

The Symbol is shown in Fig. 5.3

Varactor diodes are used in high frequency circuits.

Fig. 5.3 Varactor diode

In the case of abrupt junction,

$$C = \frac{\in A}{W}$$

But $W \alpha (V)^{1/2}$

$\therefore$ $C \alpha (V)^{-1/2}$

In the case of linearly graded junction,

$W \alpha (V)^{1/3}$

$\therefore$ $C \alpha (V)^{-1/3}$

The variation of impurity atom concentration with ω for different types of junctions is shown in Fig. 5.4.

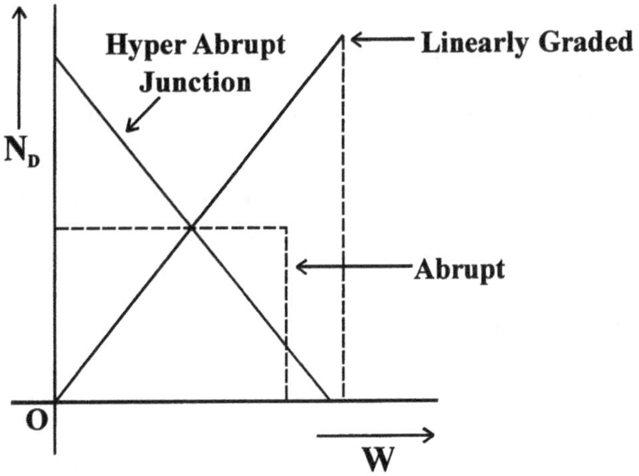

Fig. 5.4 Impurity concentration variation in the junction region.

But 'C' can be made proportional to V^2 by changing the doping profile. Such a junction is called ***Hyper Abrupt Junction***. This junction can be obtained by ***Epitaxial Process***. Therefore, C changes rapidly with voltage. Varactors are used in Reverse Bias condition, since C_T proportional to V and $C_D = \tau \times g$ No direct relation with V. So Varactors are used in Reverse Bias condition.

Capacitance, $$C = \frac{\in A}{W}$$

As the reverse bias voltage increases, width of the depletion region increases. So W increases $C \alpha \frac{1}{W}$. Hence C decreases. C is maximum under no reverse bias condition, since W is minimum C is minimum under maximum reverse bias, since W is maximum.

$$\text{Figure of merit} = \frac{C_{Max}}{C_{Min}}$$

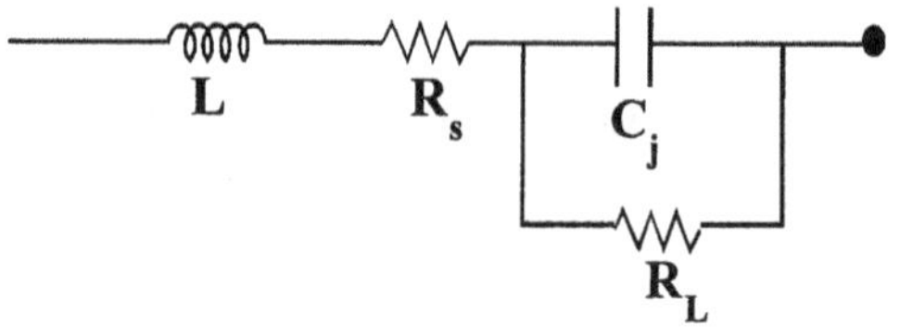

Fig. 5.5 Equivalent circuit of varactor diode

L : Lead inductance.

R_s : Series resistance of the diode due to the semiconductor material

C_j : Junction Capacitance

R_L : Leakage resistance of the diode since the capacitor can not be ideal one with out any leakage.

5.3 SCHOTTKY BARRIER DIODE

At small range frequencies, an ordinary diode turnoff when the bias changes from forward to reverse.

- As frequency increases the diode reaches a point where it cannot turn off fast as to prevent noticeable current during reverse half cycle. This phenomenon is caused the charge storage which restricts the useful frequency of ordinary rectifier diode.
- The occurance of charge storage is explained below.
- When a diode in forward bias, few carriers in depletion region not yet recombined.
- If suddenly diode is reverse-biased then the carriers can flow in the reverse for a while.
- The lifetime is greater, the longer these charges can contribute to reverse current.
- The reverse recovery time is so short in small signal diodes that its effect cannot be noticed at freqeucies below 10 MHz.
- It's significance increases above 10 MHz.

- For this solution can be found by establishing a special purpose device caused schottky diode.
- For this diode has no depletion layer eliminating the stored charges at the junction.
- Due to shortage of charge storage can switch off faster than ordinary diode.

The construction is very different from the normal P-N junction in that a metal-Semiconductor junction is developed as shown in Fig. 5.7 (a).

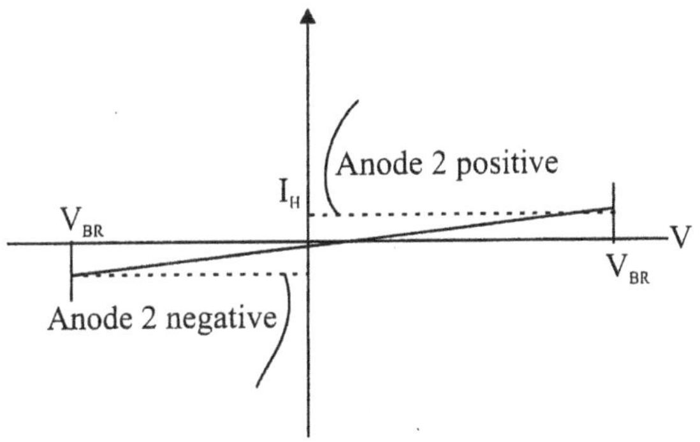

Fig. 5.6

- One side is a metal and other sides of a junction is N-type extrinsic semiconductor.
- Different construction techniques result in a different set of characteristics for the device which includes increased frequency range, lower forward bias and so on.
- Schottky diode has more uniform junction region and it is more ruggedness when compared to a point contact diode its main rival.
- In schottky diode both materials, the electrons are majority carriers.
- In the metal, the level of minority carriers is insignificant.
- When the diode is unbiased, electrons on N-side have low energy levels than electrons in the metal so the electrons cannot cross the junction barrier, caused the schottky barrier.
- When in the case forward bias, elelctrons on N-side gain energy required for crossing the junction and enter the metal.
- Since these electrons enters into the metal with very high energy, they are usually caused the hot carriers so this diode called hot-carrier diode.

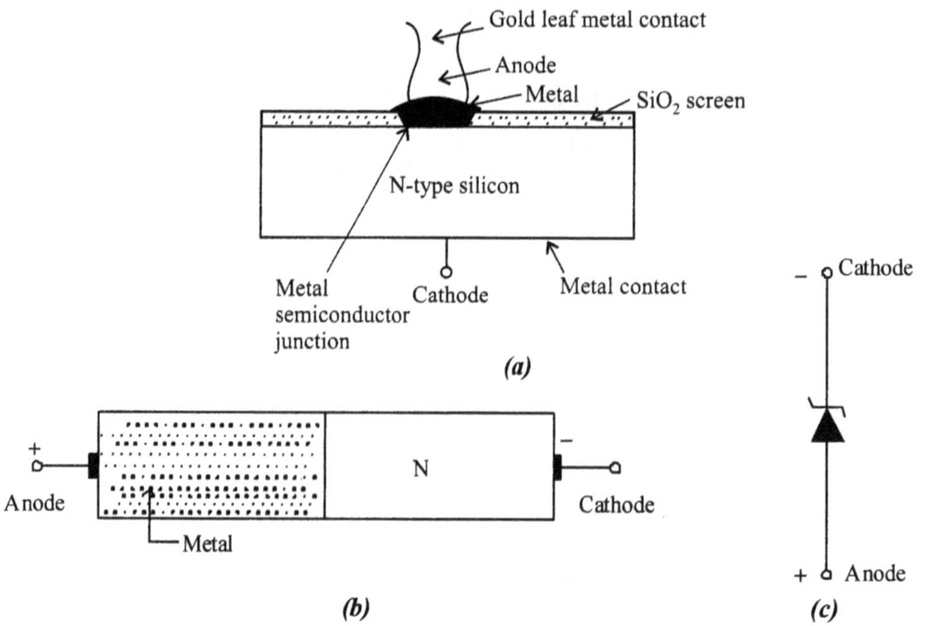

Fig. 5.7

The circuit equivalent for the device is given in Fig. 5.8(a).

- The inductance L_p and capacitance C_p are referred as package values and R_S is the body (ohmic) series resistance of the diode.
- The r is the diode resistance and C is the junction capacitance and approximate equivalent cirucuit as shown in Fig. 5.8(b).
- The comparison of V-I characteristics of PN diode, schottky diode and point contact are as shown in Fig. 5.8(c).
- The most important application of schottky diode is in digital computers.

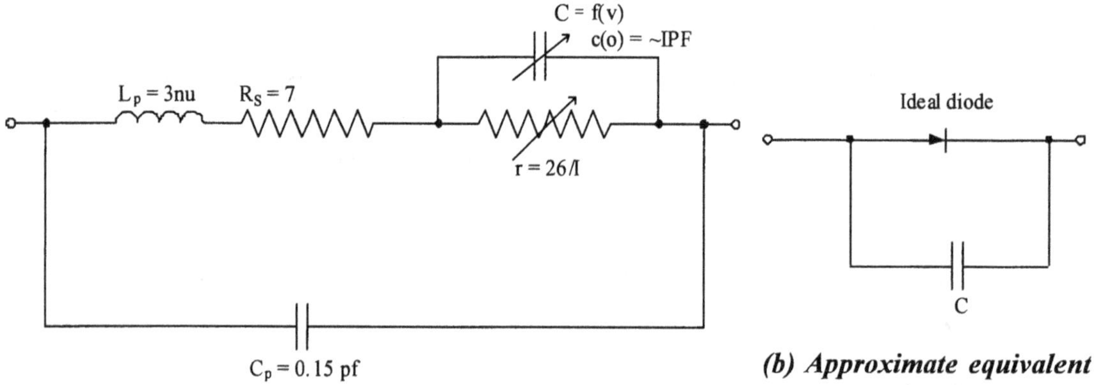

(a) Equivalent circuit

(b) Approximate equivalent circuit

Fig. 5.8

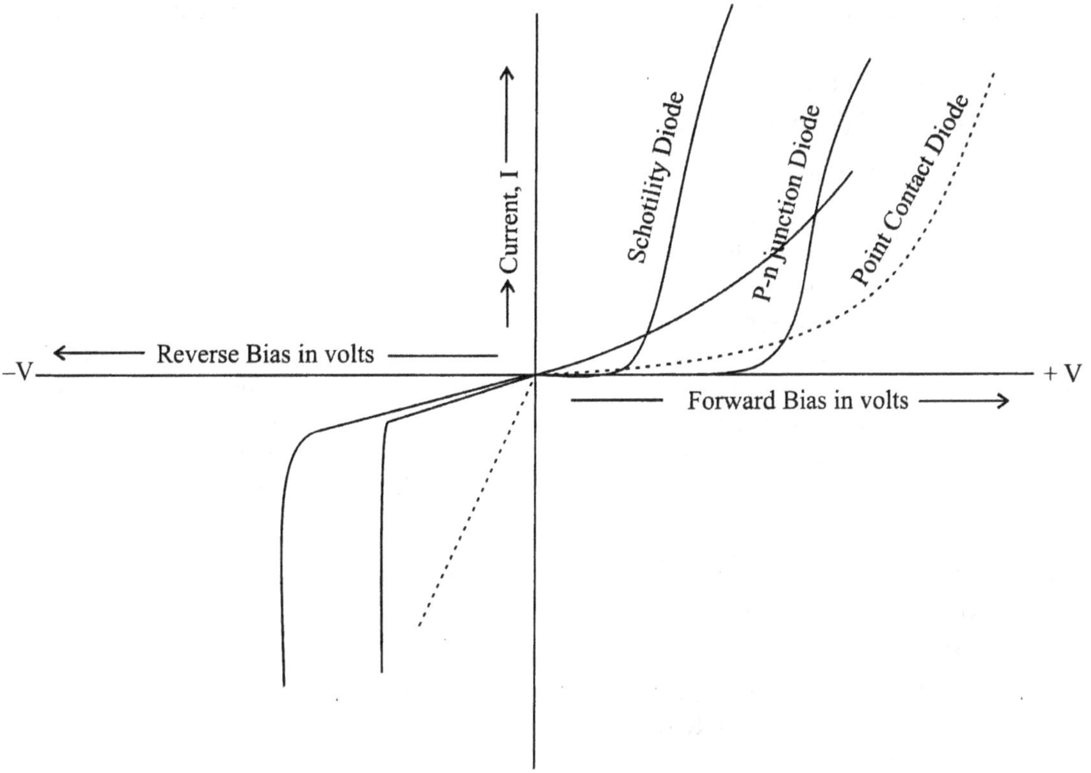

Fig. 5.8 (c) Comparision of V-I characteristics of various diodes

5.4 SILICON CONTROLLED RECTIFIER

Silicon controlled rectifier (SCR) is a 4 layer p-n-p-n device. In one type of construction, n-type material is diffused into a silicon pnp pellet to form alloy junction. From the diffused n region, the cathode connection is taken. From the p region on the other side anode is formed. Gate is taken from the p - region into which n - type impurity is diffused. So in the symbol for SCR, the gate is shown close to the cathode, projecting into it. This is known as planar construction.

5.4.1 ANNULAR TYPE OF CONSTRUCTION OF SCR

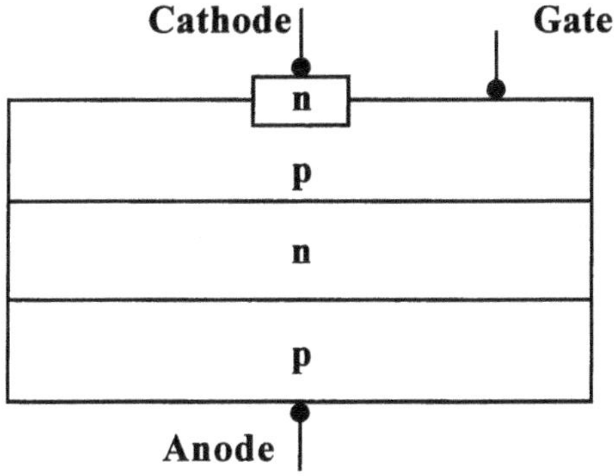

Fig. 5.9 SCR structure.

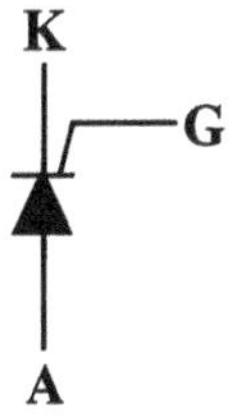

Fig. 5.10 SCR symbol.

In this type of construction gate is central to the device, with the cathode surrounding it. This is also called as shorted emitter construction. The advantages of this type construction are

(i) Fast turn on time

(ii) Improved high temperature capabilities.

5.4.2 Static Characteristics

The gate current I_G determines the necessary forward voltage to be applied between anode and cathode (Anode positive with reference to cathode) to cause conduction or to turn SCR 'ON'. So that current flows through the device.

As the gate current increases, the voltage to be applied between anode and cathode to turn the device ON decreases. If the gate is open, $I_G = 0$, the SCR is in the OFF state. This is also ***known as forward blocking region.*** The anode current that flows is only due to leakage. But if the voltage between anode and cathode is increased beyond a certain voltage, V_{BO} called as the forward breakover voltage, the SCR will turn ON and current will be limited by only the applied voltage V_{BO} called as the forward breakover voltage, the SCR will turn ON and current will be limited by only the applied voltage.

V_{BOO} : THE FORWARD BREAK OVER VOLTAGE

This is the voltage to be applied between anode and cathode to turn the SCR ON, when $I_G = 0$.

B.O : Breakover O : Zero gate current.

As the gate current is increased, the forward break over voltage decreases. If the value of I_G is very large ≈ 50 mA, the SCR will turn ON immediately when same voltage is applied between anode and cathode.

Usually, some voltage is applied between the anode and cathode of the SCR, and it is turned ON by a pulse of current in the gate.

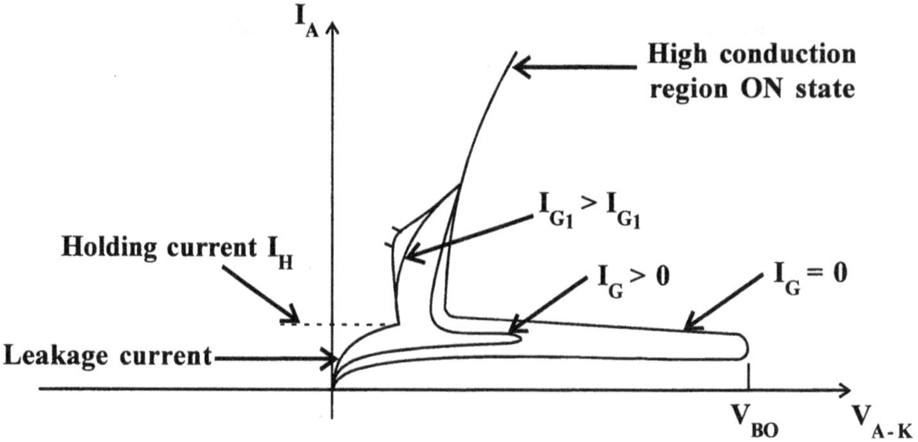

Fig. 5.11 SCR Characteristics

HOLDING CURRENT I_H

Once SCR is ON, a minimum amount of current is required to flow to keep the device ON. If the current is lowered below I_H, by increasing external circuit resistance, the SCR will switch off.

Once the SCR is turned ON, the gate looses control over the device. The gate can not be used to switch off the device. If V_{A-K} is reduced to zero (as automatically happens in rectifying application) or if the current is reduced ***below I_H, the device will turn off.***

I_H : Holding Current

Minimum Anode Current required to keep SCR in the conducting state. If I_A goes below I_H, SCR is turned off (To hold in conducting state).

LATCHING CURRENT I_L

Minimum value of I_A to be reached to keep SCR in the ON state even after the removal of gate trigger signal. (To latch on to the conducting state.)

5.4.3 ANALYSIS OF THE OPERATION OF SCR

The SCR can be regarded as two transistor connected together.

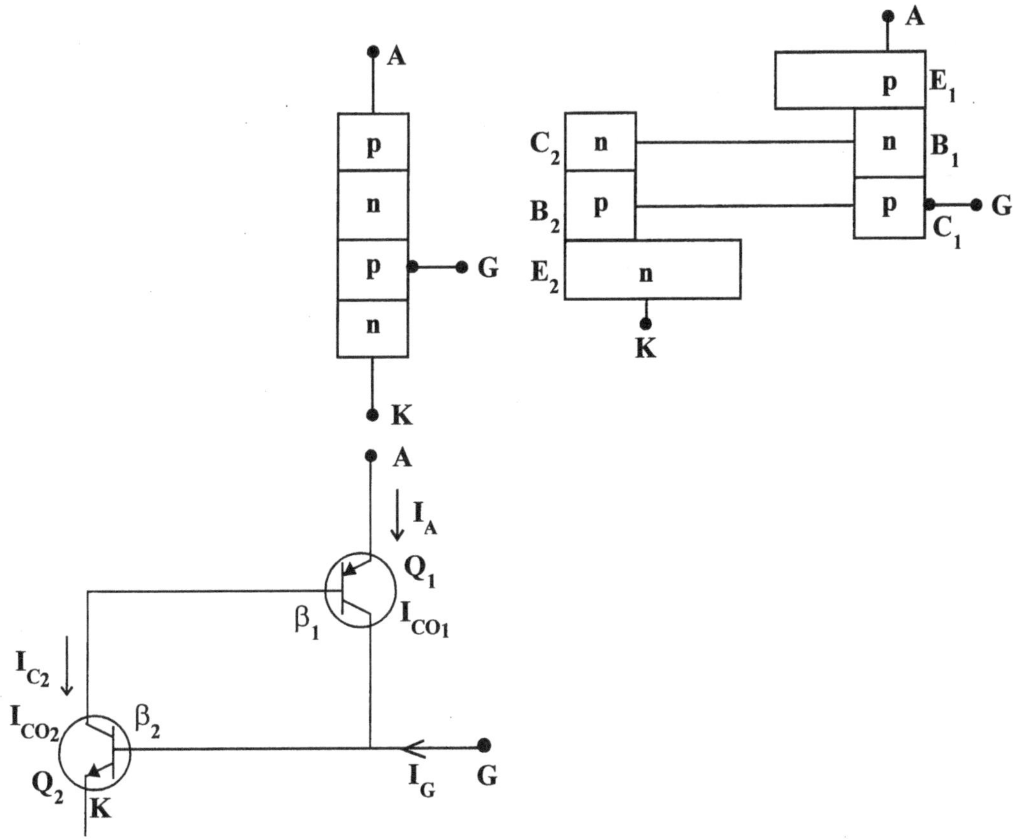

Fig. 5.12 Two transistor analogy of SCR.

The SCR can be regarded as two transistor p-n-p and n-p-n. The Base of p-n-p transistor is connected to the collector of the n-p-n transistor. The base of n-p-n transistor is connected to the collector of p-n-p transistor. The emitter of n-p-n transistor is the cathode. The emitter of p-n-p transistor is the anode.

Suppose a positive voltage is applied to the anode (p-type) since there is no base injection for both the transistors, the transistors are off. The only current is the leakage current. I_{CO1} and I_{CO2} of the two transistors. Therefore when SCR is not turned ON, the current that results is the leakage current ($I_{CO1} + I_{CO2}$).

In a transistor,

$$I_C = \beta I_B + (\beta + 1) I_{CO} \quad \text{..........(5.1)}$$

$$= \beta I_B + \beta I_{CO} + I_{CO} \quad \text{..........(5.2)}$$

$$= \beta (I_B + I_{CO}) + I_{CO} \quad \text{..........(5.3)}$$

CONSIDER TRANSISTOR Q_1

The base current for the transistor Q_1, I_{B1} is the collector current I_{C2} of Q_2.

$$I_C = \beta (I_B + I_{CO}) + I_{CO}$$

$$I_{B1} = I_{C2}$$

$$\therefore \quad I_{C1} = \beta_1 (I_{C2} + I_{CO1}) + I_{CO1} \quad \text{..........(5.4)}$$

CONSIDER TRANSISTOR Q_2

The base current of transistor Q_2 is the collector current I_{C1} of Q_1.

$$I_{C2} = \beta_2 (I_{C1} + I_{CO2}) + I_{CO2} \quad \text{..........(5.5)}$$

Anode current is $I_{E2} = I_{B2} + I_{C2}$

But $I_{B2} = I_{C1}$

$\therefore$ $I_A = I_{C1} + I_{C2}$

I_A can be written in the forms,

$$\frac{(1+\beta_1)(1+\beta_2)(I_{CO1}+I_{CO2})}{1-\beta_1 . \beta_2}$$

The two transistor will have wide base regions. So β will be small. The value of β depends upon the value of I_E. When I_E is very small, $\beta \simeq 0$.

Therefore when $\beta_1 = \beta_2 = 0$.

$$I_A = \frac{(I_{CO1}+I_{CO2})(1+0)(1+0)}{1-0.0}$$

$$\simeq I_{CO1} + I_{CO2}$$

When a negative voltage is applied from gate to cathode, to inject holes into the base of Q_1, emitter base junction of Q_1 is forward biased. So I_{C1} increases. But this is the base current for Q_2. So transistor Q_2 is tuned ON. This increases collector and emitter currents. The I_{C2} of Q_2 now becomes trigger current and so increase I_{C1}. So the action is one of internal regeneration feed back , until both the transistor are driven into saturation. Once the SCR is turned ON, removal of gate current will not stop conduction because already the transistor Q_2 is turned ON and it provides the base injection for Q_1.

SCR can be represented as a pnp and npn transistors connected together, as shown in Fig. 5.13.

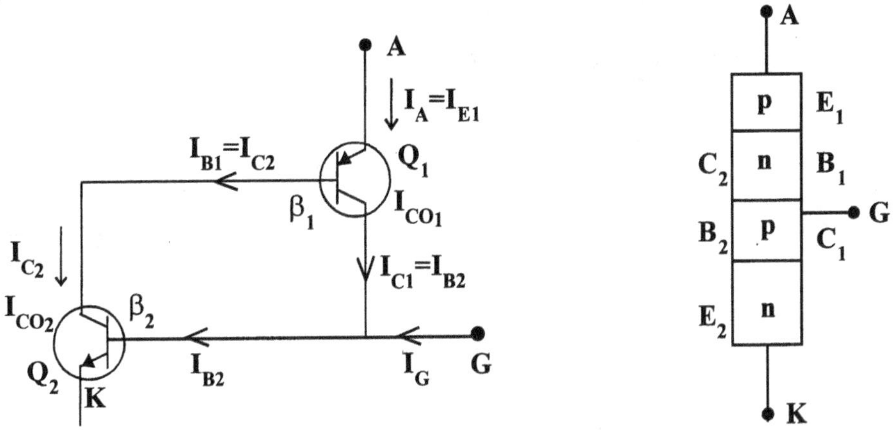

Fig. 5.13 Two transistor analogy of SCR.

$$I_{B1} = I_{C2}$$
$$I_{C1} = I_{B2}$$
Anode current $= I_A = I_{E1}$

General expression for I_C in term of β, I_B and I_{CO} is
$$I_C = \beta\, I_B + (\beta + 1)\, I_{CO}$$

Since for transistor Q_1,
$$I_{C1} = \beta_1\, I_{B1} + (\beta_1 + 1)\, I_{CO1} \qquad \text{..........(5.6)}$$

Since for transistor Q_1,
$$I_{C2} = \beta_2\, I_{B2} + (\beta_2 + 1)\, I_{CO2}$$

But $\quad I_{B2} = I_{C1}$

$\therefore \quad I_{C2} = \beta_2\, I_{C1} + (\beta_2 + 1)\, I_{Co2} \qquad \text{..........(5.7)}$

$I_A = I_{E1} = I_{C1} + I_{B1}$ But I_{C1} is given by eq. (5.6)

$\quad = \beta_1 \cdot I_{B1} + (\beta_1 + 1)\, I_{CO1} + I_{B1}$

$\therefore \quad I_{E1} = I_A = (\beta_1 + 1)(I_{B1} + I_{CO1}) \qquad \text{..........(5.8)}$

Since we must get an expression for I_{B1}.

Substitute eq. (5.6) in (5.7).

$\therefore \quad I_{C2} = \beta_2\, [\beta_1\, I_{B1} + (\beta_1 + 1)\, I_{CO1}] + (\beta_2 + 1)\, I_{CO2}$

But $\quad I_{C2} = I_{B1}$

$\therefore \quad I_{B1} = \beta_2\, \beta_1\, I_{B1} + \beta_2\, (1 + \beta_1)\, I_{CO1} + (\beta_2 + 1)\, I_{CO2}$

or $\quad (1 - \beta_1\beta_2)\, I_{B1} = \beta_2\, (1 + \beta_1)\, I_{CO1} + (1 + \beta_2)\, I_{CO2}$

or
$$I_{B1} = \frac{\beta_2(1+\beta_1)I_{CO1} + (1+\beta_2)I_{CO2}}{1-\beta_1\beta_2} \quad \text{..........(5.9)}$$

Substitute the value of I_{B1} from eq. (5.9) in (5.8).

$$\therefore \quad I_{E1} = I_A = (1+\beta_1)\left[\frac{\beta_2(1+\beta_1)I_{CO1} + (1+\beta_2)I_{CO2}}{1-\beta_1\beta_2} + I_{CO1}\right]$$

$$= (1+\beta_1)\left\{\frac{\beta_2 I_{CO1} + \beta_1\beta_2 I_{CO1} + I_{CO2} + \beta_2 I_{CO2} + I_{CO1} - \beta_1\beta_2 I_{CO1}}{1-\beta_1\beta_2}\right\}$$

$$= (1+\beta_1)\left\{\frac{I_{CO1}(1+\beta_2) + I_{CO2}(1+\beta_2)}{1-\beta_1\beta_2}\right\}$$

$$\boxed{I_A = \frac{(1+\beta_1)(1+\beta_2)(I_{CO1}+I_{CO2})}{1-\beta_1\beta_2}}$$

5.4.4 Anode to Cathode Voltage - Current Characteristics

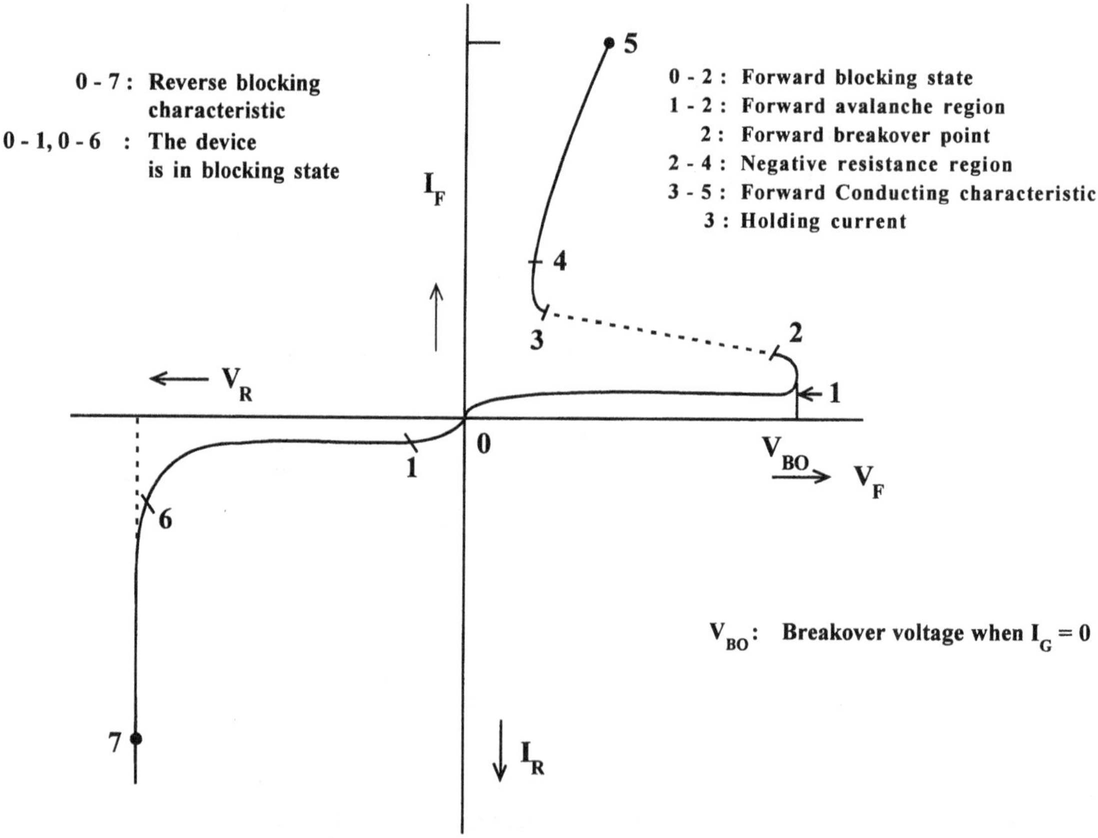

Fig 5.14 SCR characteristics.

5.5 DIAC

***Definition*:** The name DIAC indicates two (Di) – terminal parallel inverse combination of semiconductor layers that permits triggering in two directions.

***Construction and Operation*:** The semiconductor layers of the diac can be basically arranged as shown in Fig. 5.15.

- Here it is important to remember that neither terminal is referred to as cathode.
- Here they are two anodes referred as anode 1 and anode 2.
- If anode 1 is positive with respect to anode 2 then the semiconductor layers forms p_1 n_2 p_2 and n_3 particularly.
- In similar way if anode 2 is positive with respect to anode 1 then the semiconductor layers forms p_2 n_2 p_1 and n_1.

5.5.1 DIAGRAM FOR BASIC CONSTRUCTION AND SYMBOL

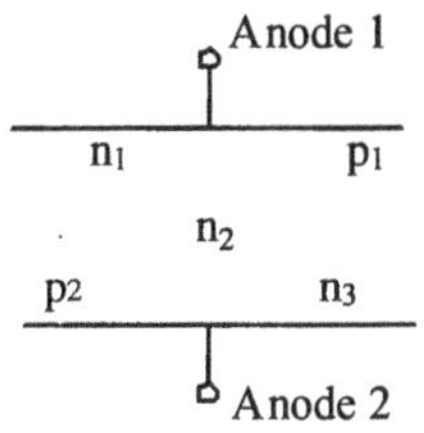

Fig. 5.15 Basic construction

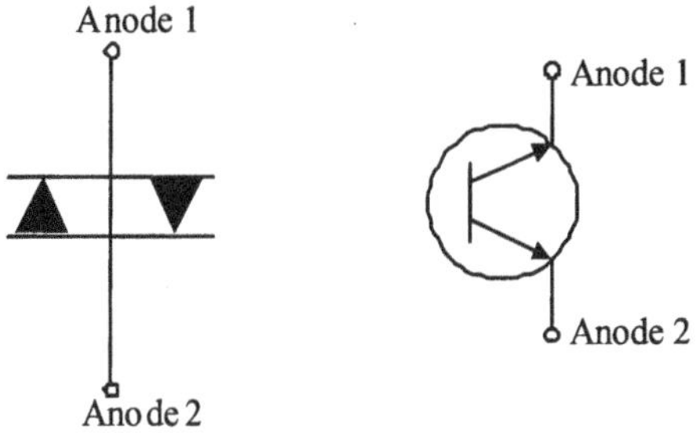

Fig. 5.16 Symbols

5.5.2 CHARACTERISTICS of DIAC

The special characteristics of this device is it offers break over voltage in two directions this indicates possibility of an 'on' condition in either direction this is the main advantage in AC applications.

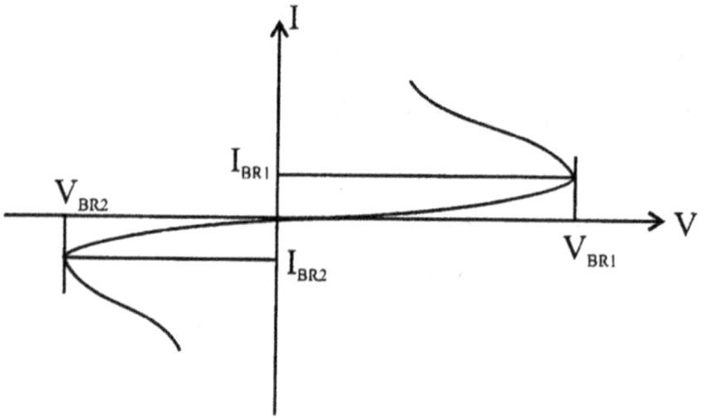

Fig. 5.17 Characteristics

Here the breakdown voltages are related by the following equation since here DIAC has two breakdown voltage in two directions which are mention in specification sheet is given by,

$$V_{BR_1} = V_{BR_2} \pm 0.1\ V_{BR_2} \quad \text{....(5.10)}$$

- The breakdown voltages are very close in magnitude that means approximately equal in magnitude but varies from 28 V (min) to 42 V (max).
- Here the current levels $\left(I_{BR_1} \text{and } I_{BR_2}\right)$ are also approximately equal in magnitude for each device whose value is 200 μA.

5.6 TRIAC

TRIAC is formed from DIAC for which gate terminal is connected which is used for controlling the device in either direction for turning on.

- In other way, the action of the device in either direction can be controlled by gate current.

***Symbol*:**

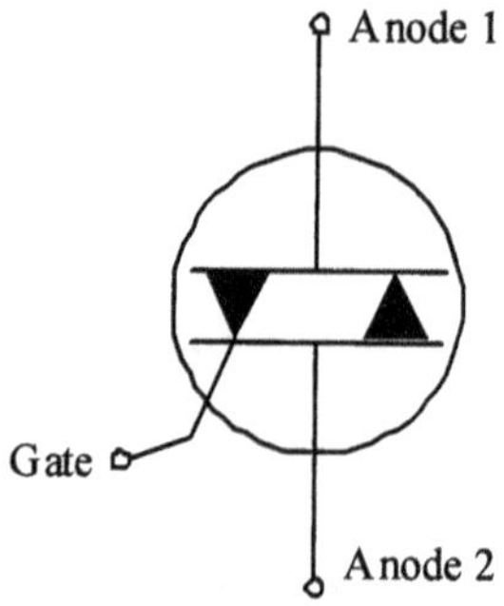

Fig. 5.18

5.6.1 Construction

The semiconductor layers of the diac can be basically arranged as shown in Fig. 5.19.

- Here they are two anodes and gate terminal.

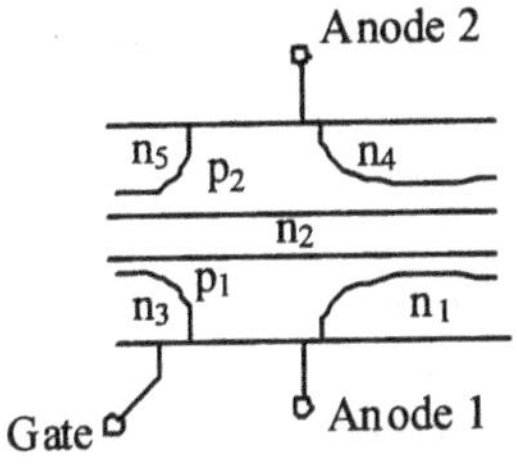

Fig. 5.19

Operation: Here for every possible direction of conduction, input signal is applied to the gate terminal can control the combination of semiconductor layers.

5.6.2 Characteristics

Here in TRIAC the characteristics are different from DIAC. When we observe characteristics from Fig. 5.20 the first and third quadrants are different from DIAC.

- Mainly the holding current not exist in each direction in the characteristics of DIAC.

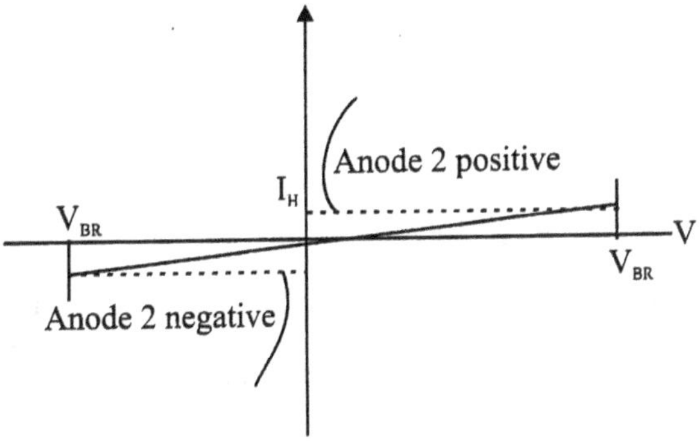

Fig. 5.20 Characteristics

5.7 UNIJUNCTION TRANSISTOR (UJT)

This device has only one *p-n* junction and three leads like transistor. Hence it is called as unijunction transistor (UJT).

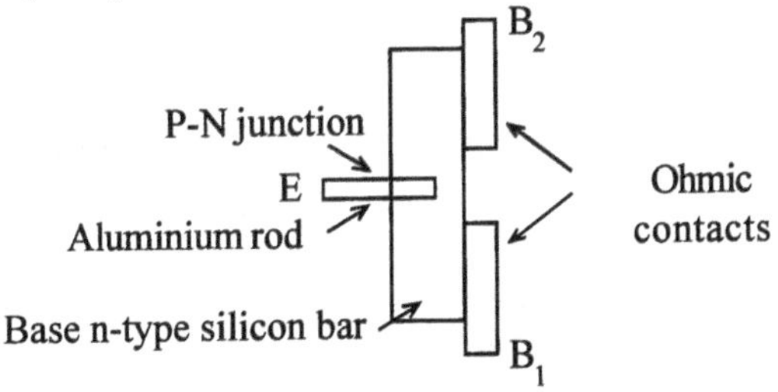

Fig. 5.21 (a) UJT structure

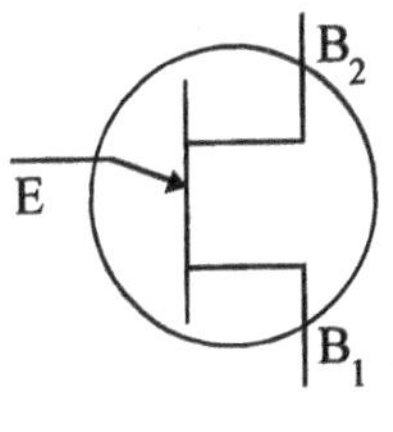

Fig. 5.21 (b) Symbol

The construction of this device is as shown in Fig. 5.21.

A bar of high resistivity **n-type silicon of typical dimensions** $8 \times 10 \times 35$ mils called base B, is taken and two ohmic contacts are attached to it at the two ends to form the base leads B_2 and B_1. A 3 mil aluminium wire, called emitter E is alloyed to the base close to B_2 to form a *p-n* rectifying junction. This device was originally described in the literature as the **double-base diode**. But now it is called as **UJT** ($\because$ it has only one *p-n* junction and two base leads).

5.7.1 UJT CONSTRUCTION AND OPERATION

In UJT, the alloy is formed by diffusing Alluminium into *n* type silicon close to the B_2 lead as shown in Fig. 5.22. The doping and construction is such that, when E - B_1, junction is forward biased, holes are injected into the *n*-silicon bar from the *p*-region towards the B_1 lead. Holes will not travel upwards because B_2 is at positive potential. The doping concentration of *p*-region is large. So there is sudden increase in the number of holes in the region close to B_1. To recombine with these holes, the free electrons from B_2 region will move towards the B_1 lead. Thus there is large increase in the number of holes and electrons in the region corresponding to R_{B_1}.

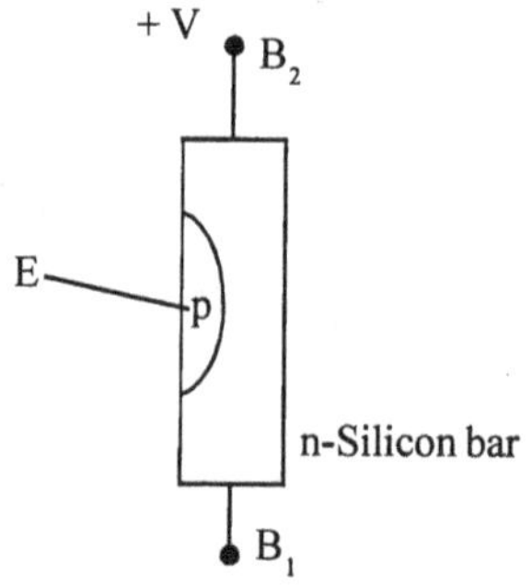

Fig. 5.22 UJT structure.

$\therefore$ $\sigma = n e \, \mu_n + p \, e \, \mu_p$. *p* will be almost comparable with *n*. Because of increase in the values of *n* and *p*, σ increases, R_B, decreases. When the E-B, junction is forward biased, even after recombination of holes and electrons, still there will be large number of electrons and holes in the region close to B_1. So conductivity σ increases.

R_{B_1} and R_{B_2} have the same temperature coefficient. Therefore material is the same (*n* type silicon bar). With temperature, because of the increase in the number of free carriers, R_B decreases. The net effect is *n* decreases slightly with the temperature. The change is $\simeq 4\%$ for 100^0 C rise in temperature. $V_P = (\eta V_{BB} + V_V)$. V_V decreases with increase in temperature. Because threshold voltage decreases with increase in temperature. Therefore V_P decreases with temperature. Now R_2 has positive temperature coefficient. If R_2 is chosen such that it increases by the same amount as to decrease in R_{B1} and η and V_V, V_P will remain constant. Thus, R_2 will provide temperature compensation.

UJT basically consists of a bar of *n*-type silicon with one *p*-type emitter providing *p-n* junction. The emitter will be close to the base two (B_2) than base one (B_1) (see Fig. 5.23). So UJT consists of emitter and two bases with the emitter close to the second base. The emitter is p-type and the two bases are *n*-type. The symbol for the voltage between emitter and base 1 (B_1 is *n*-type). Polarity is as shown in Fig. 5.23. Suppose V_{BB} is the voltage from B_2 to B_1. Now if $V_{BB} = 0$, and positive voltage V_E is applied. The resulting current I_E gives the emitter to base B_1

diode characteristic. With V_{BB} = 10V or 20V, there is leakage current I_{EO} from B_2 to emitter. ($\because$ B_2 is *n*-type and E is *p*-type when B_2 is at a higher potential, the minority carries from B_2 and E will flow which results in leakage current). It takes $\simeq$ 7 volts from E to B_1, to reduce this current to zero and then cause current to flow in the opposite direction, reaching peak point V_P. After this point I_E increases suddenly and V_E drops. This is the unstable resistance region. This lasts until the valley voltage is reached and the device saturates.

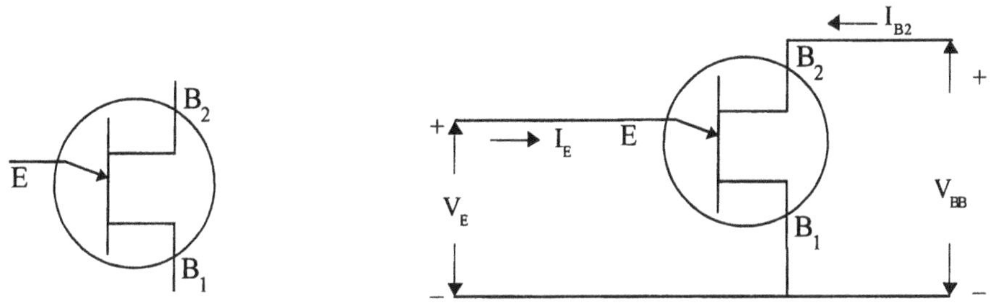

(a) Symbol ***(b) V-I components***

Fig. 5.23

The equivalent circuit for UJT is as shown in Fig. 5.24. The resistance of the silicon bar is represented by two series resistors. R_{B_2} is the resistance of base two portion, R_{B_1} is variable resistance. Its value depends upon the bias voltage V_0. The p-n junction formed due to emitter is shown as a diode.

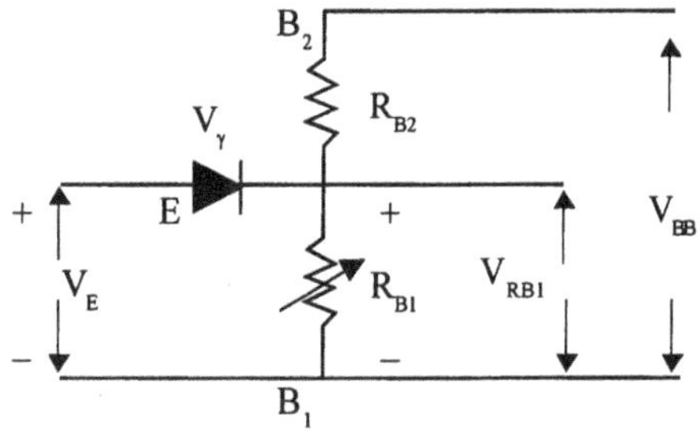

Fig. 5.24 Equivalent circuit of UJT.

With no voltage applied to the UJT, $R_{BB} = R_{B_1} + R_{B_2}$. Its value varies from 4.7KΩ to 10KΩ for (2N1671). With emitter open, some voltage is applied at V_{BB}. It gets divided across R_{B_2} and R_{B_1}. (Fig. 5.24). The fraction of voltage appearing across R_{B_1} is

$$V_{RB1} = \eta V_{BB} = \frac{R_{B1}}{R_{B1} + R_{B2}} \times V_{BB},$$

where,
$$\eta = \frac{R_{B1}}{R_{B1} + R_{B2}} = \textbf{Intrinsic stand off ratio.}$$

Its value lies between 0.5 and 0.82. Since no voltage is applied at the emitter, the diode is reverse biased, because the cathode voltage is ηV_{BB} and anode current is 0. Now as V_E is increased, from zero and when V_E is > ηV_{BB}, the diode will be forward biased. So the resistivity between E and B_1 decreases since holes are injected from emitter into B_1. Therefore R_{B1} decreases or V_E decreases and I_E increases. So negative resistance region results. Peak voltage $V_P = \eta V_{BB} + V_\gamma$. $V_\gamma = 0.5V$.

5.7.2 V-I Characteristics of UJT

Suppose some voltage V_{BB} is applied between B_2 and B_1. If V_{BB} is small, $I_E = 0$. Then the silicon bar can be considered as an ohmic resistance R_{BB} between the leads B_2 and B_1. R_{BB} will be in the range 5 to 10 kΩ. $R_{BB} = (R_{B1} + R_{B2})$. When $I_E = 0$, the voltage across $R_{B1} = \eta V_{BB}$. Where,

$\eta = \dfrac{R_{B1}}{R_{B1} + R_{B2}}$. η is called is ***intrinsic stand off ratio*** and its value lies between ***0.5 and 0.75***.

If V_E is < $\eta\ V_{BB}$ the p-n junction diode or the E-B junction of UJT is reverse biased and the current I_E is negative. It is the reverse saturation current I_{E0} which is of the order of 10 R_B.

If V_{EE} is increased beyond ηV_{BB}, input diode becomes forward biased and becomes positive. So holes are injected from the emitter into the base region B_1. As V_{EE} increases, I_E increases. So the holes injected into B_1 region also increases ($\sigma = pe\ \mu_p$).

∴ As the hole concentration in the base region increases, its resistivity decreases. So the voltage across R_B, ηV_{BB} decreases. This is the voltage V_E between E and B_1. ***So as I_E increases, V_E decreases and the device exhibits negative resistance region. (Fig. 5.25(b))***

When I_E is very large the current I_{B2} flowing from V_{BB} into B_2 load can be neglected. For current greater than the valley current, the resistance becomes positive.

$$V_P = \eta\ V_{BB} + V_V$$
$$V_P = \text{Peak Voltage}$$
$$V_V = \text{Valley Voltage}$$

For current greater than the valley current the resistance becomes positive. (Fig. 5.25.)

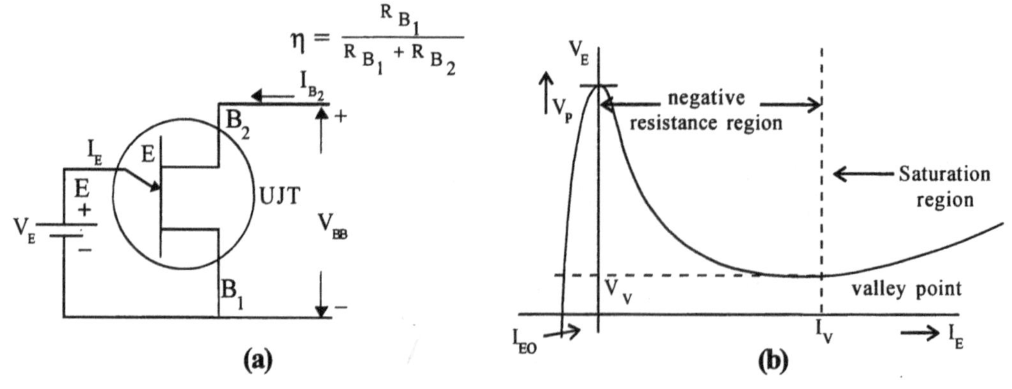

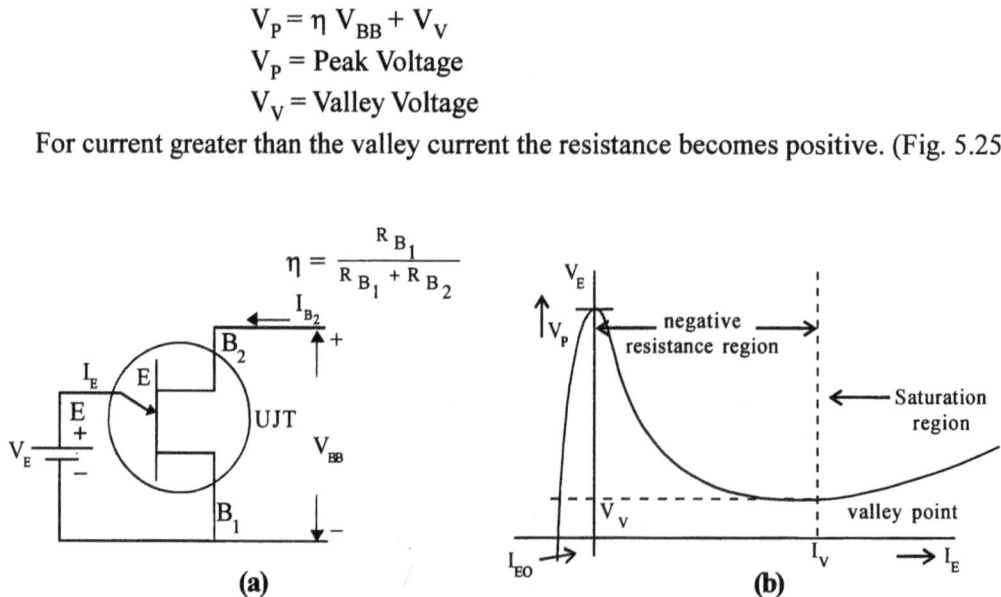

Fig. 5.25 (a) UJT circuit (b) V-I Characterisitic of UJT

5.7.3 Applications of UJT

1. In Relaxation oscillator circuits to produce sawtooth waveforms.
2. In triggering SCR circuits.
3. In thyristor circuits.

5.8 SEMICONDUCTOR PHOTO DEVICES

5.8.1 Light Dependent Resistors (LDR)

Generally the conduction in semiconductor materials occurs when free charge carriers are available in the material when an electric field is applied.

- In some semiconductors, light energy fall on them release correct order of magnitude which release charge carries which in flow of current produce by an voltage applied.
- Increase in current results in increase of light intensity decreases the resistance of semiconductors of constant Voltage applied.
- So these semiconductors are named as LDR or photoconductive cells or photoresistors in which resistance is varied according to light.
- The mostly used photoconductive semiconductor material are cadmium sulphide (cds) and cadmium selenide (cdse)
- The photo conductive cells are made up of chemically sintering here in this cds or cds powders are required which are made into tables of request shape and they are enclosed in a protective envelope of glass or plastic.
- Here electrodes are deposited on the tablet surface and are made of materials which give in ohmic contact here gold is typically used.
- Here the electrodes are usually inter-digital i.e., they are in the form of interlocked fingers or combs as shown in Fig. 5.26.

Characteristics of Photoconductive Cells: Characteristics of photoconductive cells vary according to the type of materials used mention in Table 5.1.

Table 5.1: Characteristics of Photoconductive Cells

Photoconductor	Time constant	Spectral Band
Cds	100 ms	$0.47-0.71\ \mu m$
cdse	100 ms	$0.6-0.77\ \mu m$
pbs	$400\ \mu m$	$1-3\ \mu m$
pbse	$10\ \mu m$	$1.5-4\ \mu m$

- When the cell is kept in darkness its resistance is refered as dark resistance be as high as $10 \times 10^{12}\,\Omega$.
- If the cell is illuminated then its resistance decreases.
- Resistance depends on physical character of the photoconductive layer as well as dimensions current depends upon d.c voltage applied.
- The spectral response characteristics of two commercial cells are shown in Fig. 5.26 .
- From the characteristics we observe that there is almost no response to a wavelength shorter than 300 nm.
- Cadmium sulphide cells have a peak response in the green part of spectrum at 510 nm.
- The maximum response of cadmium sulphoselenide is in the yellow orange at 615 nm.
- When using this device the two important parameters are proper dark resistance as well as suitably sensitivity.
- The sensitivity of photoresistive transducer is

$$S = \frac{\Delta R}{\Delta H}\ \Omega/W - m^{-2}$$

- The major disadvantages is temperature changes leads to cause substantial resistors changes for a particular light intensity so this type of device used in analog applications.

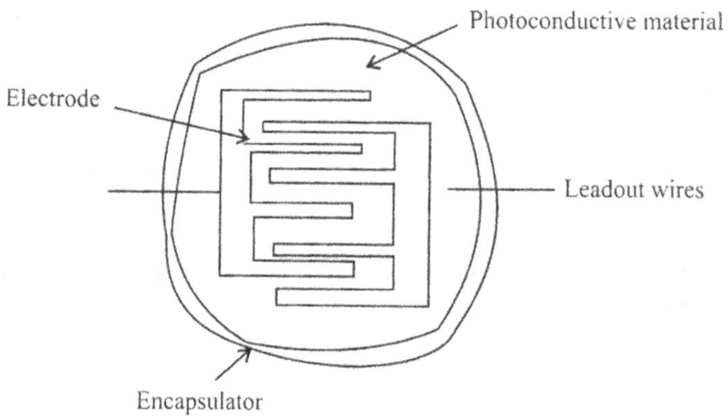

Fig. 5.26 LDR

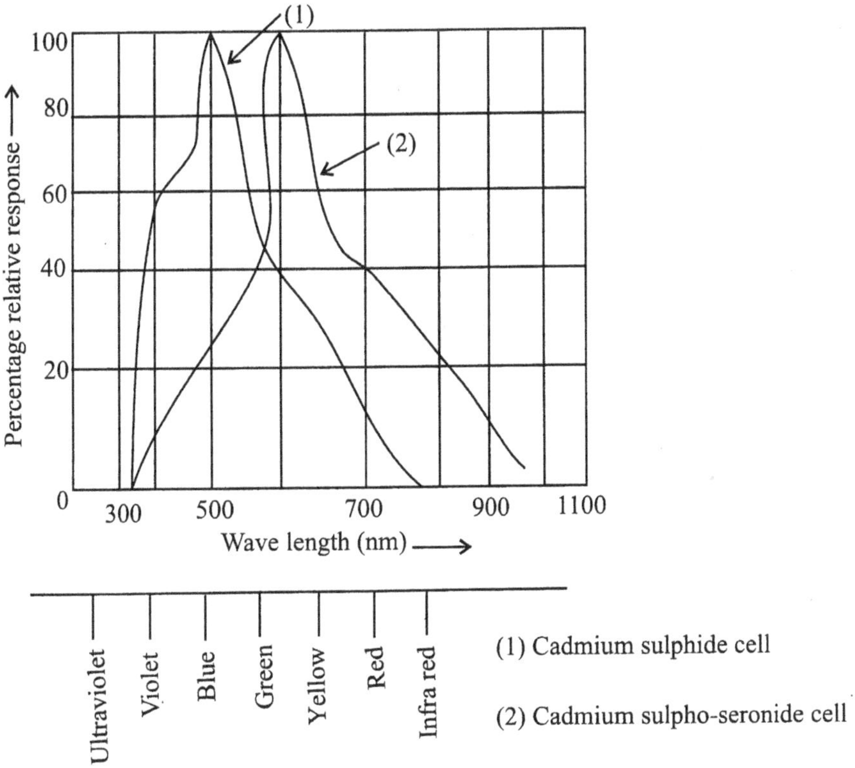

Fig. 5.27 Spectrcal response characteristics of photoconductive cells

5.8.2 LIGHT EMITTING DIODE (LED)

The Light Emitting Diode (LED) is also a p-n junction diode which is capable of emitting light when the diode is forward biased. It is a known fact that p-n junction Diode (any semiconductor device) which is made from semiconductors such as silicon and Germinium generates a small amount of heat during the flow of current across the junction. Similarly there are semiconductor materials other than Silicon (Si) and Germinium (Ge) such as Gallium Arsenide (GaS), Gallium Phosphide GaP), etc., which generates light (photons) when the diode junction is forward biased (i.e., generates light when the current flows across the junction. Hence the light emitting diode (LED) is formed by choosing semiconductor materials such as Gallium Arsenide (Gas), Gallium phosphide (GaP), that are capable of emitting the light (photons) when the junction of the diode is forward biased. In LED the Gallium Arsenide (GaS), Gallium phosphorous etc., are chosen as semiconductor material and then the corresponding pentavalent and triavalent impurities added through processes of doping to get corresponding 'n' and 'p' type materials and they are diffused to form the Light Emitting Diode (LED).

Principle of Operation

Forward Bias: The Fig. 5.28 shows the p-n junction of light emitting diodes during its forward bias. It is already known that p-side contains excess positive charge (rich in holes) and n-side contains excess negative charge (rich in electrons). When we apply forward bias voltage (V_f), the charge carriers (both holes and electrons) crosses the junction and recombines with each other. This recombination of electrons with holes generates light (photon) across the junction (The electrons will be in conduction band (CB) and holes will be in valance band (V_B). Hence the electrons will be having higher energy (E_1) and Holes will be having lesser Energy (E_2). Thus when an electron recombine with hole some energy (E_1-E_2) must be dissipated, here in LED it is dissipated as light).

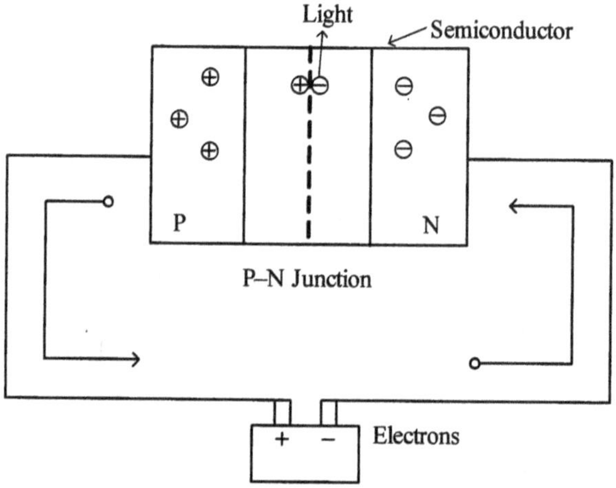

Fig. 5.28 Forward biased p-n junction of LED

Reverse Bias: During Reverse bias, the Light Emitting Diode (LED) will not produce any light. Further if we try to operate it in reverse bias condition, it may damage Light Emitting Diode (LED).

Construction of LED

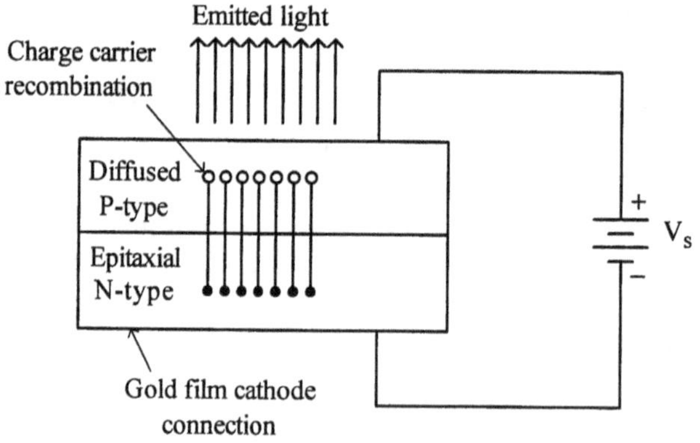

Fig. 5.29 Illustration of construction of LED

The Fig. 5.29 illustrates the construction of LED and as shown in Fig.5.29. An n-type expitaxial layer is grown on the substrate and p-region is formed through the process of diffusion. The p-region includes recombination of electrons with holes i.e., region where the light (photon) is generated. Hence for better visibility and collectability and light this p-region is chosen as surface of LED. In order to allow more surface area for emitting the light, the metal connections are taken at the outer edges of the p-region (p-layer). And in order to allow entire light to be reflected towards the surface of LED, a thin gold film is applied bottom of LED and further this gold film can also be used for cathode connection.

After manufacturing the LED, the entire LED is protected by providing encasing to the device. The nature of the light being emitted by LED depends on material used.

The following lists out the different types of materials that can be employed for generating different colors of lights.

Sl. No.	Colour of Light	Material
1.	Infra Red	Gallium arsenide (GaAs) Aluminium gallium arsenide (AlGaAs)
2.	Red	Aluminium gallium arsenide (AlGaAs) Gallium arsenide phosphide (GaAsP) Aluminium gallium indium phosphide (AlGaInP) Gallium (III) phosphide (GaP)
3.	Orange	Gallium arsenide phosphide (GaAsP) Aluminium gallium indium phosphide (AlGaInP) Gallium (III) Phosphide (GaP)
4.	Yellow	Gallium arsenide phosphide (GaAsP) Aluminium gallium indium phosphide (AlGaInP) Gallium (III) phosphite (GaP)
5.	Green	Gallium (III) phosphide (GaP) Aluminium gallium indium phosphide (AlGaInP) Aluminium gallium phosphide (AlGaP) Indium gallium nitride (InGaN)/ Gallium (III) nitride (GaN)
6.	Blue	Zinc selenide (ZnSe) Indium gallium nitride (InGaN) Silicon carbide (SiC)
7.	Violet	Indium gallium nitride (InGaN)
8.	Purple	Dual blue/red LEDS
9.	Ultra violet	Aluminium nitride (AlN) Aluminium gallium nitride (AlGaN) Aluminium gallium Indium nitride (AlGaInN)

V-I Characteristics of LED

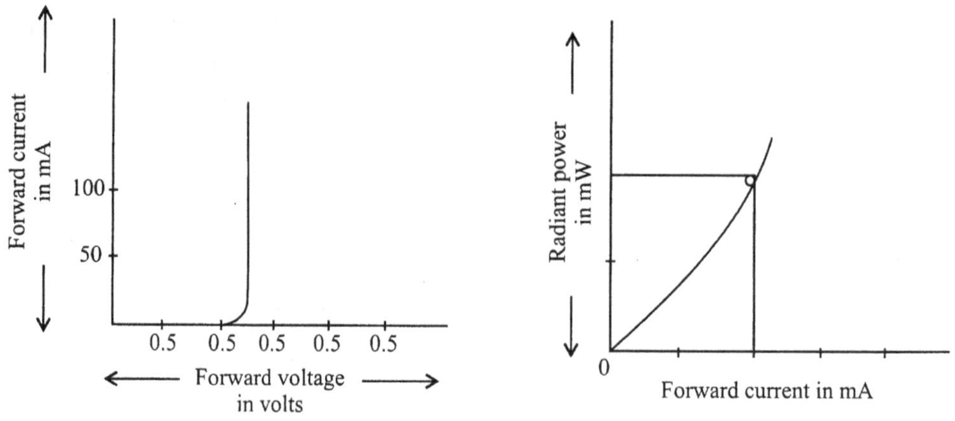

(a) Forward Bias V-I curve *(b) Output characteristics curve*

Fig. 5.30 V-I characteristics of LED

V-I Characteristics of LED (Input)

The Fig.5.30 shows V-I characteristics (Input and Output) characteristics of LED. The Fig. 5.30(a) shows input V-I characteristics of LED. The input V-I characteristics of a LED is a plot drawn between applied voltage and corresponding current through the junction. And as shown in Fig. 5.30(a) of the LED has a knee voltage of 0.5 V and hence if we increase input voltage slightly greater than 0.5 V, the LED starts conducting and allows heavy flow of current through the junction. And if the applied voltage is less than 0.5 V i.e., less than knee (or) threshold voltage of LED, no current flows through the junction. But in practical a forward bias voltage of 1 V is required to get significant forward current.

V-I Characteristics of LED (Output)

The Fig. 5.30(b) shows output V-I characteristics of LED. The output V-I characteristics of LED is a plot drawn between Radiant Power (on X-axis) and forward current (on Y-axis). It is apparent from Fig. 5.30(b) that the output characteristics of LED are linearly exponential.

Circuit Symbol of LED

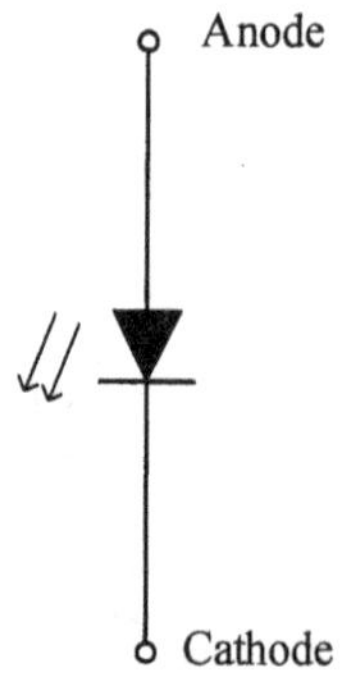

Advantages of LED

1. A very low voltage and current are enough to drive LED.
2. The response time of LED is very less about ions.
3. The LED does not require any warm up time.
4. Small in size.
5. Les in weight.
6. It has rugged construction and hence it can withstand vibrations and stocks.
7. LED's are manufactured with different colors and hence we can choose appropriate LED to get required color of light.
8. LED has more life span (about 20 years).

Disadvantages

1. Excess voltage (applied) can damage LED.
2. Temperature dependence of output power.

Applications

1. The LED's are popularly employed as indicators.
2. The LED are sucessfully employed in manufacturing Telivisions.

5.8.3 PHOTO DIODES

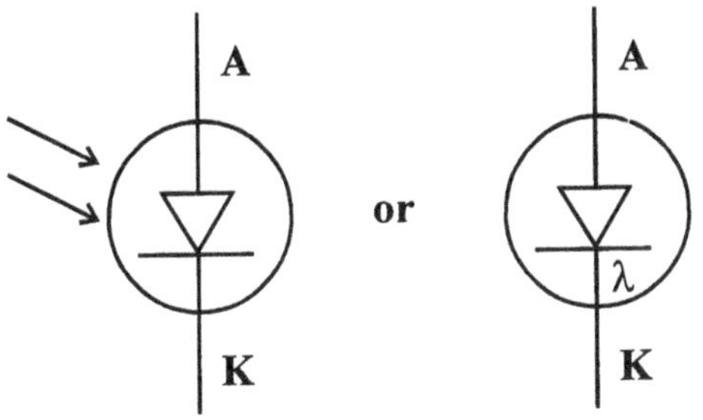

Fig 5.31 Symbols for photo diode.

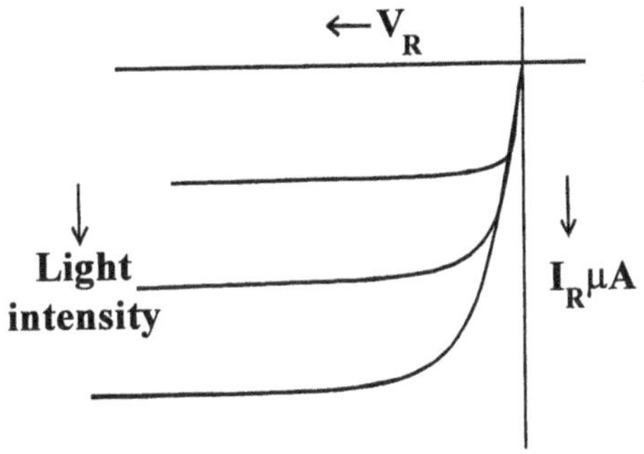

Fig 5.32 Photo diode reverse characteristics.

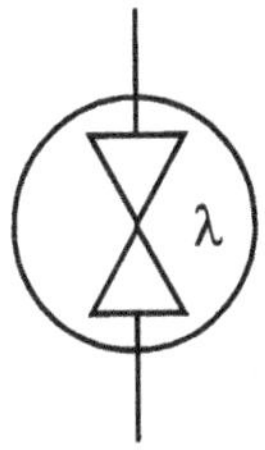

Fig. 5.33 Symbol for photo duo diode.

Instead of photoconductive cells, if a reverse biased p-n junctions diode is used the current sensitivity to radiation can be increased enormously. The mechanism of current control through radiation is similar to that of a photo conductive cell. Photons create electron hole pairs on both sides of the junction. When no light is applied the current is the reverse saturation current due to minority carriers. When light falls on the device photo induced electrons and holes cross the junction and thus increasing the current through the device. For Photo diodes also dark current will be specified. This is the current through the device with no light falling on the device. It is typically 30 - 70 n A at $V_R = 30$ V.

The active diameter of these devices is only 0.1''. They are mounted in standard To - 5 package with a window.

They can operate at frequencies of the order of 1 MHz. These are used in optical communication and in encoders.

5.8.4 Photo Transistors

In a photodiode, the current sensitivity is much larger compared to a photoconductive cell. But this can be further increased by taking advantage of the inherent current multiplication found in a transistor. Photo transistors have a lens, to focus the radiation or light on to the common base junction of the transistor. Photo induced current of the junction serves as the base current of the transistor (I_λ).

Therefore collector current $I_C = (1 + h_{fe}) I_\lambda$

The base lead may be left floating or used to bias the transistor into some area of operation.

Symbol for photo transistor shown in Fig. 5.34(a).

If the base lead in left open, the device is called photo duo diode.

Symbol is shown in Fig. 5.34(b).

It has two diodes pn and np. So it is called as photo duo diode.

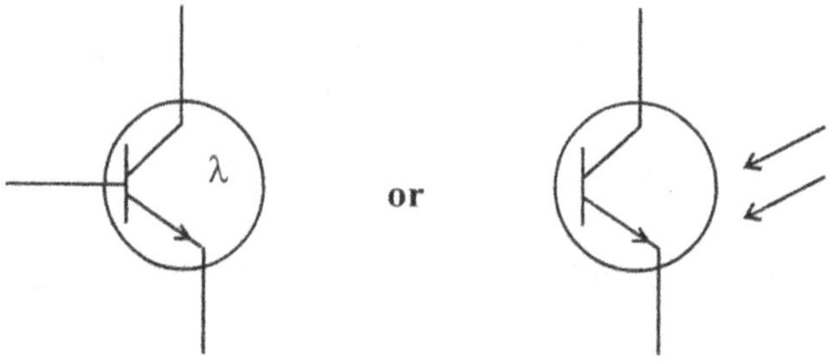

(a) *(b)*

Fig. 5.34 Symbols for photo transistors

SUMMARY

- Tunnel diode exhibits negative resistance characteristics
- Varactor diodes are operated in reverse bias and their junction capacitance varies with the voltage.
- Zener diode is also operated in reverse bias for voltage regulation.

ESSAY TYPE QUESTIONS

1. Write notes on Varactor Diode.
2. Derive the expression for the diffusion capacitance C_D in the case of *p-n junction* diode.
3. Give the constructional features of UJT.
4. Qualitatively explain the static V-I characteristics of UJT.
5. What is the significance of negative resistance region? Explain how UJT exhibits this characteristics ?

MULTIPLE CHOICE QUESTIONS

1. The symbol shown  **is that of a**

(a) LED (b) Schottky diode

(c) Varactor diode (d) Gunn diode

2. The symbol shown is  **is that of a .....**

(a) DIAC (b) GTO (c) SCR (d) TRIAC

3. DIAC is a ..... layer device

(a) 4 layer (b) 3 layer (c) 6 layer (d) Two layers

CONCEPTUAL QUESTIONS (Interview Questions)

1. Why tunnel diode device exhibits negative resistance characteristic?

Ans : The current flowing through a device is proportional to the number of free electrons. The higher the number, the more the current. If current through a device 'I' decreases as the voltage across it is being increased, the device exhibits negative resistance.

In the case of a tunnel diode, there are two conditions to be satisfied for tunnelling to take place.

(i) There must be empty energy states on the P-side corresponding to filled energy states on the n-side.

(ii) The barrier width at the junction must be very narrow (~ 3A^o), corresponding to doping concentration of 10^{19} carries/cm^3.

When forward voltage is applied to the tuneel diode, because of the energy gained from the electric field, electrons with higher energy levels tunnel through the barrier, to occupy the equivalent empty energy states on P-side. As the forward voltage is further increased, electrons with lesser energy also acquire sufficient energy from the increased electric field and will be able to tunnel through the barrier. But as the empty energy levels on the P-side get filled, after reaching the peak current I_P, even though forward voltage is being increased, the number of electrons reaching P-side decreases and hence tunnelling current decreases. So the device exhibits negative resistance region, as current is decreasing with increasing voltage. When the voltage is further increased, tunnelling phenomenon ceases, as the empty energy states on P - side would have been filled and normal forward current of the diode only flows.

6 Feedback Amplifiers

In this Chapter,

- The concept of Feedback is introduced.
- Effect of negative feedback on amplifier characteristics is explained. Necessary equations are derived.
- Voltage series and shunt amplifiers, current series and shunt amplifier circuits are given.

6.1 FEEDBACK AMPLIFIERS

Any system whether it is electrical, mechanical, hydraulic or pneumatic may be considered to have at least one input and one output. If the system is to perform smoothly, we must be able to measure or control output. If the input is 10mV, gain of the amplifier is 100, output will be 1V. If the input deviates to 9mV or 11mV, output will be 0.9 V or 1.1V. So there is no control over the output. But by introducing feedback between the output and input, there can be control over the output. If the input is increased, it can be made to increase by having a link between the output and input. By providing feedback, the input can be made to depend on output.

One example is, the temperature of a furnace. Suppose, inside the furnance, the temperature should be limited to 1000°C. If power is supplied on, continuously, the furnace may get over heated. Therefore we must have a thermocouple, which can measure the temperature. When the output of the thermocouple, reaches a value corresponding to 1000°C, a relay should operate which will switch off the power supply to the furnace. Then after sometime, the temperature may come down below 1000°C. Then again another relay should operate, to switch on the power. Thus the thermocouple and relay system provides the feedback between input and output.

Another example is traffic light. If the timings of red, green and yellow lights are fixed, on one side of the road even if very few vehicles are there, the green light will be on for sometime. On the other hand, if the traffic is very heavy on the other side of the road, still if the green lamp glows for the same period, the traffic will not be cleared. So in the ideal case, the timings of the red and green lamps must be proportional to the traffic on the road. If a traffic policeman is placed, he provides the feedback.

Another example is, our human mind and eyes. If we go to a library, our eyes will search for the book which we need and indicates to the mind. We take the book, which we need. If our eyes are closed, we can't choose the book we need. So eyes will provide the feedback.

Basic definitions

Ideally an amplifier should reproduce the input signal, with change in magnitude and with or without change in phase. But some of the short comings of the amplifier circuit are

1. Change in the value of the gain due to variation in supplying voltage, temperature or due to components.
2. Distortion in wave-form due to non linearities in the operating characters of the amplifying device.
3. The amplifier may introduce noise (undesired signals)

The above drawbacks can be minimizing if we introduce feedback

6.2 CLASSIFICATION OF AMPLIFIERS

Amplifiers can be classified broadly as,

1. Voltage amplifiers.
2. Current amplifiers.
3. Transconductance amplifiers.
4. Transresistance amplifiers.

This classification is with respect to the input and output impedances relative to the load and source impedances.

6.2.1 VOLTAGE AMPLIFIER

This circuit is a 2-port network and it represents an amplifier (see in Fig 6.1). Suppose $R_i >> Rs$, drop across Rs is very small.

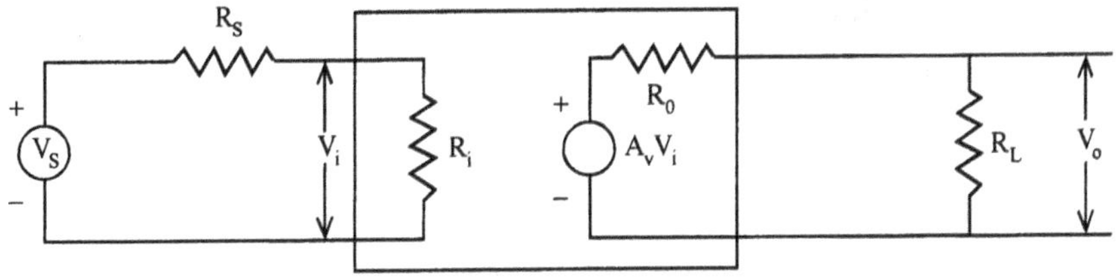

Fig 6.1 Equivalent circuit of voltage amplifiers.

$\therefore \quad V_i \simeq V_S.$

Similarly, if $R_L >> R_o$, $V_o \simeq A_V . V_i$.

But $V_i \sim V_S$.

$\therefore \quad V_o \simeq A_V . V_S.$

$\therefore$ Output voltage is proportional to input voltage.

The constant of proportionality A_V doesn't depend on the impedances. (Source or load). Such a circuit is called as ***Voltage Amplifier***.

Therefore, for ideal voltage amplifier

$$R_i = \infty .$$

$$R_o = 0.$$

$$A_V = \frac{V_o}{V_i}$$

with $\quad R_L = \infty .$

A_V represents the open circuit voltage gain. For ideal voltage amplifier, output voltage is proportional to input voltage and the constant of proportionality is independent of R_S or R_L.

6.2.2 CURRENT AMPLIFIER

An ideal current amplifier is one which gives output current proportional to input current and the proportionality factor is independent of R_S and R_L.

The equivalent circuit of current amplifier is shown in Fig.6.2.

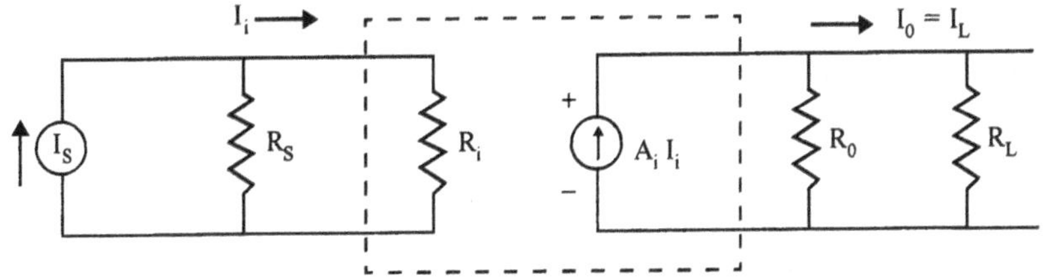

Fig 6.2 Current amplifier.

For ideal Current Amplifier,

$R_i = 0$

$R_o = \infty$.

If $R_i = 0, I_S \sim I_i$.

$\therefore$ $R_o = \infty$,

$I_L = I_o = A_i I_i = A_i I_S$.

$\therefore$ $A_I = \frac{I_L}{I_i}$ with $R_L = 0$.

$\therefore$ A_I represents the short circuit current amplification.

6.2.3 Transconductance Amplifier

Ideal Transconductance amplifier supplies output current which is proportional to input voltage independently of the magnitude of R_S and R_L.

Ideal Transconductance amplifier will have

$R_i = \infty$.

$R_o = \infty$.

In the equivalent circuit, on the input side, it is the Thevenins' equivalent circuit. A voltage source comes in series with resistance. On the output side, it is Norton's equivalent circuit with a current source in parallel with resistance. (See Fig. 6.3).

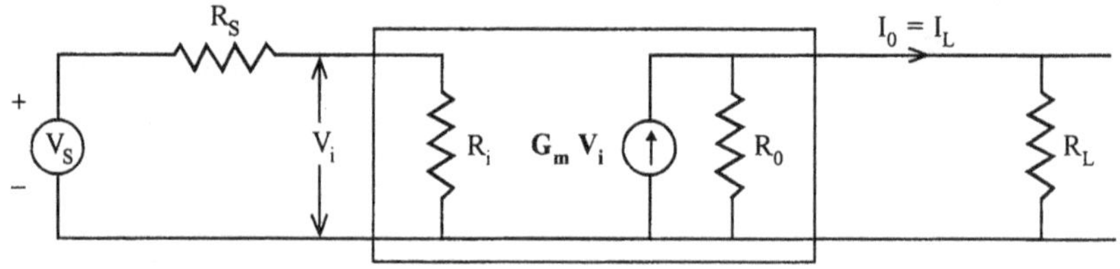

Fig 6.3 Equivalent circuit of Transconductance amplifier.

6.2.4 Trans resistance Amplifier

It gives output voltage V_o proportional to I_s, independent of $R_s \propto R_L$. For ***ideal amplifiers***

$R_i = 0, R_o = 0$

Equivalent circuit

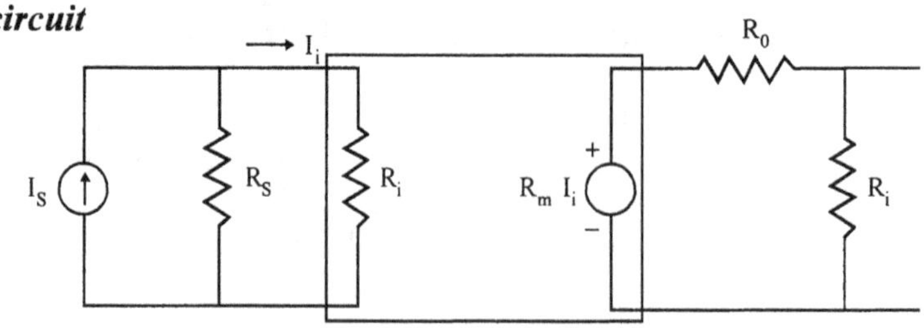

Fig 6.4 Trans resistance amplifier.

Norton equivalent circuit on the R_i input side.

Thevenins' equivalent circuit on the output side.

6.3 FEEDBACK CONCEPT

A sampling network samples the output voltage or current and this signal is applied to the input through a feedback two port network. The block diagram representation is as shown in Fig. 6.5.

GENERALIZED BLOCK SCHEMATIC

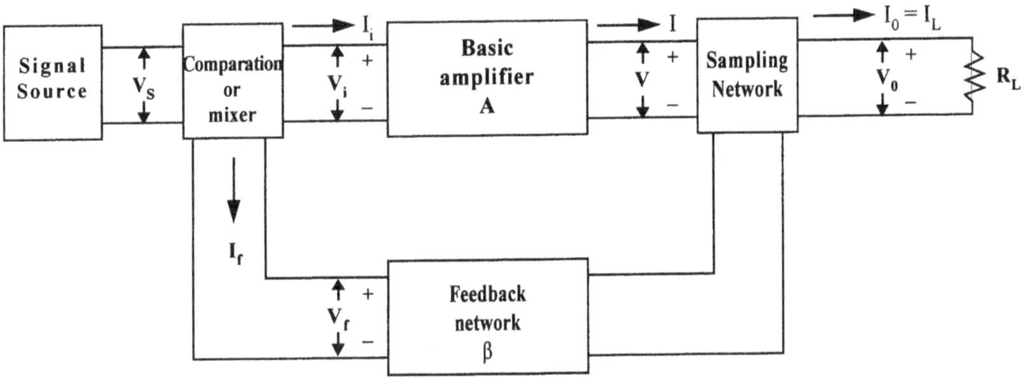

Fig 6.5 Block diagram of feedback network.

Signal Source

It can be a voltage source V_S or a current source I_S

FEEDBACK NETWORK

It is a passive two port network. It may contain resistors, capacitors or inductors. But usually a resistance is used as the feedback element. Here the output current is sampled and feedback. The feedback network is connected in series with the output. This is called as ***Current Sampling or Loop Sampling***.

A voltage feedback is distinguished in this way from current feedback. For voltage feedback, the feedback element (resistor) will be in parallel with the output. For current feedback the element will be in series.

COMPARATOR OR MIXER NETWORK

This is usually a differential amplifier. It has two inputs and gives a single output which is the difference of the two inputs.

V = Output voltage of the basic amplifier *before sampling* [see the block diagram of feedback]

V_i = Input voltage to the basic amplifier

A_V = Voltage amplification = V/V_i

A_I = Current amplification = I/I_i

G_M = Transconductance of basic amplifier = I/V_i

R_M = Transresistance = V/I_i.

All these four quantities, A_V, A_I, G_M and R_M represent the transfer gains (Though G_M and R_M are not actually gains) of the basic amplifier without feedback. So the symbol 'A' is used to represent these quantities.

A_f is used to represent the ratio of the output to the input ***with feedback***. This is called as the transfer gain of the amplifier with feedback.

$$\therefore \quad A_{Vf} = \frac{V_o}{V_s}$$

$$A_{If} = \frac{I_o}{I_s}$$

$$G_{Mf} = \frac{I_o}{V_s}$$

$$R_{Mf} = \frac{V_o}{I_s}$$

Feedback amplifiers are classified as shown below.

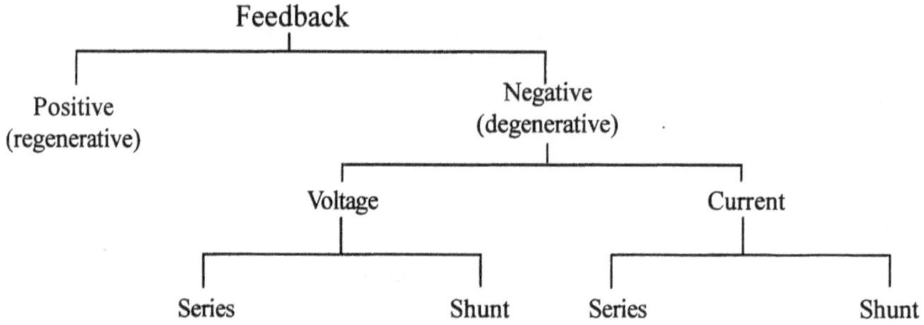

$$A = \text{voltage gain} = \frac{V_0}{V_i}$$

Feedback Factor, $\beta = \dfrac{V_f}{V_o}$ (This β is different from β used in BJTs)

$$(\beta)\,(A)\,(-1) = -\beta A$$

Loop gain / Return Ratio

– 1 is due to phase-shift of 180^0 between input and output in Common Emitter Amplifier. Since $\text{Sin}(180^0) = -1$.

Return difference $D = 1 -$ loop gain negative sign is because it is the difference

$$D = 1 - (-\beta A)$$

$$\boxed{D = 1 + \beta.A.}$$

Types of Feedback

How to determine the type of feedback ? Whether current or voltage ? If the feedback signal is proportional to voltage, it is ***Voltage Feedback.***

If the feedback signal is proportional to current, it is ***Current Feedback.***

Conditions to be satisfied

1. Input signal is transmitted to the output through amplifier A and not through feedback network β.
2. The feedback signal is transmitted to the input through feedback network and not through amplifier.
3. The reverse transmission factor β is independent of R_S and R_L.

6.4 TYPES OF FEEDBACK

Feedback means a portion of the output of the amplifier circuit is sent back or given back or feedback at the input terminals. By this mechanism the characteristics of the amplifier circuit can be changed. Hence feedback is employed in circuits.

There are two types of feedback.

1. Positive Feedback
2. Negative Feedback

Negative feedback is also called as ***degenerative feedback***. Because in negative feedback, the feedback signal opposes the input signal. So it is called as degenerative feedback. But there are many advantages with negative feedback.

Advantages of Negative Feedback

1. Input impedance can be increased.
2. Output impedance can be decreased.
3. Transfer gain A_f can be stabilized against variations in ***h-parameter*** of the transistor with temperature etc.
   i.e., stability is improved.
4. Bandwidth is increased.
5. Linearity of operation is improved.
6. Distortion is reduced.
7. Noise reduces.

6.5 EFFECT OF NEGATIVE FEEDBACK ON TRANSFER GAIN

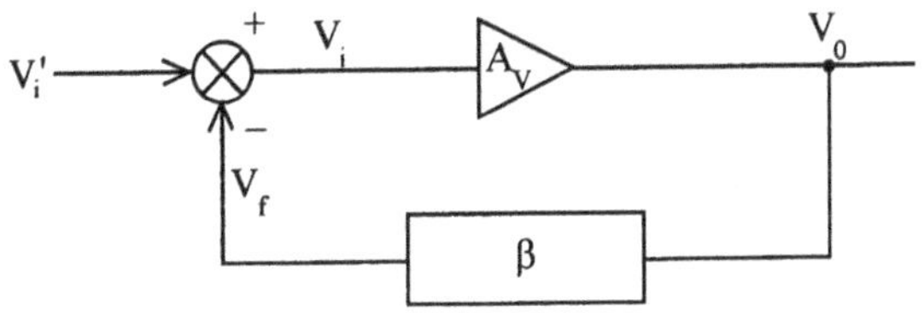

Fig 6.6 Block schematic for negative feedback.

$$A_V = \frac{V_o}{V_i}$$

$$A_V' = V_o / V_i'$$

$$V_i' = V_i - \beta V_o$$

$$V_o = A_v (V_i - \beta V_o)$$
$$V_o = A_v . V_i - A_v . \beta V_o$$
$$V_o (1 + \beta A_v) = A_v V_i$$
$$\therefore \quad A_{vf} = \frac{V_o}{V_i} = \frac{A_v}{1+\beta A_v}$$

6.5.1 Reduction in Gain

$$A_v' = \frac{A_v}{1-\beta A_v}$$

(for positive feedback)

A_v = Voltage gain without feedback. (Open loop gain).

If the feedback is negative, β is negative.

$$\therefore \quad A_v' = \frac{A_v}{1-(-\beta . A_v)}$$

For negative feedback $A_v' = \dfrac{A_v}{1+\beta A_v}$

Denominator is > 1. $\therefore \quad A_v' < A_V$

$\therefore$ There is reduction in gain.

6.5.2 Increase in Bandwidth

If, f_H is upper cutoff frequency.

f_L is lower cut off frequency.

f is any frequency.

Expression for A_v (voltage gain at any frequency f) is,

$$A_v = \frac{A_v(\text{mid})}{1 + j.\frac{f}{f_H}} \qquad \text{..... (6.1)}$$

A_v (mid) = Mid frequency gain

$$A_v = \frac{A_v (\text{mid})}{1+\beta A_v} \text{ for negative feedback.} \qquad \text{.....(6.2)}$$

Substituting equation (6.1) in (6.2) for A_V,

$$A_v = \frac{A_v(\text{mid})/1 + j.\frac{f}{f_H}}{1+\beta\left[\frac{A_v\text{mid}}{1+jf/f_H}\right]} \quad \text{(for negative Feedback, } \beta \text{ is } V_e)$$

Simplifying, $$A_V = \frac{\frac{A_V(\text{mid})}{f_H + j.f}\left[1 + j\frac{f}{f_H}\right]}{1 + j.f + \beta\left[A_V \text{ mid}\right]}$$

$$= \frac{A_V(\text{mid})}{f_H + j.f + \beta.A_{V\text{mid}} \cdot f_H}$$

$$A'_V = \frac{A_v(\text{mid})/1 + \beta_v A_v(\text{mid})}{1 + \frac{jf}{f_H\left[1 + \beta_v A_v(\text{mid})\right]}} \qquad \text{..... (6.3)}$$

Equation (6.3) can be written as,

$$A'_V = \frac{A_v'(\text{mid})}{1 + \frac{jf}{f_H'}}$$

Where $$A'_{V(\text{mid})} = \frac{A_v(\text{mid})}{1 + \beta_v A_v(\text{mid})}$$

and $$f_H' = f_H\,(1 + \beta_v\, A_{V(\text{mid})})$$

$\because$ β is negative for negative feedback, $f_H' > f_H$.

$\therefore$ ***Negative feedback, increases bandwidth.***

Similarly, $$f_L' = \frac{f_L}{1 + \beta_v A_{v(\text{mid})}} \qquad A_v = \frac{A_{v(\text{mid})}}{1 - j\left(\frac{f_L}{f}\right)}$$

or $$A_v' = \frac{A_{V(\text{mid})}/1 + \beta A_{V(\text{mid})}}{1 + j\frac{f_L}{f(1 + \beta.A_V)}}$$

$$A_v' = \frac{A_V}{1 + j\frac{f_L'}{f}}$$

6.5.3 Reduction in Distortion

Suppose, the amplifier, in addition to voltage amplification is also producing distortion D.

$$V_o = A_V \,.\, V_i + D.$$

Where, $V_i = V_i' - \beta_V.\, V_o$ (for negative feedback)

$V_i' = V_i - \beta\, V_o$ and β is negative for negative feedback,

$\therefore$ $$V_o' = A_V\left[V_i' - \beta_v V_o\right] + D$$

$$V_o\,[1 + A_V\ \beta_V] = V_i'.A_V + D$$

$$\therefore \quad V_o = \frac{V_i'.A_v}{(1 + A_V.\beta_V)} + \frac{D}{(1 + \beta_V A_V)}$$

For negative feedback, β is negative therefore denominator is > 1

$\therefore$ The distortion in the output is reduced.

$$= \frac{D}{1 + \beta_v A_v} \text{ is} < D$$

The physical explanation is, suppose the input is pure sinusoidal wave. There is distortion in the output as shown, in Fig. 6.7.

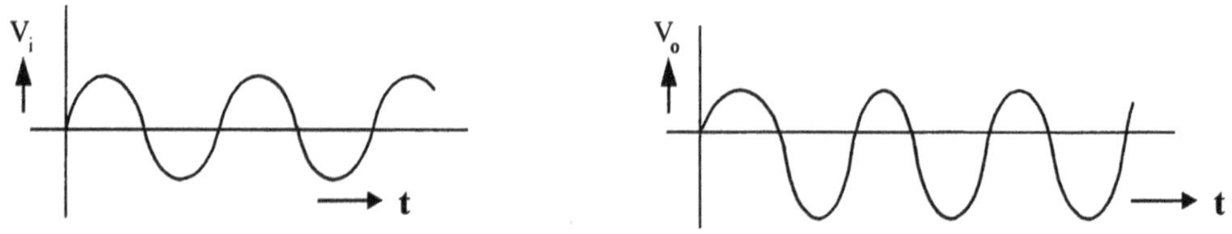

(a) Input signal *(b) Output with distortion (without negative feedback)*

Fig 6.7

Now, if a part of the distorted output is fed back to the input, so as to oppose the input, the feedback signal and input will be out of phase.

So the new input V_i will have distortion introduced init, because of mixing of distorted V_f with pure V_i. So the distortion in the output will be reduced because, the distortion introduced in the input cancels the distortion produced by the amplifier. Because these two distortions are out of phase. The feedback signal cancels the distortion produced by the amplifier, Therefore these two are out of phase.

6.5.4 Feedback to Improve Sensitivity

Suppose an amplifier of gain A_1 is required. Build an amplifier of gain A2 = DA_1 in which D is large. Feedback is now introduced to divide the gain by D. Sensitivity is improved by the same factor D, because both gain and instability are divided by D. The stability will be improved by the same factor.

6.5.5 Frequency Distortion

$$A_f = \frac{A}{1 + \beta A}$$

If the feedback network does not contain reactive elements. The overall gain is not a function of frequency. So frequency duration is less. If β depends upon frequency, with negative feedback, Q factor will be high.

6.5.6 Band Width

It increases with negative Feedback (as shown in Fig.6.2)

$$f_1' < f_1, \quad f_2' > f_2$$

$$BW' = (f_1' - f_2') : BW = (f_2 - f_1)$$

$$f_1' = \frac{f}{1+\beta A_m} \quad f_2' = f_2\ (1 + \beta A_m) \quad BW' > BW$$

$$BW = f_2 - f_1 \simeq f_2$$

$$(BW)_f = f_2' - f_1 \sim f_2'$$

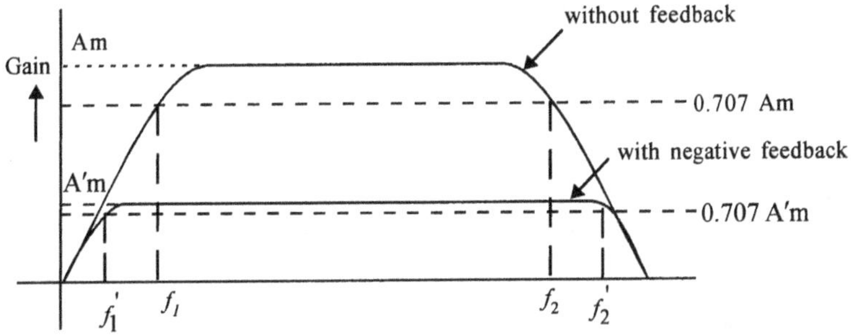

Fig 6.8 Frequency Response with and without negative feedback.

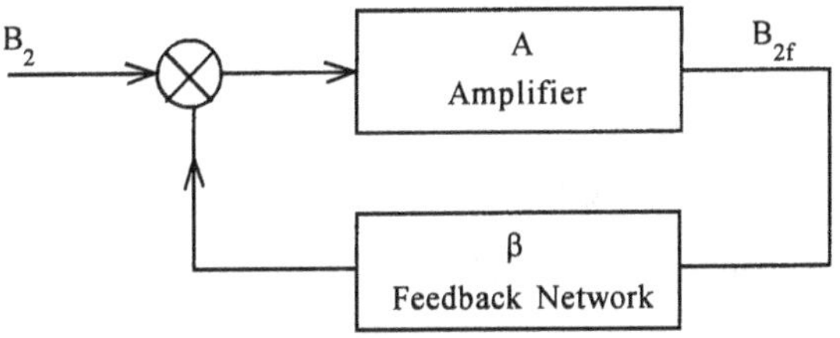

Fig 6.9 Feedback network.

$\therefore \quad (BW)_f = (1 + \beta A_m)\ BW$

6.5.7 Sensitivity of Transistor Gain

Due to aging, temperature effect etc., on circuit capacitance, transistor or FET, stability of the amplifier will be affected

Fractional change in amplification with feedback divided by the fractional change without feedback is called ***Sensitivity*** of Transistor.

$$= \frac{\left|\frac{dA_f}{A_f}\right|}{\left|\frac{dA}{A}\right|}$$

$$A_f = \frac{A}{1+\beta A}$$

Differentiating, $$\frac{dA_f}{A_f} = \frac{1}{|1+\beta A|}\left|\frac{dA}{A}\right|$$

$\therefore$ Sensitivity, $$= \frac{1}{1+\beta A}$$

Reciprocal of sensitivity is ***Densitivity*** $\boxed{D = (1+\beta A).}$

6.5.8 Reduction of Nonlinear Distortion

Suppose the input signal contains second harmonic and its value is B_2 before feedback. Because of feedback. B_{2f} appears at the output. So positive βB_{2f} is fed to the input. It is amplified to $-A\beta B_{2f}$.

$\therefore$ Output with two terms $B_2 - A\beta B_2 f$.

$\therefore$ $$B_2 - A\beta B_2 f = B_{2f}.$$

or $$B_{2f} = \frac{B_2}{1+\beta A} \quad B_{2f} < B_2$$

So it is reduced.

6.5.9 Reduction of Noise

Let N be noise constant without feedback and N_F with feedback. N_F is fed to the input and its value is βN_F. It is amplified to $-\beta AN_F$.

$\therefore$ $$N_F = N - \beta AN_F$$
$$N_F(1+\beta A) = N$$

as $$N_F = \frac{N}{1+\beta A}$$

$\therefore$ $N_F < N$. Noise is reduced with negative feedback.

6.6 TRANSFER GAIN WITH FEEDBACK

Consider the generalized feedback amplifier shown in Fig.6.10.

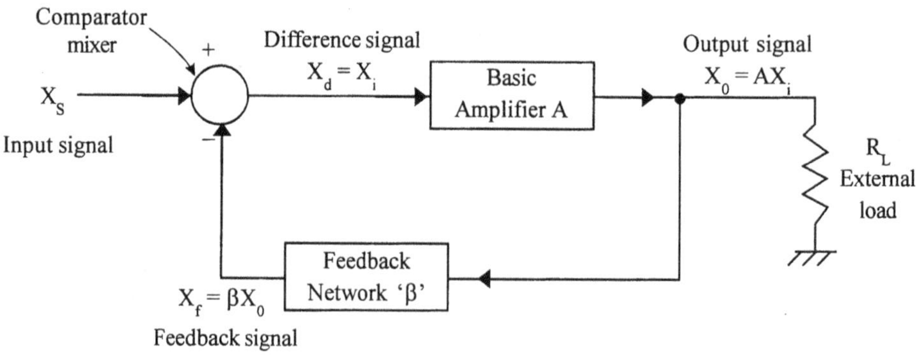

Fig 6.10

The basic amplifier shown may be a voltage amplifier or current amplifier, transconductance or transresistance amplifier.

The four different types of feedback amplifiers are,

1. Voltage series feedback.
2. Voltage shunt feedback.
3. Current series feedback.
4. Current shunt feedback.

X_s = Input signal.
X_0 = Output signal.
X_f = feedback signal.
X_d = Difference signal. [Difference between the input signal and the feedback signal].
β = Reverse transmission factor or feedback factor = X_f / X_o.

The feedback network can be a simple resistor. The mixer can be a difference amplifier. The output of the mixer is the difference between the input signal and the feedback signal.

$$X_d = X_s - X_f = X_i.$$

X_d is also called as error or comparison signal.

$$\beta = \frac{X_f}{X_o}$$

It is often a positive or negative real number.

This β should not be confused with the β of a transistor $\frac{I_C}{I_B}$

Transfer gain $$A = \frac{X_o}{X_i} = \frac{X_o}{X_d}$$

$$X_i = X_d \quad .....(6.4)$$

Gain with feedback $A_f = \frac{X_o}{X_s}$

But $$X_d = X_s - X_f = X_i \quad .....(6.5)$$

$$\beta = \frac{X_f}{X_o} \quad .....(6.6)$$

From (6.5), $X_d = X_S - X_f$.

Substitute (6.5) in (6.4).

$$\therefore \quad A = \frac{X_0}{X_s - X_f}$$

Dividing Numerator and Denominator by X_s,

$$A = \frac{X_o/X_s}{1-\frac{X_f}{X_s}} = \frac{X_o/X_s}{1-\frac{X_f}{X_o}.\frac{X_o}{X_s}}.$$

We know that $X_o / X_s = A_f$; $\frac{X_f}{X_o} = \beta$.

$$\therefore \quad A = \frac{A_f}{(1-\beta.A_f)}$$

or $$A - \beta.\ A_f.\ A = A_f.$$

$$A = A_f(1+\beta A)$$

$$\therefore \quad \boxed{A_f = \frac{A}{1+\beta A}}$$

A_f = gain with feedback.

A = transfer gain without feedback.

If $|A_f| < |A|$ the feedback is called as negative or degenerative, feedback

If $|A_f| > |A|$ the feedback is called as positive or regenerative, feedback

6.6.1 Loop Gain

In the block diagram of the feedback amplifier, the signal X_d which is the output of the comparator passes through the amplifier with gain A. So it is multiplied by A. Then it passes through a feedback network and hence gets multiplied by β, and in the comparator it gets multiplied by -1. In the process we have started from the input and after passing through the amplifier and feedback network, completed the loop. So the total product $A\beta$.

In order that ***series feedback*** is most effective, the circuit should be driven from a ***constant voltage*** source whose internal resistance R_s is small compared to R_i of the amplifier. If R_s is very large, compared with R_i, V_i will be modified not by V_f but because of the drop across R_s itself. So effect of V_f will not be there. Therefore for series feedback, the voltage source should have less resistance. (See Fig.6.11).

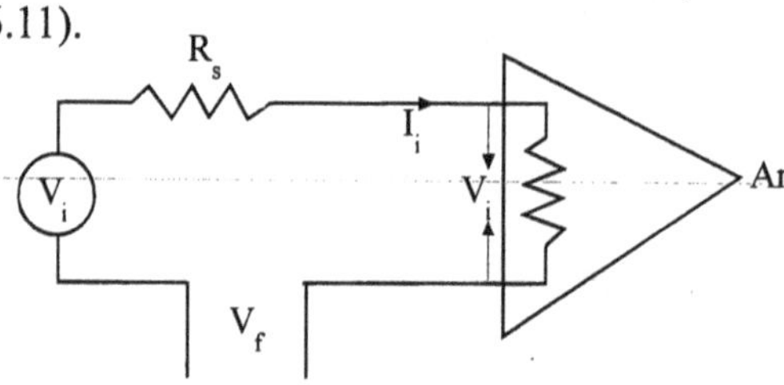

Fig 6.11 Series feedback.

In order that ***shunt feedback*** is most effective, the amplifier should be driven from a constant current source whose resistance R_s is very high ($R_s >> R_i$).

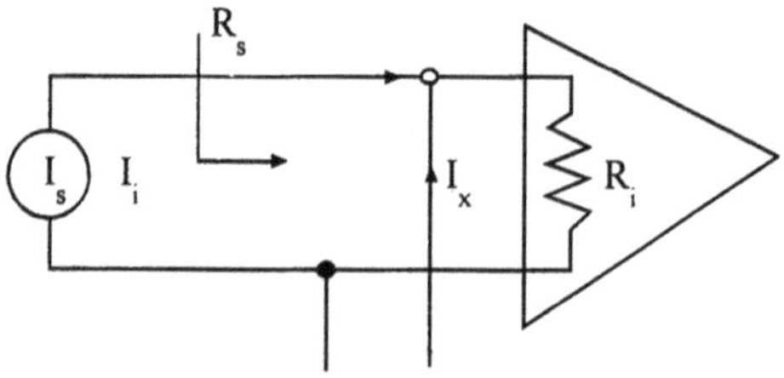

Fig 6.12 Current shunt feedback.

If the resistance of the source is very small, the feedback current will pass through the source and not through R_i. So the change in I_i will be nominal. Therefore the source resistance should be large and hence a current source should be used.

If the feedback current is same as the output current, then it is series derived feedback.

When the feedback is shunt derived, output voltage is simultaneously present across R_L and across the input to the feedback. So in this case V_f is proportional to V_o.

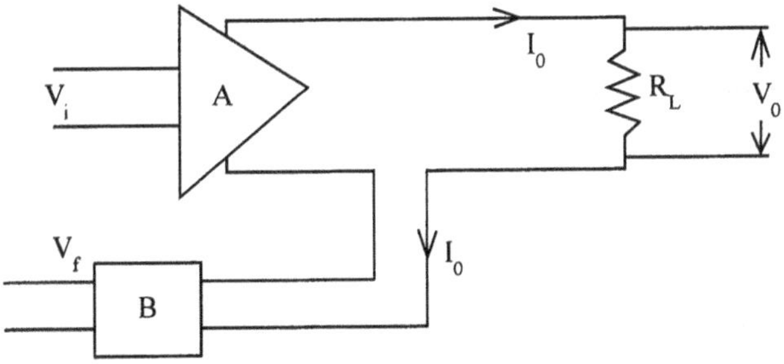

Fig 6.13(a) Series derived feedback.

An amplifier with shunt derived negative Feedback increases the output resistance. When the feedback is series derived, I_0 remains constant, so R_0 increase.

Similarly if the feedback signal is ***shunt fed*** it reduces the input resistance.

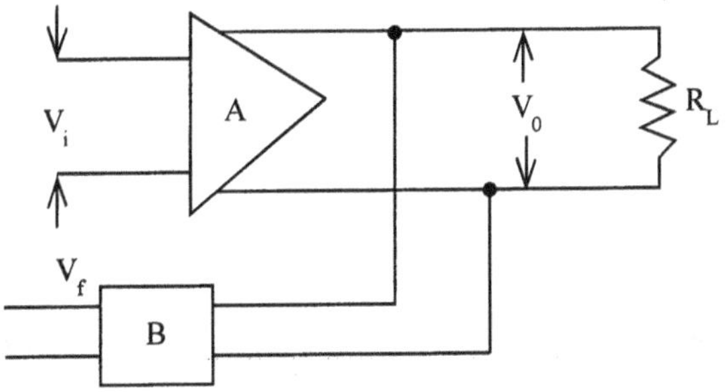

Fig. 6.13 (b) Shunt derived feedback.

If it is ***series fed***, it increases the output resistance. Therefore I_0 remains constant, even through V_0 increase, so R_0 increases.

Return Ratio

βA = Product of feedback factor β and amplification factor A is called as ***Return Ratio.***

Return Difference (D)

The difference between unity (1) and return ratio is called as ***Return difference.***

$$D = 1 - (-\beta A) = 1 + \beta A.$$

Amount of feedback introduced is expressed in decibels

$$= N \log \left(\frac{A'}{A}\right) \quad N = 20 \log \left|\frac{A_f}{A}\right| = 20 \log \left|\frac{1}{1 + A\beta}\right|$$

If feedback is negative N will be negative because $A' < A$.

6.7 CLASSIFACTION OF FEEDBACK AMPLIFIERS

There are four types of feedback,

1. Voltage series feedback.
2. Voltage shunt feedback.
3. Current shunt feedback.
4. Current series feedback.

Voltage Series Feedback

Feedback signal is taken across R_L.proportional to V_o. So it is voltage feedback. V_f is coming in series with V_i So it is Voltage series feedback.(See Fig.6.14).

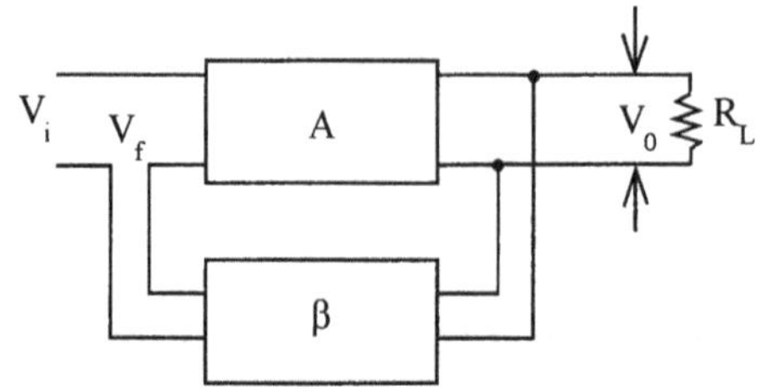

Fig 6.14 Schematic for voltage series feedback.

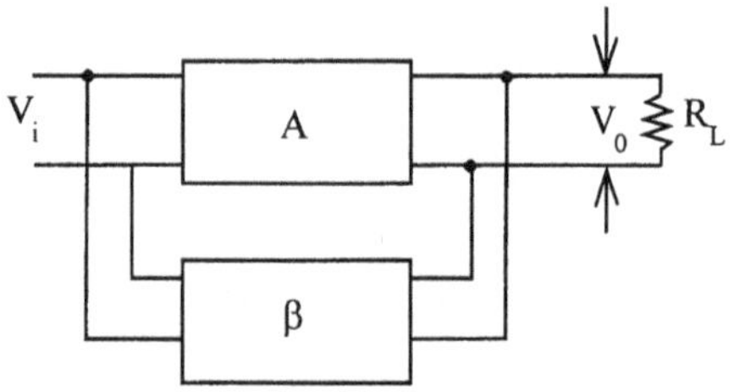

(a) Voltage shunt Feedback

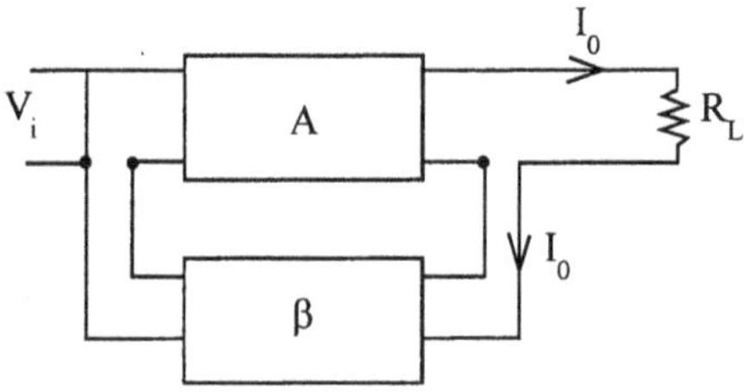

(b) Current Shunt Feedback

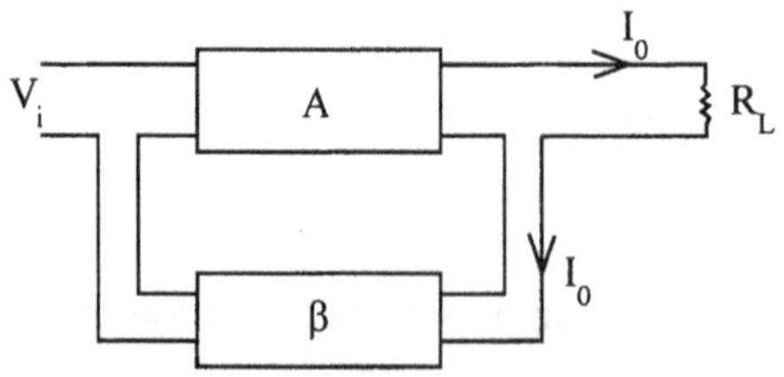

(c) Current Series Feedback

Fig 6.15

Improvement of Stability with Feedback

Stability means, the stability of the voltage gain. The voltage gain must have a stable value, with frequency. Let the change in A_v is represented by S.

$$\frac{dA_V}{A_v} = S\left(\frac{dA'_V}{A'_V}\right) = S = \frac{\left(\frac{d\,A'_v}{A'_v}\right)}{\left(\frac{d\,A_v}{A_v}\right)}$$

$$A'_V = \frac{V_0}{V_S} \text{ (for negative feedback)}$$

$$A'_V = \frac{A_V}{1 + A_V\beta_1}$$

Differentiating with respect to A_V, $\frac{dA_V}{A_V} = \frac{(1+\beta.A_V) - A_V(+\beta)}{(1+\beta.A_V)^2} = \frac{1}{(1+A_V\beta)^2}$

Dividing by A_V on both sides, of $\frac{dA'_V}{dA_V} = \frac{1}{(1+\beta A_V)^2}$

$$\frac{dA'_V}{A_V}\frac{1}{dA_v} = \frac{1}{(1+\beta A_v)}\cdot\frac{1}{A'_V}$$

But $\quad A'_V = \frac{A'_v}{1+\beta A_v}$ and $\frac{dA'_V}{A_V} = S'\ \frac{dA_v}{A_v} = S$

$$S = \frac{dA'_V}{dA_V} = \frac{1}{(1+\beta A_V)^2}$$

$$\boxed{S = \frac{dA_v}{(1+\beta A_v)}}$$

For negative Feedback, β is negative $\therefore$ denominator > 1.

$\therefore \quad S' < S$

i.e., variation in A_V, or % change in A_V is less with -negative feedback.

$\therefore$ ***Stability*** is good.

6.8 EFFECT OF FEEDBACK ON INPUT RESISTANCE

6.8.1 INPUT RESISTANCE WITH SHUNT FEEDBACK

With feedback $\quad R_{if} = \frac{V_i}{I_s}$

$$I_f = \beta_i\, I_o. \qquad \left[\because \beta_i = \frac{I_f}{I_o}\right]$$

$$I_s = I_i + \beta\, I_o$$

$$R_i' = \frac{V_i}{I_i + \beta_i I_O}$$

$$I_O = A_i . I_i$$

$$R_i' = \frac{V_i}{I_i + \beta_i A_i I_i}$$

$$R_i = \frac{V_i}{I_i (1 + \beta_i A_i)}$$

$$\boxed{R_i' = \frac{R_i}{(1 + \beta_i A_i)}}$$

∴ ***Input resistance decreases with shunt feedback.***

If the feedback signal is taken across R_L, it is α V_o or so it is ***Voltage feedback.***

If the feedback signal is taken in series with the output terminals, feedback signal is proportional to I_o. So it is ***current feedback.***

If the feedback signal is in series with the input, it is ***series feedback.***

If the feedback signal is in shunt with the input, it is ***shunt feedback.***

EXPRESSION FOR R_I WITH CURRENT SHUNT FEEDBACK

A_i = Shunt circuit current gain of the BJT

A_I = practical current gain $I_0|I_i$

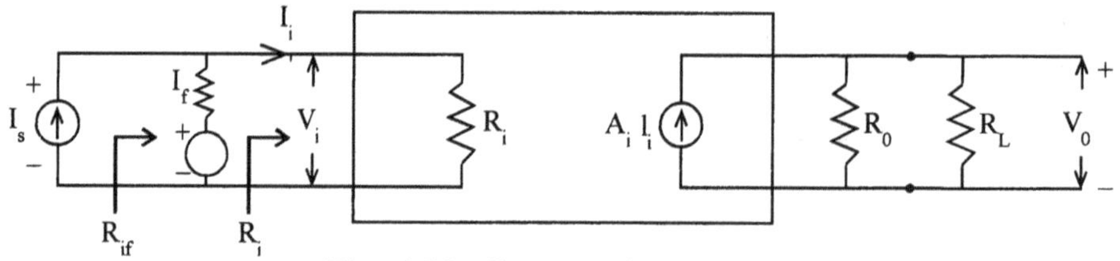

Fig. 6.16 Current shunt feedback.

A_i represents the Shunt circuit current gain taking R_s into accounts,

$$I_S = I_i + I_f = I_i + \beta\, I_0$$

and $$I_0 = \frac{A_i R_o I_i}{R_o + R_L} = A_I\, I_i \qquad \therefore \quad \text{Let } A_I = \frac{A_i R_o}{R_o + R_L}$$

$$A_I = \frac{I_o}{I_i} = \frac{A_i R_o}{R_o + R_i} = (I_i + I_f)$$

$$I_S = I_i + \beta .\, I_o = I_i \left(1 + \beta . \frac{I_o}{I_i}\right) = I_i\, (1 + \beta .\, A_I)$$

$$R_{if} = V_i \,|\, I_s \qquad R_i = V_i | \, I_i$$

$$\boxed{R_{if} = \frac{V_i}{(1 + \beta A_I) I_i} = \frac{R_i}{1 + \beta . A_I}}$$

for shunt feedback the input resistance decreases.

6.8.2 Input Impedance with Series Feedback

$$V_i' = V_i + \beta V_0 \text{ (in general case).}$$

For negative feedback, β is negative.

$$V_i = \frac{V_i'}{(1+\beta A)}$$

$$\frac{V_i}{I_i} = \frac{V_i'}{I_i(1+\beta A)}$$

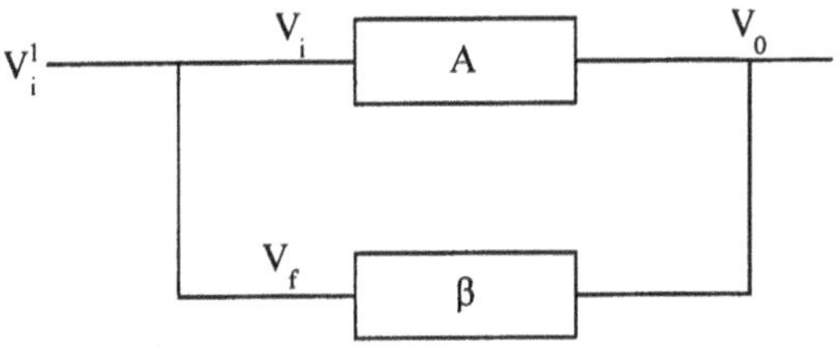

Fig 6.17 Feedback network.

In general, R_i increases,

$$V_i' = V_i + \beta V_o$$

$$V_i' = V_i + \beta.A.V_i \qquad \because \; V_0 = A.V_i$$

$$V_i' = V_i(1+\beta A)$$

But $\quad V_i' = I_i R_i$

$\therefore \quad V_i' = (1+\beta A) R_i I_i$

$$\frac{V_i'}{I_i} = \text{Input Z seen by the source} = R_i(1+\beta A)$$

$\therefore \quad R_{if} = R_i(1+\beta A)$

Expression for R_I with Voltage Series Feedback

In this circuit A_V represents the open circuit voltage gain taking R_s into account. (see Fig.6.18).

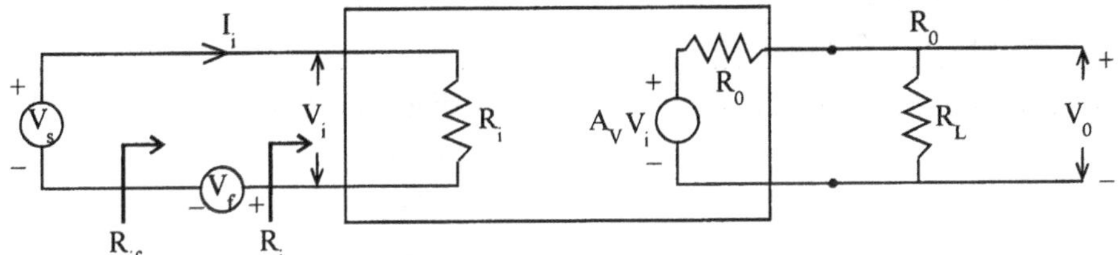

Fig. 6.18 Voltage series feedback.

$$V_i = V_S - V_f$$

or $\quad V_s = V_i + V_f$

$$R_0' = R_{if} = V_S/I_i.$$

$$V_f = I_i R_i + V_f = I_i R_i + \beta V_o$$

$$V_0 = \frac{A_v V_i R_L}{R_o + R_L} = (A_V I_i R_i)$$

Let, $\quad \frac{A_v.R_L}{R_o + R_L} = A_V, V_i = I_i.R_i.$

$$A_V = \frac{V_o}{V_i} = \frac{A_v R_L}{R_o + R_L}$$

$$V_S = I_i R_i + \beta . V_o$$

$$\boxed{\frac{V_s}{I_i} = R_i + \frac{\beta.V_o}{I_i} = R_i\left(1 + \frac{\beta.V_o}{I_i \times R_i}\right)}$$

$$R_{if} = R_i\left(1 + \beta.\frac{V_o}{V_i}\right) = R_i(1 + \beta.A_V)$$

A_V = voltage gain, without feedback.

for series feedback input resistance increases.

6.9 EFFECT OF NEGATIVE FEEDBACK ON R_O

1. Voltage feedback (series or shunt) R_o decreases.
2. Current feedback (series or shunt) R_o increases.
3. Series feedback (voltage or current) R_i increases.
4. Shunt feedback (voltage or current) R_i decreases.

Output Resistance

Negative feedback tends to decrease the input resistances. Feeding the voltage back to the input in a degenerative manner is to cause lesser increase in V_o. Hence the output voltage tends to remain constant as R_L changes because output resistance with feedback $R_{of} << R_L$.

Negative feedback, which samples the output current will tend to hold the output current constant. Hence an output current source is created ($R_{of} >> R_L$). So this type of connection increases output resistance.

For voltage sampling $R_{of} < R_o$. For current feedback $R_{of} > R_o$.

6.9.1 Voltage Series Feedback

Expression for R_{of} looking into output terminals with R_L disconnected,

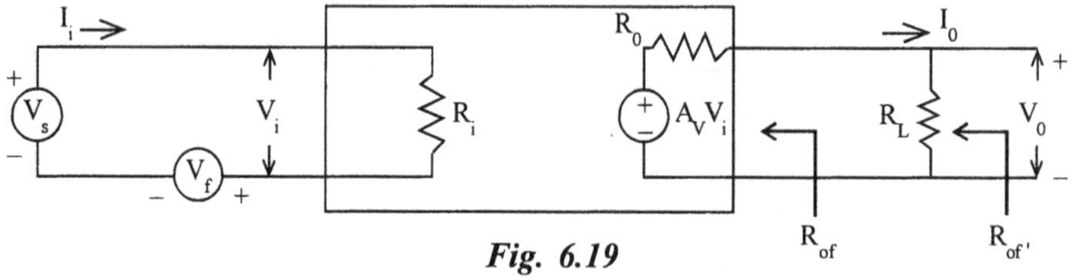

Fig. 6.19

R_0 is determined by impressing voltage 'V' at the output terminals or messing 'I', with input R'_{of} terminals shorted.

Disconnect R_L. To find R_{of}, remove external signal (set $V_s = 0$, or $I_s = 0$)

Let $R_L = \infty$.

Impress a voltage V across the output terminals and calculate the current I delivered by V. Then, R_{of} = V|I.

$$I = \frac{V_o - A_V\ V_i}{R_o} = \frac{V_o + \beta A_v V}{R_o}$$

$\because$ V_o = output voltage.

$V_i = -\beta_V$

Because with $V_S = 0,\ V_i = -V_f = -\beta_V$

Hence, $R_{of} = \frac{V_0}{I} = \frac{R_o}{1+\beta A_v}$

This expression is excluding R_L. If we consider R_L also R_{of} is in parallel with R_L.

$R_{of}' = \frac{R_{of}.R_L}{R_{of} + R_L}$ Substitute the1 value of R_{of}

$$R_{of}' = \frac{\frac{R_o}{1+\beta A_V} \text{ x } R_L}{\frac{R_0}{1+\beta A_V} + R_L} = \frac{R_o R_L}{R_o + R_L + \beta.A_v R_L}$$

6.9.2 Current Shunt Feedback

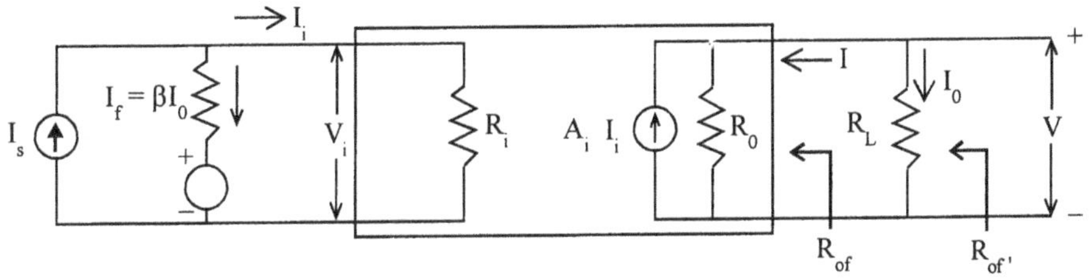

Fig. 6.20 Block schematic for current shunt feedback amplifier.

$\therefore$ $I_o = -I$

A_i = Shunt circuit current gain

A_I = Practical current gain $\left(\frac{I_o}{I_i}\right)$

$I = \frac{V}{R_o} - A_i I_i$

With, $I_S = 0,\ I_i = -I_f = -\beta I = +\beta I,$

$I = \frac{V}{R_o} - \beta A_i I$ or $I(1+\beta A_i) = \frac{V}{R_o}$

$R_{of} = \frac{V}{I} = R_o(1+\beta A_i)$

A_i = short circuit current gain.

R_{of}' includes R_L as part of the amplifier.

$$R'_{of} = \frac{R_{of}.R_L}{R_{of} + R_L}$$

$$= \frac{R_o(1+\beta A_i)R_L}{R_o(1+\beta A_i)+R_L}$$

$$= \frac{R_o R_L}{R_o + R_L} \times \frac{1+\beta A_i}{\dfrac{1+\beta A_i R_o}{R_o + R_L}}$$

$$A_I = \frac{I_o}{I_i} = \frac{A_i R_o}{R_o + R_L}$$

$$R'_o = \frac{R_o.R_L}{R_o + R_L}$$

$\therefore$ $$R'_{of} = R'_o \frac{1+\beta A_i}{1+\beta A_I}$$

If, $R_L = \infty$,

$A_I = 0,\ R'_o = R_0$

$\therefore$ $$\boxed{R'_{of} = R_0\,(1+\beta A_i)}$$

Problem 6.1

The following information is available for the generalized feedback network. Open loop voltage amplification $(A_v) = -100$. Input voltage to the system $(V_i') = 1\text{mV}$. Determine the closed loop voltage amplification, the output voltage, feedback voltage, input voltage to the amplifier, and type of feed back for (a) $\beta = 0.01$, (b) $\beta = -0.005$ (c) $\beta = 0$ (d) $\beta = 0.01$.

Solution

$$A'_v = \text{closed loop voltage amplification} = \frac{A_v}{1-\beta_v A_v}$$

Sign must be considered,

$$A'_V = \frac{-100}{1-0.01(-100)} = -50.$$

$\because$ It is positive feedback, A'_v is less negative $\therefore$ there is increase in gain.

V_o Output voltage $= V_i A_V$

$= -50 \times 10^{-3}$ mV

$= -50$ mV.

V_x Feedback voltage $= \beta V_o = 0.01\,(-50 \times 10^{-3})$

$= -0.5$ m V.

$V_i = V_i' + \beta_v.\ V_o$

$= 10^{-3} + (-0.5 \times 10^{-3})$

$= 0.5$ m V.

This is negative feedback because, $V_i < V_i'$.

$\therefore \quad |A'_V| < |A_v|$.

Problem 6.2

In the above problem determine the % variation in A'_V resulting from 100 % increase in A_v when $\beta_v = 0.01$.

When $A_v = -100$, $A'_V = -50$.

Solution

If A_v increases by 100 %, Then new value of $A_v = -200$.

$$A'_V = \frac{A_v}{1-\beta_v A_v}$$

$$A'_V = \frac{A_v}{1-\beta_v A_v}$$

Now, $$A'_V = \frac{-200}{1-0.01(-200)} = -66.7.$$

Change in A'_V is $-66.7 - 50 = 16.7$

$$\therefore \quad \text{Variation} = \frac{16.7}{50} \times 100 = 33.3\%$$

Problem 6.3

An amplifier with open loop voltage gain $A_v = 1000 \pm 100$ is available. It is necessary to have an amplifier where voltage gain varies by not more than $\pm$ 0.1 %

(a) Find the reverse transmission factor β of the feedback network used

(b) Find the gain with feedback.

Solution

(a) $$\frac{dA_f}{A_f} = \frac{1}{1+\beta A} \cdot \frac{dA}{A}$$

or $$\frac{0.1}{100} = \frac{1}{(1+\beta A)} \cdot \frac{100}{1000}$$

$\therefore \quad (1+\beta A) = 100$ or $\beta A = (100-1) = 99$.

Hence, $$\beta = \frac{99}{1000} = 0.099$$

(b) $$A_f = \frac{A}{1+\beta A} = \frac{1000}{1+99} = 10$$

Problem 6.4

A gain variation of + 10 % is expected for an amplifier with closed loop gain of 100. How can this variation be reduced to + 1 %.

Solution

$$S' = \frac{S}{1-\beta A_v}.$$

for positive feedback,

$$S' = \frac{S}{1-\beta A_v}$$

$$S' = 0.01, \quad S = 0.1$$

for negative Feedback, $(1+\beta A_V) = \frac{0.1}{0.01} = 10 = \frac{S}{S'}$

$$A'_V = 100 = \frac{A_v}{1+\beta A_v} = \frac{A_v}{10}$$

$\therefore$ $$A'_V = 100 \times 10 = 1000.$$

$$1 + \beta A_v = 10\,; \qquad \therefore \quad \beta.A_v = 10 - 1 = 9$$

$\therefore$ $$\beta = \frac{9}{A_v} = \frac{9}{1000} = 0.009$$

$\therefore$ By providing negative feedback, with $\beta = 0.009$, we can improve the stability to 1%.

Problem 6.5

An amplifier with $A_v = -500$, produces 5% harmonic distortion at full output. What value of β is required to reduce the distortion to 0.1% ? What is the overall gain?

Solution

$$D' = \frac{D}{1+\beta A_v} \quad \text{(for negative feedback)}$$

$$D = 5; \quad D' = 0.1$$

$$N' = \frac{N}{(1+\beta A_v)}$$

$$0.1 = \frac{5}{1+\beta\,[500]}$$

or $$\beta = \frac{4.9}{50} = 0.098.$$

$$A'_V = \frac{-A_v}{1+\beta A_v} = -\frac{500}{50} = -10.$$

$$\frac{D'}{D} = \frac{A'_v}{A_v}.$$

$$\therefore \quad A'_V = A_v\left(\frac{0.1}{5}\right) = -10$$

Problem 6.6

An amplifier has voltage gain of 10,000 with ground plate supply of 150 V and voltage gain of 8000 at reduced rate supply of 130 V. On application of negative feedback. The voltage gain at normal plate supply is reduced by a factor of 80. Calculate the 1) voltage gain of the amplifier with feedback for two values of plate supply voltage. 2) Percentage reduction in voltage gain with reduction in plate supply voltage for both condition, with and without feedback.

Solution

$$A = \frac{V_o}{V_s}. \text{ When } V_s = 150 \text{ V and } A = 10{,}000.$$

$$\therefore \quad V_o = 10{,}000 \times 150 = 15 \times 10^5 \text{ V. (at normal supply voltages)}$$

$$\beta = \frac{V_f}{V_o}. \ A_{fb} = \frac{A}{1+\beta A}$$

But $$A_{fb} = \frac{10{,}000}{80}$$

$\because$ it is reduced by a factor of 80 with feedback. $\left(\frac{1}{80}^{Th} \text{ of } 10{,}000\right)$

$$\therefore \quad \frac{10{,}000}{80} = \frac{10{,}000}{1+\beta\times 10000}$$

$$1 + \beta\,(10{,}000) = 80$$

$$\therefore \quad \beta = +\frac{79}{10^4}$$

$$= 0.0079$$

V_o when $V_s = 130$ V is gain $\times\ V_s = 8000 \times 130 = 104 \times 10^4$ V

Voltage gain of amplifier with feedback when $V_s = 150$V,

$$A_{fb_1} = \frac{10{,}000}{80} = 125.$$

$$V_s = 130 \text{ V.}$$

$$A_{fb_2} = \frac{8000}{1+0.0079\times 8000} = 124.7$$

$$\% \text{ stability of gain without feedback } = \frac{10{,}000 - 8000}{10{,}000} \times 100 = 20\%$$

$$\% \text{ stability of gain with feedback } = \frac{125 - 124.7}{125} \times 100 = 0.24\%$$

$\therefore$ With negative feedback stability is improved.

6.10 ANALYSIS OF FEEDBACK AMPLIFIERS

When an amplifier circuit is given, separate the basic amplifier block and the feedback network. Determine the gain of the basic amplifier A. Determine the feedback factor β. Knowing A and β of the feedback amplifier, the characteristics of the amplifiers R_i, R_o, noise figure, A_f etc., can be determined .

Complete analysis of the feedback amplifier is done by the following steps :

Step I : First identify whether the feedback signal X_f is a voltage feedback signal or current feedback signal. If the feedback signal X_f is applied in shunt with the external signal, ***it is shunt feedback***. If the feedback signal is applied in series, it is ***series feedback***.

Then determine whether the sampled signal X_o is a voltage signal or current signal. If the sampled signal X_o is taken between the output node and ground, it is voltage feedback.

If the sampled signal is taken from the output loop, it is current feedback.

Step II :

2. Draw the basic amplifier without feedback
3. Replace the active device (BJT or FET) by proper model (hybrid – π equivalent circuit or ***h-parameter*** model)
4. Indicate V_f and V_0 in the circuit.
5. Calculate $\beta = V_f / V_0$
6. Calculate 'A' of basic amplifier.
7. From A and β, calculate $A_f = \dfrac{A}{1+\beta A}$; R_{if} and R_{of}.

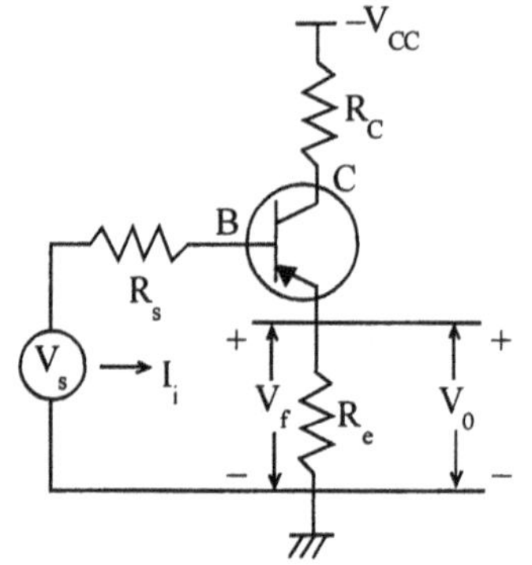

Fig. 6.21 Circuit for voltage series feedback.

6.10.1 VOLTAGE SERIES FEEDBACK

Amplifier circuit is shown in Fig. 6.22.

This is emitter follower circuit because output is taken across R_e. The feedback signal is also across R_e. So $V_f = V_o$. because feedback signal is proportional to output voltage. If V_o increases V_f also increase and if V_o, V_f also decreases because $V_f \alpha V_o$. So it is ***Voltage feedback***.

Now the drop across R_e, i.e. V_f opposes the input voltage. It reverse biases the feedback in V_f coming in series with V_{BE} and it opposes it. So it is negative ***series voltage feedback***.

In the current series feedback, V_f is the voltage across R_e but, output is taken across R_c or R_L and not Re. So in that case $V_f \propto I_o$ or I_c or not V_o. But in this case, it is emitter follower circuit. Output is taken across R_e and that itself is the feedback signal V_f.

Let us draw the base amplifier without feedback.

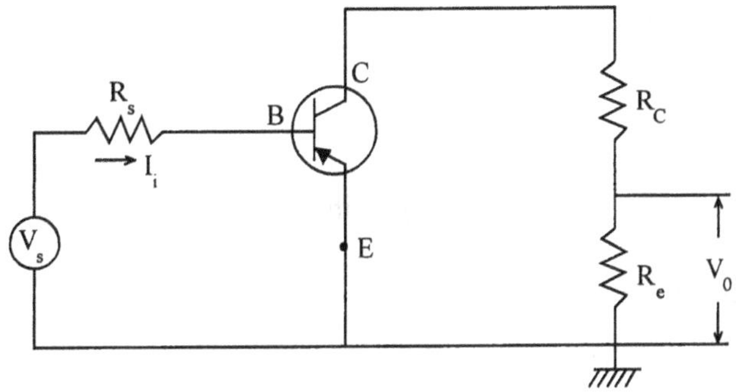

Fig. 6.22 Simplified circuit.

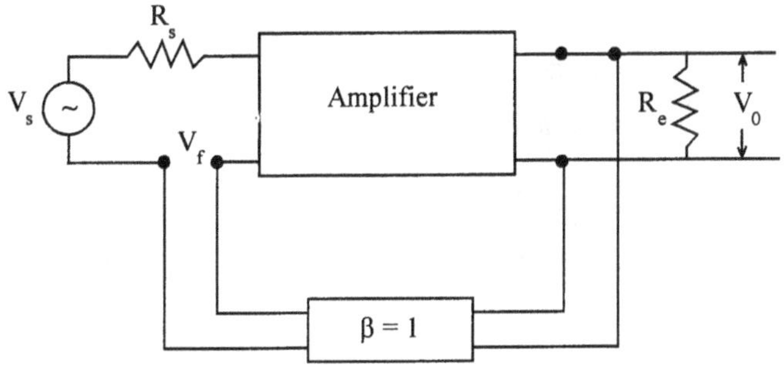

Fig. 6.23 Block schematic.

EQUIVALENT CIRCUIT

Now replace the transistor by its low frequency h-parameter equivalent circuit.

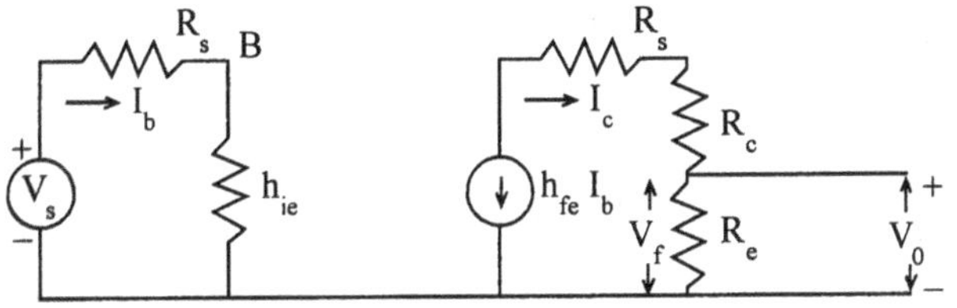

Fig. 6.24 Block Schematic

$$V_o = V_f$$

$$\therefore \quad \boxed{\beta = \frac{V_o}{V_f} = 1.}$$

V_S is considered as part of the amplifier. So $V_i = V_S$.

$$A_v = \frac{V_o}{V_i} \cdot V_o = h_{fe} \cdot I_b \cdot R_e$$

$I_C \simeq I_e$; $I_C = h_{fe} \cdot I_b$ ($\because$ It is voltage source, R_s in services with h_{ie})

$V_i = (R_s + h_{ie})\, I_i$

$I_i = I_b$

$$A_v = \frac{h_{fe}\, I_b\, R_e}{I_b (R_s + h_{ie})}$$

$$\therefore \quad A_v = \frac{h_{fe} . R_e}{R_s + h_{ie}} = \text{Voltage gain without feedback}$$

A'_V : Voltage gain with feedback.

$$A'_V = \frac{A_v}{1 + \beta A_v}$$

$\beta = 1$

$$A_V = \frac{h_{fe} . R_e}{R_s + h_{ie}}$$

$$= \frac{\dfrac{h_{fe} \cdot R_e}{R_S + h_{ie}}}{1 + \dfrac{1.h_{fe} \cdot R_e}{R_S + h_{ie}}}$$

$$A'_V = \frac{h_{fe} . R_e}{R_s + h_{ie} + h_{fe} R_e}$$

$h_{fe}\, R_e >> (R_s + h_{ie})$

$\therefore$ $A_{Vf} \cong 1.$ or <1, which is true for emitter follower.

R'_i = Input resistance without feedback is $R_s + h_{ie}$.

R'_i = Input resistance with feedback

$R'_i = R_i\,(1+\beta\, A_v)$ (voltage Feedback increase input resistance)

$\beta = 1,$

$$A_V = \frac{h_{fe} . R_e}{R_s + h_{ie}}$$

$$\therefore \quad R'_i = (R_s + h_{ie}) \left(1 + \frac{h_{fe} . R_e . 1}{R_s + h_{ie}}\right)$$

$$R_i' = \frac{(R_S + h_{ie})(R_S + h_{ie} + h_{fe}R_e)}{(R_S + h_{ie})}$$

$$\boxed{R_i' = R_s + h_{ie} + h_{fe} R_e}$$

R_i' : Output resistance with feedback

Output resistance of the circuit is R_e. Because output taken across R_e and not R_c and ground. Load resistance is also R_e.

∴ Considering load resistance,

$$R_0' = \frac{R_o'}{1+\beta A_v}$$

$$R_0' = \frac{R_e}{1+\dfrac{1.h_{fe}.R_e}{R_s + h_{ie}}}$$

Problem 6.7

Calculate A_{Vf}, R_{of} and R_{if} for the voltage series feedback for the circuit shown in Fig. 6.25.

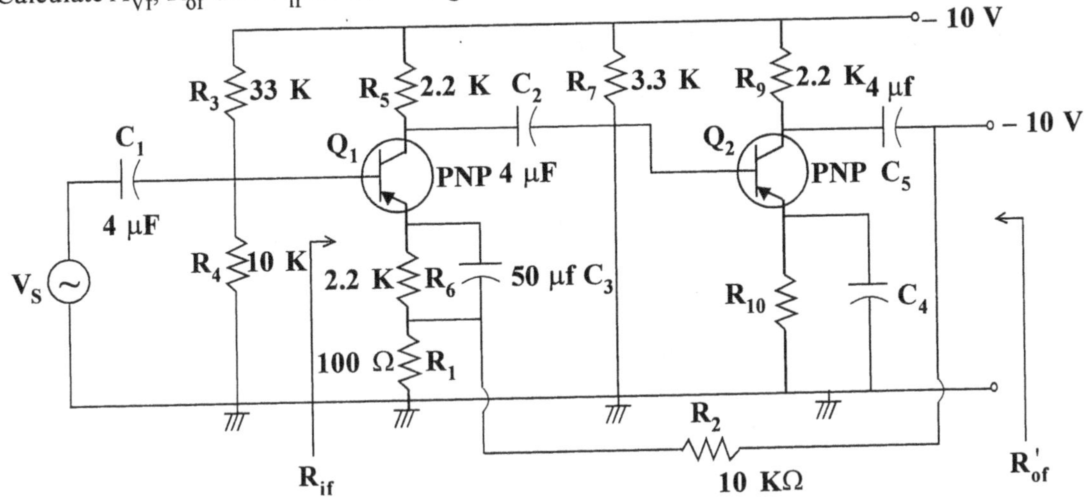

Fig. 6.25 For Problem 6.7.

Assume, $R_s = 0$, $h_{fe} = 50$,
$h_{ie} = 1.1$ kh_{re}
$= h_{oe} = 0$ and identical transistor.

The second collector is connected to the first emitter through the voltage divider R_1 R_2. Capacitor C_1, C_2, C_5 are DC blocking capacitors. C_3 and C_4 are bypass capacitors for the emitter resistors. All these capacitances represent negligible reactance at high frequencies.

Solution

Voltage gain without feedback $A_v = A_{v1} \times A_{v2}$.

Load resistance, R'_{L1}

R_5 in parallel with R_7 ($\because$ at high frequency X_{C2} is negligible), in parallel with

R_8, and h_{ie2} $R'_{L1} = 2.2 \parallel 33 \parallel 4.3 \parallel 1.1\ k\Omega = 980\Omega.$

$$R'_L = R_5 \parallel R_7 \parallel R_8 \parallel h_{ie_2}$$

Effective load resistance R'_{L2} of transistor Q_2 is its collector resistance R_9 in parallel with

$(R_1 + R_2)$; $R'_{L2} = R_9 \parallel (R_1 + R_2)$

$\therefore$ $R'_{L2} = 2.2\ k\Omega \,||^{le}$ with $10.1 k\Omega \simeq 2\ k\Omega$

Effective emitter impedance R_e of Q_1 is R_1 parallel with R_6.

$R_{e1} = R_1 \,||^{le} R_6$ $\because$ C_3 is short for AC

$R_e = 0.1 \,||^{le}$ with $2.2\ k\Omega = 0.098 k\Omega = 98\Omega$

The voltage gain A_{V_1} of Q_1 for a common emitter transistor with emitter resistance

is
$$A_{V_1} = \frac{V_1}{V_i} = \frac{-h_{fe}R'_L}{h_{ie} + (1 + h_{fe})R_e}$$

For Q_1 emitter is not at GROUND potential. So for A_{V_1} this formula must be used.

$$= \frac{-50 \times 0.980}{1.1 + (51) \times 0.098} \simeq -7.78$$

Voltage gain A_{v2} of transistor Q_2,

$$A_{v_2} = A_I \cdot \frac{R_L}{R_i} = -\frac{h_{fe}R_L}{h_{ie}}$$

$$A_{v_2} = -\frac{h_{fe} \cdot R_{L_2}}{h_{ie}} = -\frac{50 \times 2}{1.1} = -91$$

For Q_2 emitter is bypassed. So $R_e = 0$. $\therefore$ For A_v this formula is used.

Voltage gain A_v of the two stages is cascade without feedback

$$A_v = \frac{V_o}{V_i} = A_{V1} \times A_{V2} = 7.78 \times 91 \simeq 728$$

$$\beta = \frac{R_1}{R_1 + R_2} = \frac{100}{100 + 10{,}000} = \frac{100}{10100} = \frac{1}{101} \simeq 0.01$$

$A_v \times \beta = 728 \times 0.01 = 7.28$

$D = 1 + \beta A_v = 8.28$.

$$A_{Vf} = \frac{A}{1 + \beta A} = \frac{A_v}{D} = \frac{728}{8.28} \simeq 90$$

$$A_{vf} = \frac{A}{1 + \beta A} = \frac{A_v}{D} = \frac{728}{8.28} \simeq 90$$

Input resistance without external feedback

$$R_i = h_{ie} + (1+h_{fe})\,R_e = 1.1 + (1+50)\;0.098 \simeq 6.1\text{K}\,\Omega$$

$$R_i' = R_{if} = R_i(1+\beta A) = 6.1\;(1+7.28) = 50.5 \approx 51$$

Output resistance without feedback $R' = R'_{L_2} = 2\text{ k}\Omega$

Output resistance without feedback $R'_{of} = \dfrac{R_o}{1+\beta A} = \dfrac{2}{8.28} = 0.24\text{ K}\,\Omega$

Equivalent Circuit

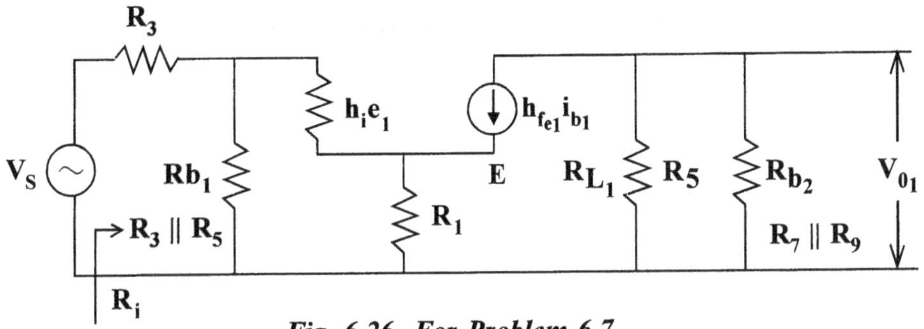

Fig. 6.26 For Problem 6.7.

BJT will not behave like a fixed resistor with h_{ie}, value . So circuit analysis is not simply parallel or series combination.

$$A_V = \frac{V_o}{V_i} = \frac{R_L}{R_s + r_{bb}'}$$

$$R_o = \frac{V_o}{R_L} = \frac{V_i}{R_s + r_{bb}'}$$

$$A_I = \frac{I_o}{I_i} = \frac{R_s}{R_s + r_{bb}'}$$

6.10.2 Current Shunt Feedback

The circuit is shown in Fig. 6.27.

Amplifier circuit parameters in terms of transistor h-parameters.

$$h_{ie} = R_i$$
$$h_{fe} = \beta$$
$$h_{oe} = R_o$$

The circuit shows two transistors Q_1 and Q_2 in cascade(See Fig.6.30). Feedback is provided from the emitter of Q_2 to the base of Q_1. This is negative feedback because, V_{i2} the input voltage to Q_2 is >> $V_{i1}.V_{i2}$ is out of phase with V_{i1}. V_{i2} >> V_{i1} because Q_1 is in Common emitter configuration. A_V is large. Also, V_{i2} is 180^0 out of phase with V_{i1}. Q_2 is emitter follower because emitter is not at ground potential. Voltage is taken across R_e. [This voltage follows the collector voltage. So it is emitter follower]. So $A_V < 1. \simeq 0.99$

∴ V_{e2} is slightly less than $V_{i2,}$ and there is no phase shift. [because emitter follows action]

∴ V_{e2} is in phase with V_{i2}.

V_{i2} is out of phase with V_{i1}.

∴ V_{e2} is out of phase with V_{i1} (180^0). So it is negative feedback

$V_{i2} >> V_{i1}$. $V_{e2} \simeq V_{i2}$.

$V_{e2} >> V_{i1}$.

If the input signal V_s increases, I_S' the input current from source also increases. If I_S' increases, I_f also increases ($\because$ V_{e2} increases as I_S' increases)

$I_i = I_S' - I_f$. (I_i is the base current for the transistor Q_1. So it is negative feedback)

This is current shunt feedback because,

$$I_f = \frac{V_{e_2} - V_i}{R'}$$

But $V_{i1} << V_{e2}$.

$$I_f \simeq + \frac{V_{e_2}}{R'}.$$

I_0 = collector current of Q_2. $V_{e2} = (I_0 - I_f) R_e$

$\simeq$ emitter current of Q_2.

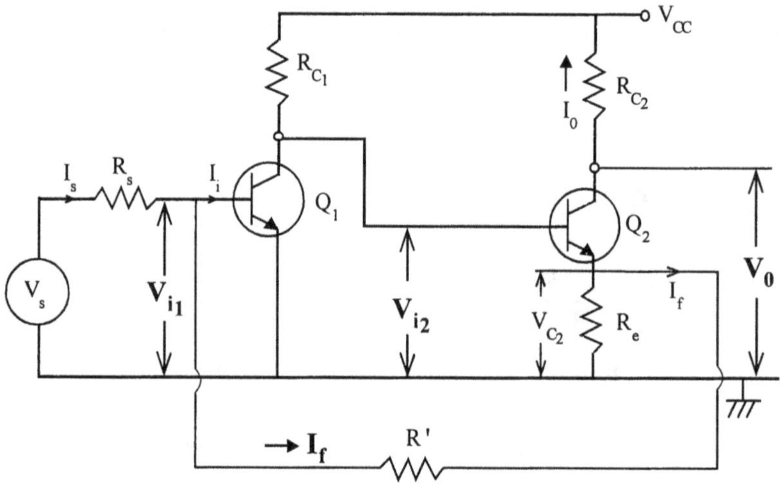

Fig. 6.27 Current shunt feedback.

Dividing by R'

$$\frac{V_{e2}}{R'} = I_f = \frac{(I_o - I_f)R_e}{R'}.$$

∴
$$I_f = + \frac{(I_o - I_f)R_e}{R'};$$

$$I_f + \frac{I_f.R_e}{R'} = \frac{I_o.R_e}{R'}$$

$$I_f\left(1+\frac{R_e}{R'}\right) = \frac{I_o.R_e}{R'}$$

$$I_f = \frac{R'.I_0.R_e}{(R'+R_e)R'} = \frac{R_e}{(R'+R_e)} I_0$$

$$\boxed{\beta = \frac{I_f}{I_o} = \frac{R_e}{R'+R_e}}$$

$$I_f \propto I_0$$

∴ This is current feedback

A_I' : Current gain with feedback.

$$A_I' = \frac{I_o}{I_s'}$$

$$I_S' = I_i + I_f \text{ which is small } \mu A \; I_i = I_b$$

∴ $$I_S' \simeq I_f$$

∴ $$A_I' = \frac{I_o}{I_f} = \frac{1}{\beta} \qquad \because \beta = \frac{I_f}{I_o}$$

∴ $$\boxed{A_I' = \frac{R'+R_e}{R_e}}$$

A_V' : Voltage gain with feedback.

$$\frac{V_o}{V_s} \text{ (with feedback)}$$

$$V_o = I_o . R_{c2}$$

$$V_s = I_s . R_s.$$

$$\frac{V_o}{V_s} = \frac{I_o.R_{c2}}{I_s R_s} \qquad I_S \simeq I_f$$

$$A_V' \simeq \frac{I_o}{I_f}.\frac{R_{c2}}{R_s} \qquad \frac{I_0}{I_f} = \frac{1}{\beta}$$

$$= \frac{1}{\beta}.\frac{R_{c2}}{R_s} \qquad \beta = \frac{R_e}{R'+R_e}$$

$$\boxed{A_V' = \frac{R'+R_e}{R_e}.\frac{R_{c2}}{R_s}}$$

We shall determine R_i', R_o' taking numerical values. The actual circuit, without feedback, considering R' can be drawn as shown in Fig. 6.28 To represent R' on the input side, with output terminals open circuited ($\therefore$ it is emitter follower configuration, output is taken across emitter and ground). When E_2 is left open (To make $I_o = 0$). R' is in series with R_{e2} and thus total resistance is between base B_1, and ground. Hence on the input side, R' is in series with R_{e2}.

To find the output resistance, the base of Q_1 is shorted to ground because one terminal of R' is connected to (input shorted looking the high output terminals) B_1 is at ground. R' or R_{e2} are in parallel.

$\therefore$ The circuit is as shown below, (Fig. 6.28)

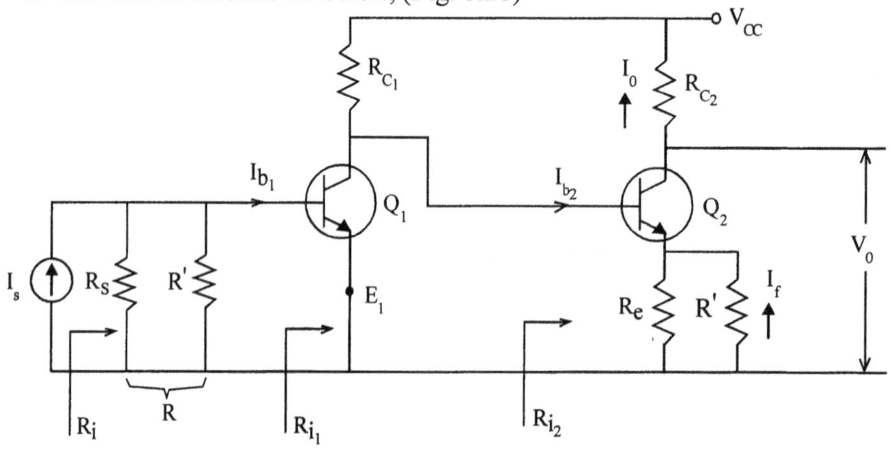

Fig. 6.28 Current shunt feedback without R'.

Because it is shunt feedback, we have shown this as a current source with R_s in parallel with I_S.

Problem 6.8

To find A_V', R_i', R_o'. A_I' for the circuit shown in Fig. 6.31.

$$R_{c1} = 3\text{ k}\Omega, \qquad R_{c2} = 500\Omega; \quad R_{e2} = 50\Omega$$

$$R' = R_s = 1.2\text{ k}\Omega \quad h_{fe} = 50.$$

$$h_{ie} = 1.1\text{ k}\Omega \qquad h_{re} = h_{re} = 0.$$

Solution

In this problem, output is taken at the collector 2 and not emitter 2. So the formulae derived earlier can be used directly.

$$A_I = \frac{-I_{c_2}}{I_s} = \frac{-I_{c2}}{I_{b2}} \cdot \frac{I_{b2}}{I_{c1}} \cdot \frac{I_{c1}}{I_{b1}} \cdot \frac{I_{b1}}{I_s}$$

(Multiplying and dividing by I_{b2}, I_{c1} and I_{b1})

$$\frac{-I_{c_2}}{I_{b_2}} = -h_{fe} = -50; \text{ (Emitter follower)}$$

$$\frac{I_{c_1}}{I_{b_1}} = + h_{fe} = +50 \text{ (common emitter configuration)}$$

$I_{b_2} \# I_{c_1}$,

$\because$ The current gets divided in the ratio of the resistances, current gets divided depending upon R_{c_1} and R_{i_2} of transistor Q_2.

$$\frac{I_{b2}}{I_{c1}} = \frac{-R_{c_1}}{R_{c_1} + R_{i_2}}$$

$$= \frac{-3}{3 + 3.55} = -0.457.$$

$$R_{i2} = h_{ie} + (1+h_{fe})(R_{e_2} \parallel R').$$

(For emitter follower configuration this is the input resistance).

$$= 1.1 + (50 + 1)\left[\frac{0.05 \times 1.20}{1.25}\right] \quad (R_{e2} = 0.05 \text{ k}\Omega)$$

$$= 3.55 \text{ k}\Omega$$

Let $\quad R = R_S$ in parallel with $(R' + R_{e_2})$

$$= \frac{1.2 \times 1.25}{1.2 + 1.25} = 0.612 \text{ k}\Omega$$

$$= \frac{I_{b_1}}{I_S} = \frac{R}{R + h_{ie}}$$

$$= \frac{0.61}{0.61 + 1.1} = 0.358\ \Omega$$

$\therefore$

$$A_I = (-50)(-0.457)(50)(0.358)$$

$$= +406$$

$$\beta = \frac{R_{e_2}}{R_{e_2} + R'}$$

$$= \frac{50}{1{,}250} = 0.04.$$

$$(1 + \beta A_i) = 1 + (0.040)(406) = 17.2$$

$$A_i' = \frac{A_I}{1 + \beta A_i} = \frac{406}{17.2} = 23.6$$

$$A_V' = \frac{V_o}{V_s} = \frac{-I_{c_2}.R_{c_2}}{I_s.R_s} = \frac{A_I'.R_{c_2}}{R_s}$$

$$= \frac{(23.6)(0.5)}{1.2} = 9.83$$

$$[A_V' \text{ is also} = \frac{R_{c_2}}{\beta R_s} = \frac{0.5}{(0.040)(1.2)} = 10.4]$$

R_i = Input resistance without feedback

= R in parallel with h_{ie}

$$= \frac{(0.61)(1.1)}{1.71} = 0.394 \text{ k}\Omega$$

$$R_i' = \frac{R_i}{1+\beta A_i} = \frac{394}{17.2} = 23.0\Omega$$

R_{oL} = Output resistance considering R_L,

$$= R_o \text{ in parallel with } R_{c2} = R_{c2} \quad \because R_o \text{ is large } \frac{1}{h_{ie}} = \alpha.$$

R_L' = with feedback considering R_L,

$$= \frac{R_{0L}(1+\beta A_i)}{1+\beta A_i} = R_0' = R_{c2} = 500$$

When we represent R on the input side and output side and calculate the value of A_I, it is not the current gain with feedback. Because, R' is represented on the input side leaving E_2 terminal open and on the output side shorting B_1 to ground (to make I_o and $I_s = 0$) . So thus A_I will not be the same if R' is actually connected between E_2 and B_1. We are taking into account the effect of R' and not the feedback effect. In practice for shunt or series feedback, the signal generator will act as a current source or voltage source. Therefore it is capable of supplying current or voltage required .In theory we assume ideal voltage and current sources. Therefore for shunt feedback we must have a current sources irrespective of input voltage. The current source will supply sufficient current to drive the resistance.

R_i = (1+9) 20 kΩ = 200 kΩ

R_s is negligible compared to 200 kΩ.

$$R_0 = \frac{r_o}{1+\beta A_o} = \frac{20\Omega}{10} = 2\Omega$$

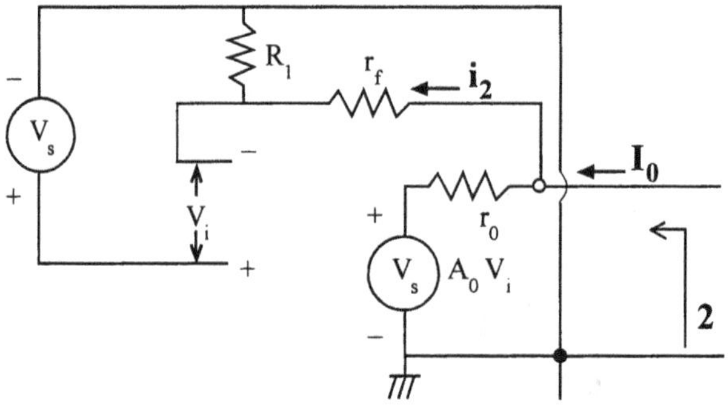

Fig. 6.29 Equivalent circuit.

6.10.3 Current Series Feedback

INPUT RESISTANCE (R_I')

$$V_i = V_i' - V_f$$

$$R_i' = \frac{V_i'}{I_i} = \frac{V_i + V_f}{I_i} = \frac{V_i}{I_i} + \frac{V_f}{I_i} \qquad \therefore \qquad V_i' = V_i + V_f.$$

$$\frac{V_i}{I_i} = R_i;$$

$V_f = I_e R_e$; $I_i = I_e - I_c$ [∵ negative current series feedback] (I_b)

$$\therefore \qquad R_i' = R_i + \frac{I_e R_e}{I_e - I_c}$$

$$R_i' = R_i + \frac{R_e}{1 - \frac{I_c}{I_e}}$$

Dividing II term by I_e.

$$\frac{I_C}{I_e} = \alpha = \frac{\beta}{1+\beta}$$

$$\frac{I_C}{I_e} = \frac{I_C}{I_C + I_b}$$

$$= \frac{I_c | I_b}{\frac{I_c}{I_b} + 1} = \frac{h_{fe}}{1 + h_{fe}}.$$

$$\therefore \qquad R_i' = R_i + \frac{R_e}{1 - \frac{h_{fe}}{1 + h_{fe}}}$$

$$R_i' = R_i + (1 + h_{fe}).R_e$$

But $\qquad R_i = h_{ie}$ for the transistor

$$\boxed{\therefore \; R_i' = h_{ie} + (1 + h_{fe})\, R_e}$$

If we consider the bias resistor also, R_1 and R_2 will come in parallel with R_i'

$$= R_i \parallel R_1 \parallel R_2$$

VOLTAGE GAIN (A_V')

$$A_v = \frac{A_i.R_L}{R_i}$$ [This is the general expression for A_V interms of A_i]

$$\therefore \qquad A_V' = \frac{A_i.R_L}{R_i'}$$

[With feedback, A_i will not change, R_L will not change, but R_i will be R_i']

$$A_i \simeq -h_{fe}$$

$$\therefore \quad A'_V = \frac{-h_{fe}.R_L}{h_{ie}+(1+h_{fe})R_e}$$

$$\therefore \quad A'_V < A_v$$

OUTPUT RESISTANCE (R'_o)

R without feedback $\simeq \dfrac{1+h_{fe}}{h_{oe}}$

This will be very large.

Taking into effect, the feedback,

$$R'_o = R_o \parallel R_L$$

$$\frac{1}{R_o} = h_{oe} - \frac{h_{fe}h_{re}}{h_{ie}+R_s}$$

ACTUAL EXPRESSION FOR β THE FEEDBACK FACTOR :

With negative feedback, $R'_i = R_i(1+\beta.A_v)$

$$A_v = \frac{-h_{fe}.R_L}{R_i};$$

$$\therefore \quad R'_i = R_i\left[1-\frac{h_{fe}.R_L}{R_i}.\beta\right]$$

$$= R_i - h_{fe}.R_L.\beta \qquad .....(6.7)$$

The expression we get for

$$R'_i = R_i + (1+h_{fe}).R_e \qquad .....(6.8)$$

Comparing (6.7) and (6.8), we find that

$$\beta = \frac{(1+h_{fe})}{h_{fe}} \cdot \frac{R_e}{R_L}$$

This is the actual expression for β

$$\frac{1+h_{fe}}{h_{fe}} \simeq 1.$$

$$\therefore \quad \beta \simeq \frac{R_e}{R_L}$$

6.10.4 CURRENT SERIES FEEDBACK (TRANS CONDUCTANCE AMPLIFIER)

Consider the circuit shown in Fig. 6.30 output is taken between collector and ground. The drop across R_e is the feedback signal V_f. The sampled signal is the load current I_o and not V_o. This is current series feedback. Because V_f is in series with V_i. It is current series feedback.

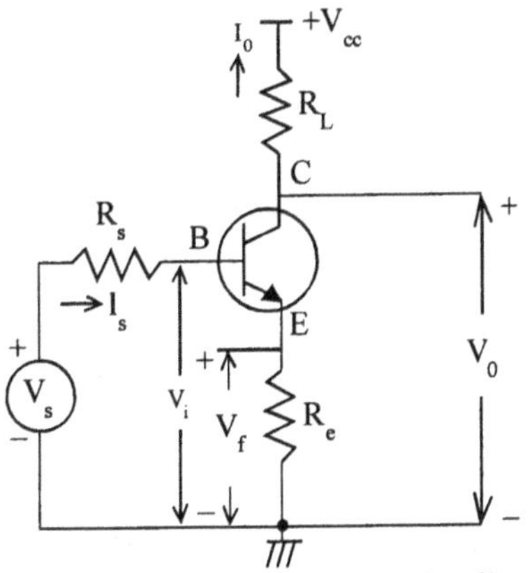

Fig. 6.30 Current series feedback.

$$V_f = I_e R_e.$$
$$I_e \simeq I_c = I_o.$$
$\therefore$ $$V_f = I_o R_e.$$
$\because$ R_e is constant,
$$V_f \propto I_o.$$
[β must be independent of R_s or R_L]

$$\beta = \frac{V_f}{V_o} = \frac{-I_o.R_e}{I_o.R_L} = -\frac{R_e}{R_L}$$

The basic amplifier circuit without feedback is shown in Fig.6.31 below,

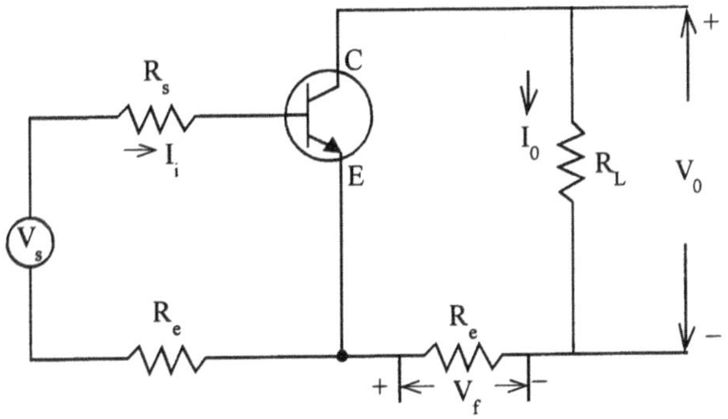

Fig.6.31 Circuit without feedback.

The equivalent circuit when the active device is replaced by ***h-parameter*** circuit is shown in Fig.6.32.

This has to be considered as Transconductance Amplifier, Since, β is taken as $\frac{V_f}{V_o}$.

It depends, on R_L, because for this circuit, G_M is considered.

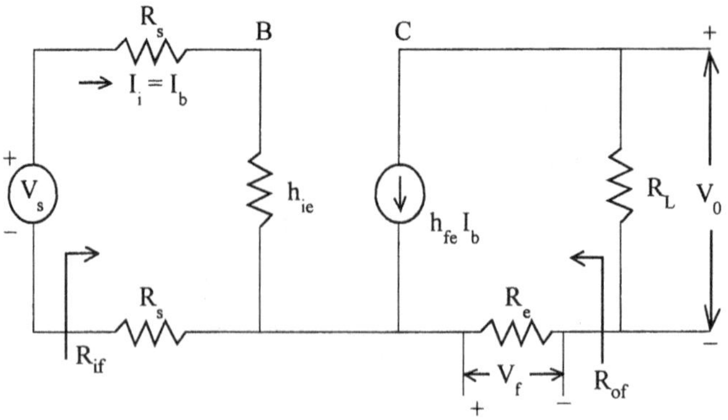

Fig. 6.32 h - parameter equivalent circuit.

$$\beta = \frac{V_f}{I_o} = \frac{-I_o.R_e}{I_o} = -R_e.$$

($\because$ the sampled signal is I_o and not V_o.)

$$G_M = \frac{I_o}{V_i} \quad I_0 = -h_{fe}.I_b \text{ (Input current gain)}$$

Since, $V_i = I_b(R_s + h_{ie} + R_e)$

$\therefore$
$$G_m = \frac{-h_{fe}.I_b}{I_b(R_s + h_{ie} + R_e)}$$

$$= \frac{-h_{fe}}{R_s + h_{ie} + R_e}$$

$$D = 1 + \beta.G_M = 1 + \frac{h_{fe}.R_e}{R_s + h_{ie} + R_e}$$

$$= \frac{R_s + h_{ie} + (1 + h_{fe})R_e}{R_s + h_{ie} + R_e}$$

$$G_{Mf} = \frac{G_M}{D} = \frac{-h_{fe}}{R_s + h_{ie} + (1 + h_{fe})R_e}$$

Voltage gain
$$A_{vf} = \frac{I_o.R_L}{V_s} = \frac{I_o}{V_s} = G_{Mf}$$

$\therefore$
$$A_{vf} = G_{Mf}.R_L$$

$$= \frac{-h_{fe}.R_L}{R_s + h_{ie} + (1 + h_{fe})R_e}$$

$$(1+h_{fe})R_e >> R_s + h_{ie}.$$

Since h_{fe} is a large quantity.

$\therefore$ Denominator $\simeq (1+h_{fe})R_e$.

$$h_{fe} >> 1.$$

$\therefore$ Denominator $\simeq h_{fe} \cdot R_e$.

$$\therefore \quad \boxed{A_{vf} = \frac{-h_{fe}R_L}{h_{fe}.R_e} = -\frac{R_L}{R_e}}$$

$R_i = R_s + h_{ie} + R_e$ (without feedback).

$R_{if} = R_i\,D = R_s + h_{ie} + (1 + h_{fe})\,R_e$.

6.10.5 VOLTAGE SHUNT FEEDBACK (TRANS RESISTANCE AMPLIFIER)

This is ***voltage shunt feedback*** because, the feedback current through R_B is proportional to the output voltage or it is in shunt with the input. So it is voltage shunt feedback. [We are not interested whether voltage is fed or current is fed. But the feedback signal is proportional to output voltage. ***So it is voltage feedback***]

The circuit can be written as, shown in Fig. 6.33.

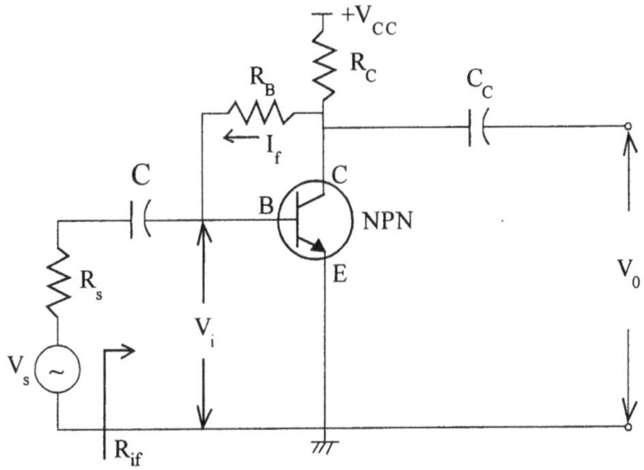

Fig. 6.33 Circuit for voltage shunt feedback.

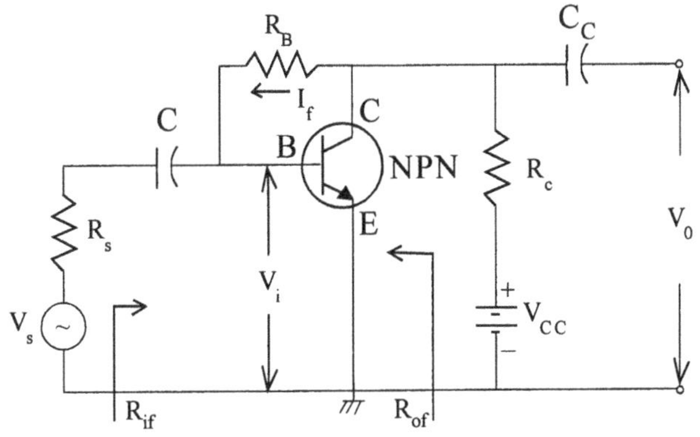

Fig. 6.34 Redrawn circuit for voltage shunt feedback.

The low frequency ***h-parameter*** equivalent circuit is, shown in Fig. 6.35.

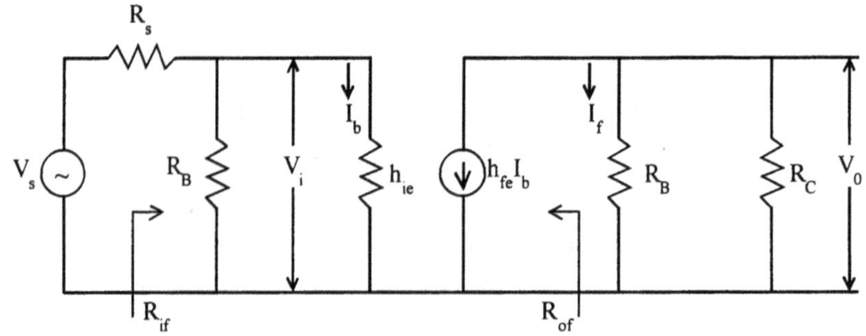

Fig. 6.35 h-parameter equivalent circuit.

$V_o >> V_i$ $\quad \because$ Amplification is there.

$$I_f \simeq \frac{-V_o}{R_B},$$

$$\therefore \quad I_f \simeq \frac{-V_o}{R_B};$$

$$\therefore \quad \beta = \frac{-I_f}{V_o}$$

$$\beta = \frac{-1}{R_B}$$

$$A_V = \frac{V_o}{V_i} = \frac{-h_{fe}.R_L'}{h_{ie}} \text{ without feedback} \left(\frac{A_I R_L}{R_v}\right)$$

$$R_L' = R_B \parallel R_C.$$

$$A_V' = \frac{V_o}{V_s} \text{ (with feedback)}$$

$$\boxed{A_V' = A_V \cdot \frac{R_i'}{R_s + R_i'}}$$

R_i' = Input resistance with feedback

$$A_{Vf} = \frac{V_o}{V_s}$$

$$= \frac{V_o}{I_s.R_s} \simeq \frac{1}{\beta R_s} = \frac{-R_B}{R_s}$$

R_i' :

$$I_s = I_f + I_b,\ I_b \text{ is negligible}$$

$$\therefore \quad I_s \simeq I_f \quad \text{or} \quad \beta = \frac{-1}{R_B}.$$

$$R_i' = h_{ie} \text{ in parallel with } \frac{R_B}{1 - A_v}$$

Transresistance $R_M = \dfrac{V_o}{I_s} = \dfrac{-I_o.R_c}{I_s} = \dfrac{-h_{fe}.I_b.R_c}{I_s}$

$$R_C' = R_c \parallel R_B;\ I_s = \frac{(R + h_{ie})}{R} I_b. \qquad \text{Where } R = R_s \parallel R_B.$$

$$\therefore \quad R_M = \frac{-h_{fe}.R_c'.R}{(R + h_{ie})};$$

$$I_b = \frac{-h_{fe}.R_e'.R}{(R + h_{ie})}$$

$$R_M' = \frac{R_M}{1 + \beta R_M}.$$

$$R_i = \frac{R \times h_{ie}}{R + h_{ie}}$$

$$R_i' = \frac{R_i}{1 + \beta R_M}$$

$$R_o' = \frac{R_B}{R_s} \times \frac{R_s + h_{ie}}{h_{fe}}$$

Problem 6.9

For the circuit shown, $R_c = 4\ k\Omega$, $R_B = 40$ k, $R_s = 10$ k, $h_{ie} = 1.1\ k\Omega$, $h_{fe} = 50$, $h_{re} = h_{oe} = 0$. Find A_{vf}, R_{if}.

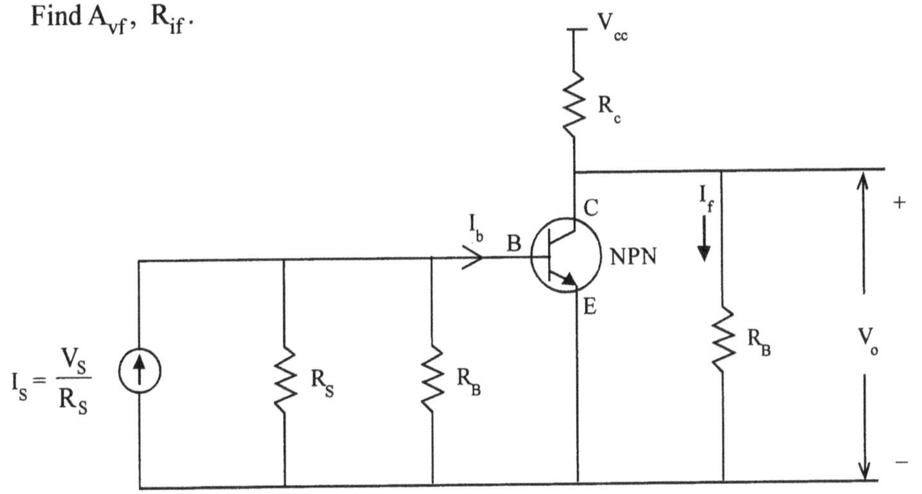

Fig.6.36 For Problem 6.9.

Solution

$$R'_C = R_c \,||^{le}\, R_B = \frac{4 \times 40}{4 + 40} = 3.64 \text{ k}\Omega$$

$$R = R_s \,||^{le}\, R_B = \frac{10 \times 40}{10 + 40} = 8 \text{ k}\Omega$$

$$\text{Transresistance}\, R_M = \frac{V_o}{I_s} = \frac{-I_C.R_c'}{I_s} = \frac{-h_{fe}.I_b.R_c'}{I_s} \qquad \because\ I_c = h_{fe} \cdot I_b$$

$$I_s = \frac{(R + h_{ie})}{R} I_b; \qquad \therefore\ R_M = \frac{-h_{fe}.I_b\, R'_C \,.\, R}{(R + h_{ie}).I_b}$$

$$R_M = -\,160 \text{ k}$$

$$\therefore \qquad \beta = -\frac{-1}{R_B} = -\,0.025 \text{ mA|V}$$

$$R'_M = \frac{R_M}{1 + \beta R_M} = \frac{-160}{5.00} = -\,32.0 \text{ k}\Omega$$

$$A'_V = \frac{V_o}{V_s} = \frac{V_o'}{I_s.R_s}$$

$$= \frac{V_o'}{I_s} = R'_M$$

$$\therefore \qquad A'_V = \frac{R_M'}{R_s} = -\frac{32}{10} = -\,3.2\,.$$

$$R_i = \frac{R \times h_{ie}}{R + h_{ie}} = \frac{8 \times 1.1}{8 + 1.1} = 0.968 \text{ k}\Omega;$$

$$R'_i = \frac{R_i}{1 + \beta R_M} = \frac{968}{5.0} = 193\ \Omega$$

SUMMARY

- An Amplifier circuit is to provide voltage gain or current gain or both in the form of power gain. But the other desirable characteristics of the amplifier circuits are high Z_i, Low Z_o, Large B.W, low distortion, low noise and high stability. To achieve these characteristics a part of output signal is feedback and coupled to input, to oppose in phase with (V_i). So it is negative feedback. Though gain reduces due to negative feedback, it is employed in amplifier circuits to get the other advantages.
- Feedback factor $\beta = V_f/V_0$. The product βA is called return ratio. $(1 + \beta A)$ is called return difference D or desensitivity factor.
- The different types of feedback are (i) voltage series (ii) current series (iii) voltage shunt and (iv) current shunt.
- With voltage feedback (series or shunt) output resistance R_0 decreases.
- With current feedback (series or shunt) R_0 increases.
- With series feedback (current or voltage) R_i increases.
- With shunt feedback (current or voltage) R_0 decreases.
- With negative feedback, distortion, noise, gain reduce by a factor $(1 + \beta A)$. Bandwidth, f_2, stability improve by $(1 + \beta A)$.
- If the feedback signal is proportional to voltage, it is voltage feedback. If the feedback signal is porportional to current, it is current feedback. If the feedback signal V_f is coming in series with input signal V_i, it is series feedback. If V_f is in parallel with V_i it is shunt feedback.
- If the feedback signal V_f is out of phase with V_i, opposing it is negative feedback. If V_f is in phase with V_i, adding to it or aiding V_i, it is positive feedback.

OBJECTIVE TYPE QUESTIONS

1. The disadvantage of negative feedback is _______________.
2. The expression for sensitivity of an amplifier with negative feedback is _______________.
3. The expression for Desentivity of negative feedback amplifier is D = _______________.
4. The relation between bandwidth of an amplifier without feedback and with negative feedback is _______________.
5. Relation between upper cut-off frequency f_2' with negative feedback and f_2 without negative feedback is f_2' = _______________.
6. Negative feedback is also called as _______________.
7. For Ideal transconductance amplifier R_i = _______________. R_0 = _______________.

8. For practical transresistance amplifier, it is desirable that R_i is ______________ and R_o is ______________.
9. Voltage sampling is also known as ______________.
10. Characteristics of ideal voltage amplifier are : ______________.
11. Desirable characteristics of practical current amplifier are = ______________.
12. βA in feedback amplifier circuits is called ______________.
13. With voltage feedback, (series or shunt), output resistance R_o of an amplifier ______________.
14. With series feedback, (voltage or current), input resistance R_i of an amplifier ______________.
15. Expression for reverse transmission factor or feedback factor β = ______________.
16. Identify the type of feedback.

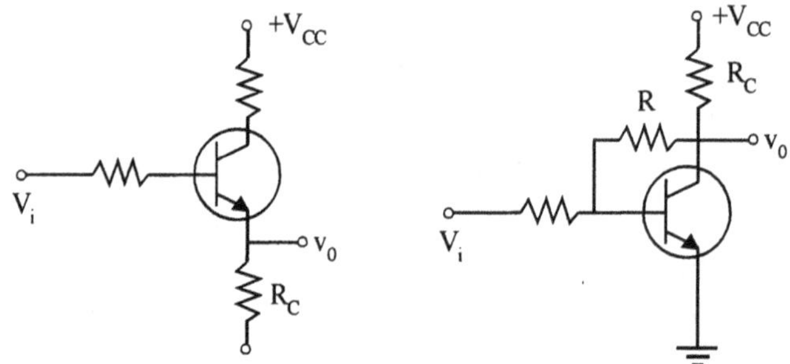

(i) (ii)

17. In the case of Voltage - series feedback, expression for output impedance with feed back z_{of} is ....
18. In the case of Voltage shunt feedback amplifier, expression for input impedance with feedback is Z_{if} = ...
19. In the case of current series feedback amplifier, expression for output impedance with feedback z_{of} is ....
20. Expression for z_{if} with current shunt feedback is ....

ESSAY TYPE QUESTIONS

1. Explain the concept of feedback as applied to electronic amplifier circuits. What are the advantages and disadvantages of positive and negative feedback ?
2. With the help of a general block schematic diagram explain the term feedback.
3. What type of feedback is used in electronic amplifiers ? What are the advantages of this type of feedback ? Prove each one mathematically.

4. Define the terms Return Ratio, Return Difference feedback factor, closed loop voltage gain and open loop voltage gain. Why negative feedback is used in electronic amplifiers eventhough closed loop voltage gain decreases with this type of feedback ?
5. Give the equivalent circuits, and characteristics of ideal and practical amplifiers of the following types (i) Voltage amplifier, (ii) Current amplifiers, (iii) Transresistance amplifier, (iv) Transconductance amplifier.
6. Derive the expression for the input resistance with feedback R_{if} (or R_i') and output resistance with feedback R_{of} (or R_i') in the case of
   (a) Voltage series feedback amplifier.
   (b) Voltage shunt feedback amplifier.
   (c) Current series feedback amplifier.
   (d) Current shunt feedback amplifier.
7. In which type of amplifier the input impedance increases and the output impedance decreases with negative impedance ? Prove the same drawing equivalent circuit.
8. Draw the circuit for Voltage series amplifier and justify the type of feedback. Derive the expressions for A_V', β, R_i' and R_o' for the circuit.
9. Draw the circuit for Current series amplifier and justify the type of feedback. Derive the expressions for A_V', β, R_i' and R_o' for the circuit.
10. Draw the circuit for Voltage shunt amplifier and justify the type of feedback. Derive the expressions for A_V', β, R_i' and R_0' for the circuit.
11. Draw the circuit for Current shunt amplifier and justify the type of feedback. Derive the expressions for A_V', β, R_i' and R_o' for the circuit.

MULTIPLE CHOICE QUESTIONS

1. Loop sampling is also known as

(a) Voltage sampling (b) Current sampling
(c) Power sampling (d) Node sampling

2. Positive feedback is also known as

(a) regenerative feedback (b) degenerative feedback
(c) Loop feedback (d) return feedback

3. Equation for feedback factor 'β' is ....

(a) $\frac{V_o}{V_f}$ (b) $\frac{V_o'}{V_f}$ (c) $\frac{V_f}{V_o'}$ (d) $\frac{V_f}{V_o}$

4. With negative feedback, the linearity of operation of the amplifier circuit

(a) deteriorates (b) improves (c) remian same (d) none of these

5. Expression for distortion D' with negative feedback, with usual notation is D' =

(a) $\frac{D}{1-\beta A}$ (b) D(1–bA) (c) $\frac{D}{1+\beta A}$ (d) D(1+bA)

6. Expression for densitivity of an amplifier circuit with negative feedback is,

(a) $D=(1-\beta A)$ (b) $D=\frac{1}{\beta A}$ (c) $D=(1+\beta A)$ (d) $D=\frac{1}{(1+\beta A)}$

7. Characteristics of ideal volt age amplifier are ...

(a) $R_i=\infty, R_o=0$ (b) $R_i=0, R_o=0$

(c) $R_i=0, R_o=\infty$ (d) $R_i=\infty, R_o=\infty$

8. The product of feedback factor 'β' and amplification factor 'A' is called ...

(a) return difference (b) return ratio

(c) series ratio (d) shunt ratio

9. Negative series feedback voltage or current ... input resistance

(a) decreases (b) no effect (c) can't be said (d) increases

10. With voltage feedback series or shunt, output ressitance R_0 ....

(a) decreases (b) increases (c) remains same (d) non of these

CONCEPTUAL QUESTIONS (Interview Questions)

1. Give physical example for a feed back systems.

Ans : A person searching for a book in Library is one example. The book seen by eyes and signal sent to brain act as closed loop system. If the eyes are closed, it is like open loop system, as no feed back can be given to the brains.

2. Give a practical example of a feed back system.

Ans : Controlling the temperature of a furnace is an example of feedback system. The temperature is sensed by a thermo couple and the signal is fed back to a controlling circuit. When set temperature is reached, the electric supply is turned off. When the temperature goes below the set value, electric supply is again turned on.

7 Oscillators

In this Chapter,

- Basic principle of oscillator circuits is explained. Generation of sinusoidal waveforms by the oscillator circuits without external A.C. input is explained.
- Barkhausen criteria to be satisfied for generation of oscillations is given.
- R-C phase shift oscillator circuit, Hartley oscillator, Colpitts oscillator, crystal oscillator, Resonant oscillator circuits are given.

7.1 OSCILLATORS

Oscillator is a source of AC voltage or current. We get A.C.output from the oscillator circuit. In alternators (AC generators) the thermal energy is converted to electric energy at 50Hz.

In the oscillator circuits that we are describing now, the electric energy in the form of DC is converted into electric energy in the form of AC. 'Invertors' in electrical engineering convert DC to AC, but there, only output power is the criterian and not the actual shape of the wave form.

An amplifier is different from oscillator in the sense that an amplifier requires some A.C. input which will be amplified. But an oscillator doesn't need any external AC signal. This is shown in Fig. 7.1 below :

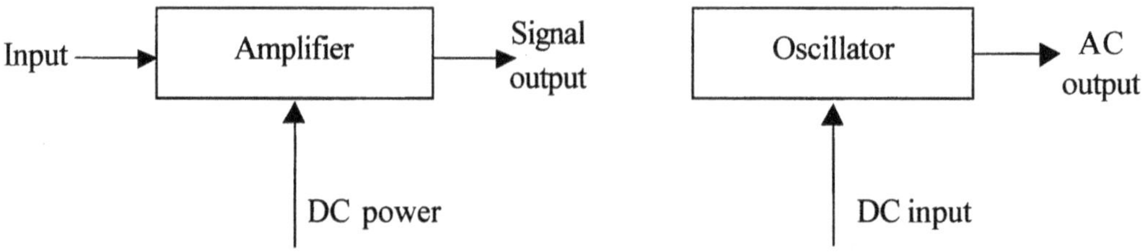

(a) Amplifier circuit *(b) Oscillator circuit*

Fig. 7.1

For an amplifier, the additional power due to amplification is derived from the DC bias supply. So an amplifier effectively converts DC to AC. But it needs AC input. Without AC input, there is no AC output. In the oscillator circuits also DC power is converted to AC. But there is no AC input signal. So the difference between amplifier and oscillator is in amplifiers circuits, the DC power conversion to AC is controlled by the AC input signal. But in oscillators, it is not so.

There are two types of oscillators circuits :

1. Harmonic Oscillators
2. Relaxation Oscillators.

Harmonic Oscillators produce sine waves. Relaxation Oscillators produce sawtooth and square waves etc. Oscillator circuits employ both active and passive devices. Active devices convert the DC power to AC. Passive components determine the frequency of oscillators.

7.1.1 Performance Measures of Oscillator Circuits

1. ***Stability :*** This is determined by the passive components. R,C and L determine frequency of oscillations. If R changes with T, f changes so stability is affected. Capacitors should be of high quantity with low leakage. So silver mica and ceramic capacitors are widely used.
2. ***Amplitude stability :*** To get large output voltage, amplification is to be done.
3. ***Output Power :*** Class A, B and C operations can be done. Class C gives largest output power but harmonics are more.

   Class A gives less output power but hormonics are low.
4. ***Harmonics :*** Undesirable frequency components are harmonics. An elementary sinusoidal oscillator circuit is shown in Fig.7.2.

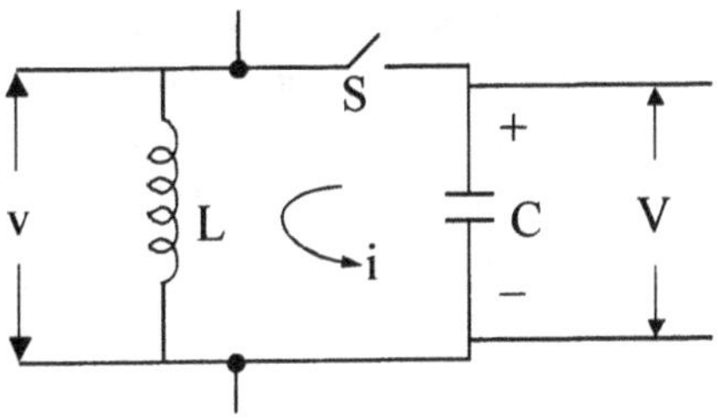

Fig. 7.2 LC Tank circuit.

L and C are reactive elements. They can store energy. The capacitor stores energy whenever there is voltage across its plates. Inductor stores energy in its magnetic field whenever current flows. Both C and L are lossless, ideal devices. So quality factor is infinity ∞. Energy is introduced into the circuit by charging capacitor to 'V' Volts. If switch is open, C cannot discharges because there is no path for discharge current to flow.

Suppose at $t = t_0$, switch 'S' is closed. Then current flows. Voltage across L will be V. At $t = t_0$, the Voltage across C is V volts. When switch is closed, current flows. So the charge across Capacitor C decreases. Voltage across C decreases, as shown in the waveform. So as the energy stored in Capacitor decreases, the energy stored in inductor L increases, because current is flowing through L Thus total energy in the circuit remains the same as before. When V across C becomes 0, current through the inductor is maximum. When the energy in C is 0, energy in L is maximum. Then the current in L starts charging C in the opposite directions. So at $t = t_1$ current in L is maximum and for $t > t_1$, the current starts charging C in the opposite direction. So V across C becomes negative as shown in the wavefrom. Thus we get sinusoidal oscillations from LC circuit.

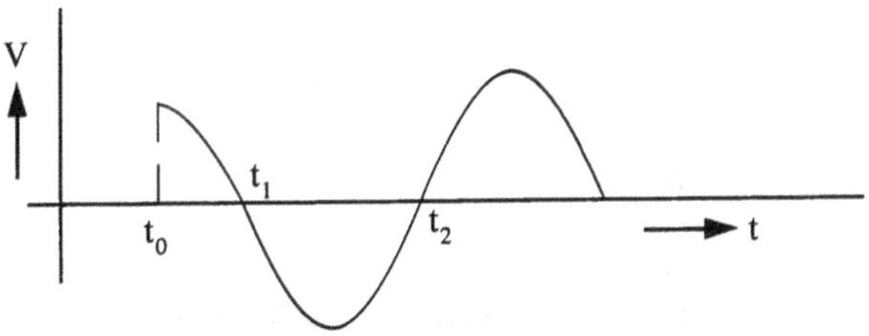

Fig. 7.3 Waveform.

Thus we are getting sinusoidal variations without giving any one input. What we have done is depositing some charge on C, so that the circuit operates on its own. But why should we get sinusoidal wave and not triangular or square wave ? Sinusoidal function is the only function that satisfies the conditions governing the exchange of the energy in the circuit.

But the above circuit is not a practical circuit. Because we have to take output from the circuit; i.e. energy has to be extracted, from the circuit. As we draw energy from the circuit, the energy stored in C and L decreases. So output also decreases or the voltage across C and L decreases so we get damped oscillation as shown below.

Fig. 7.4 Damped oscillations.

The energy lost by the elements must be replenished, so that oscillations are obtained continuously. If a negative resistance R is connected in the circuit, it replenishes, whatever energy that is lost in the circuit. Certain devices like tunnel diode, UJT, etc., exhibit -ve resistance. The energy supplied by the -ve resistance to the circuit actually comes from the DC bias supply.

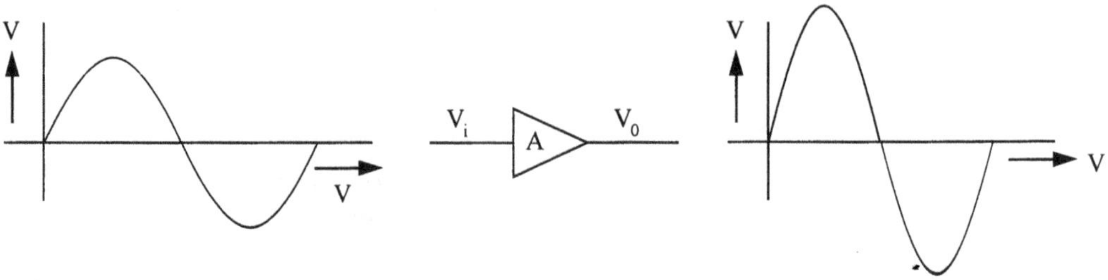

Fig. 7.5 Amplifier response.

Another method of producing sinusoidal oscillations is :

Suppose we have an amplifier with gain A_V and phase shift 180^0, $A_V >> 1$.

If we connect V_0 through a feedback network to V_i as shown in Fig.7.6 so that after feedback, the feedback signal $V_X = V_i$, and also, the feedback network provides 180^0, phase shift, after feedback the feedback signal V_X will be removed. Thus, without any input we get sinusoidal output.

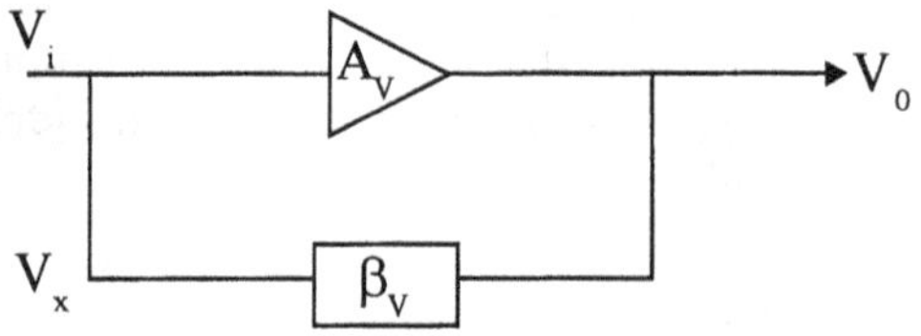

Fig. 7.6 Feedback network.

V_i is provided from the feedback signal of V_0 itself. To get initially Vo, the noise signal of transistor after switching itself is sufficient. Thus without external AC input we get sinusoidal oscillators.

$$\beta_v = \frac{V_x}{V_o}.$$

$$A_v = \frac{V_o}{V_i}.$$

$$V_X = V_i$$

$$\therefore \quad \beta . A_v = \frac{V_x}{V_o} \times \frac{V_o}{V_i}$$

$$= \frac{V_x}{V_i} \quad \text{or} \quad V_x = V_i$$

$$\therefore \quad \beta A_V = 1$$

Total phase shift = 360° (180 + 180). Therefore, to get sustained oscillations,

1. The loop gain must be unit 1.
2. Total Loop phase shift must be 0° or 360°. (Amplifier circuit produces 180^0 phase shift and feedback network another 180°.

7.2 SINUSOIDAL OSCILLATORS

Figure 7.7 shows an amplifier, a feedback network and input mixing circuit not yet connected to form a closed loop. The amplifier provides an output signal X_0, as a consequence of the signal X_i applied directly to the amplifier input terminal. Output of feedback network

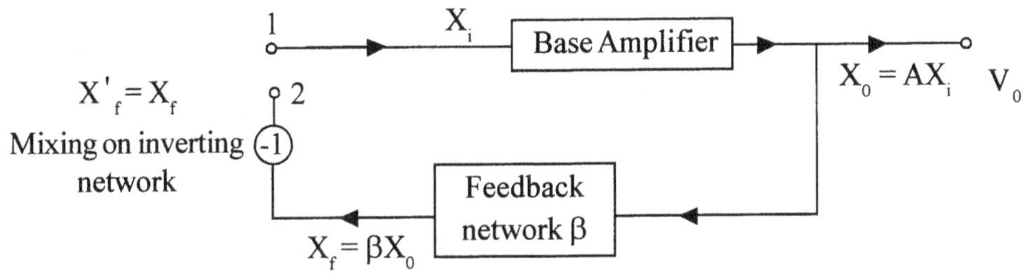

Fig. 7.7 Block schematic.

is $X_f = \beta X_o = A\beta X_i$ and the output of the mixing circuit,

is $X_f' = -X_f = -A\beta X_i$

$$\text{Loop gain,} \quad = \frac{X_f'}{X_i} = \frac{-A\beta X_i}{X_i} = -\beta A.$$

If X_f' were to be identically equal to X_i, input signal, $-\beta A$ should be = 1, so the output will be X_0, even if the input source is removed. The condition that $-\beta A = 1$ means, the loop gain must be 1.

7.3 BARKHAUSEN CRITERION

We make an asssumption that the circuit operates only in the linear region, and the amplifier feedback network contains reactive elements. For a sinusoidal wave form, if $X_i = X_f'$, the amplitude, phase and frequency of X_i and X_f' be identical. ***The frequency of a sinusoidal oscillator is determined by the condition that loop gain, Phase shift is zero at that frequency.***

For oscillator circuits positive feedback must be there i.e., V_f must be in phase with V_i to get added to V_i. When active device BJT or FET gives 180^0 phase shift, the feedback network must produce another 180^0 phase shift so that net phase shift is 0^0 or 360^0 and V_f is in phase with V_i to make it positive feedback.

Oscillations will not be sustained if, at the oscillator frequency the magnitude of the product of the transfer gain of the amplifier and of β are less than unity.

The conditions $-A\beta = 1$ is called ***Barkhausen criterion***

i.e. $|\beta A| = 1$ and phase of $-A\beta = 0$.

But $A_f = \dfrac{A}{1+\beta A}$. If $\beta A = -1$, than $A_f \to \infty$.

which implies that, there exists an output voltage even in the absence of an externally applied voltage.

A 1-V signal appearing initially at the input terminals will, after a trip around the loop and back to the input terminals, appear there, with an amplitude larger than 1V. This larger voltage will then reappear as a still larger voltage and so on. So if $|\beta A|$ is larger than1, the amplitude of oscillations will continue to increase without limit. But practically, to limit the increase of amplitude of oscillations, nonlinearity ability of the circuit will be set in. Though a circuit has been designed for $|\beta A| = 1$, since circuit components and transistor change characteristics with age, temperature, voltage etc. $|\beta A|$ will become larger or smaller than 1. If $\beta A < 1$, the oscillations will stop. If $\beta A >$ 1, the amplitude practically will increase. So, to achieve a sinusoidal osciallation practically $|\beta A| >$ 1 to 5%, so that with incidental variations in transistor circuit parameters, $|\beta A|$ shall not fall below unity.

The type of noise in electronic circuits and the causes are :

1. ***Johnson noise or Thermal Noise :*** Due to temperature.
2. ***Schottky noise or Shot noise :*** Because of variation is concentration in the Semiconductor Devices.

7.4 R – C PHASE-SHIFT OSCILLATOR (USING JFET)

This is voltage series feedback. FET amplifier is followed by three cascaded arrangements of a capacitor C, resistor R. The output of the last RC combination is returned to the gate. This forms the feedback connection. The FET amplifier shifts the phase of voltage appearing at the gate by 180^0. The RC network shifts the phase by additional amount. At some frequency, the phase-shift introduced by this network will be exactly 180^0. The total phase-shift at this frequency, from the gate around the circuit and back to the gate is $+180 - 180 = 0^0$. At this particular frequency, the circuit will oscillate. (Fig. 7.8).

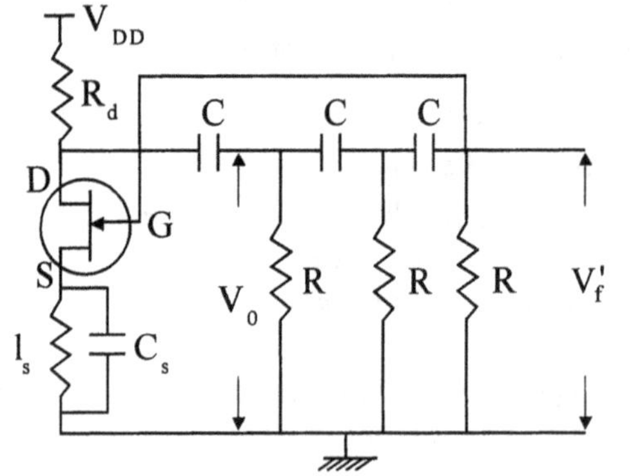

Fig. 7.8 JFET R – C phaseshift oscillator circuit.

$\therefore$ At that frequency, $V_f' = -V_f$ $\quad\because$ 180^0 out of phase.

The equivalent circuit is, show below in Fig. 7.9.

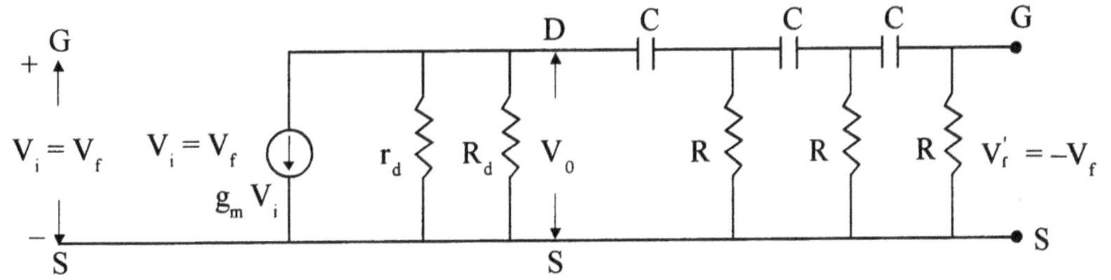

Fig. 7.9 Equivalent circuit.

The transformation of RC net work is $\frac{V_f}{V_o} = -\beta$; $\therefore$ $\beta = \frac{V_f}{V_0}$ and $V_f' = -V_f$

$$\therefore \qquad -\beta = \frac{V_f'}{V_o} = \frac{1}{1-5\alpha^2 - j(6\alpha - \alpha^3)}.$$

where $\alpha = \frac{1}{\omega RC}$.

The phase-shift of $\frac{V_f'}{V_o}$ is 180°, for $\alpha^2 = 6$ or

β must be real. Therefore $j(6\alpha - \alpha^3) = 0$.

$$\therefore \qquad f = \frac{1}{2\pi RC\sqrt{6}}. \quad \because \quad 6 = \frac{1}{\omega^2 R^2 C^2}.$$

$$\omega^2 = \frac{1}{6R^2C^2}.$$

$$\therefore \qquad \boxed{f = \frac{1}{2\pi RC\sqrt{6}}.}$$

when $\alpha^2 = 6, -\beta = \frac{1}{1-5\times 6 - 5(6\alpha - 6\alpha)} = \frac{-1}{29}$

or $\boxed{\beta = \frac{1}{29}.}$

In order that $|\beta A|$ is not les than unity, A should be atleast $|29|$. So select FET whose μ is atleast 29.

7.4.1 To find the β of the RC Phase-shift Network (JFET)

Each RC network introduces a phase-shift of 60°. Therefore, total phase-shift = 180°. (See Fig.7.10).

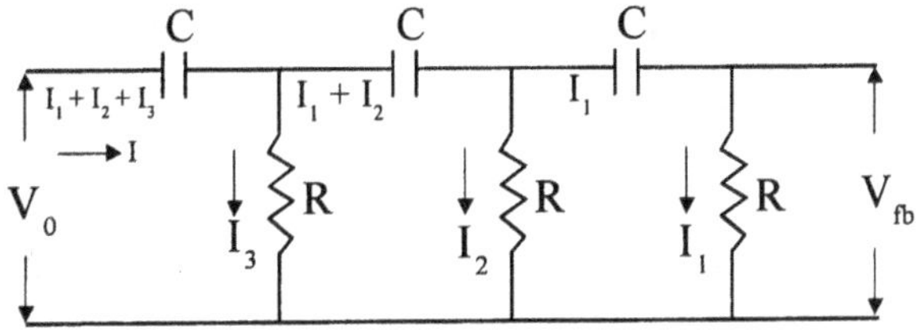

Fig. 7.10 R C phase shift network.

$$V_{fb} = I_L . R \qquad ....(7.1)$$

$$I_2 .R = \frac{I_1}{j\omega C} + I_1 . R \qquad .....(7.2)$$

$$I_3.R = \frac{(I_2 + I_1)}{j\omega C} + I_2.R \qquad .....(7.3)$$

$$V_o = \frac{(I_1 + I_2 + I_3)}{j\omega C} + I_3.R \qquad .....(a)$$

But,
$$I_3.R = \frac{(I_2 + I_1)}{j\omega C} + I_2.R$$

$$I_2 .R = \frac{I_1}{j\omega C} + I_1.R$$

$$I_1 R = V_{FB}$$

$$\therefore \quad I_3.R = \frac{(I_2 + I_1)}{j\omega c} + \frac{I_1}{j\omega c} + V_{fb}.$$

By substituting eq. (7.1), (7.2), and (7.3) in (a) we get,

$$V_o = \frac{(I_1 + I_2 + I_3)}{j\omega c} + \frac{(I_2 + I_1)}{j\omega c} + \frac{I_1}{j\omega c} + V_{fb}.$$

$$V_o = \frac{3I_1 + 2I_2 + I_3}{j\omega c} + V_{fb}. \qquad .....(b)$$

To eliminate I_1, I_2, and I_3,

$$I_1 = \frac{V_{fb}}{R} \qquad .....(c)$$

$$I_2 = \frac{I_1}{j\omega cR} + I_1 \quad = \frac{V_{fb}}{j\omega cR^2} + \frac{V_{fb}}{R}$$

$$I_2 = V_{fb}\left(\frac{1}{R} + \frac{1}{J\omega CR^2}\right) \qquad .....(d)$$

$$I_3 = \frac{(I_2 + I_1)}{j\omega CR} + I_2$$

$$I_3 = \frac{V_{fb}}{j\omega cR}\left(\frac{2}{R} + \frac{1}{j\omega cR^2}\right) + V_{fb}\left(\frac{1}{R} + \frac{1}{j\omega CR^2}\right) \qquad .....(e)$$

Substitute c, d, and e equation in (b) and simplify

$$V_o = \frac{3V_{fb}}{R(j\omega c)} + \frac{2V_{fb}}{j\omega c}\left(\frac{1}{R} + \frac{1}{j\omega cR^2}\right) + \frac{V_{fb}}{(j\omega c)^2 R}\left(\frac{2}{R} + \frac{1}{j\omega cR^2}\right)$$

$$+ \frac{V_{fb}}{j\omega c}\left(\frac{1}{R} + \frac{1}{j\omega cR^2}\right) + V_{fb}$$

$$\frac{V_o}{V_{fb}} = \frac{1}{\beta} = 1 - \frac{5}{\omega^2 c^2 R^2} + j\left(\frac{1}{\omega^3 C^3 R^3} - \frac{6}{\omega CR}\right)$$

Let, $\alpha = \dfrac{1}{\omega CR}$

$\therefore$ $\dfrac{1}{\beta} = 1 - 5\alpha^2 + j(\alpha^3 - 6\alpha)$

or $\beta = \dfrac{1}{1 - 5\alpha^2 - j(6\alpha - \alpha^3)}$

$\beta A = 1$

But β is a complex number. Therefore equating imaginary part to zero.

$$6\alpha - \alpha^3 = 0.$$

$$6\alpha = \alpha^3. \quad \text{or} \quad \alpha = \sqrt{6}.$$

But $\alpha = \dfrac{1}{\omega CR}.$

$\therefore$ $\dfrac{1}{\omega CR} = \sqrt{6}.$

or $\omega = \dfrac{1}{\sqrt{6}CR}$

or $\boxed{f = \dfrac{1}{2\pi CR\sqrt{6}}}$

The gain A, corresponding to this frequency will be 29.

7.5 TRANSISTOR RC PHASE-SHIFT OSCILLATOR

For transistor circuit, voltage shunt feedback is employed, because the input impedance of transistor is small. If voltage series feedback is employed, the resistance of feedback network will be shunted by the low 'R' of the transistor. (Fig. 7.11(a) and Fig.7.11(b)).

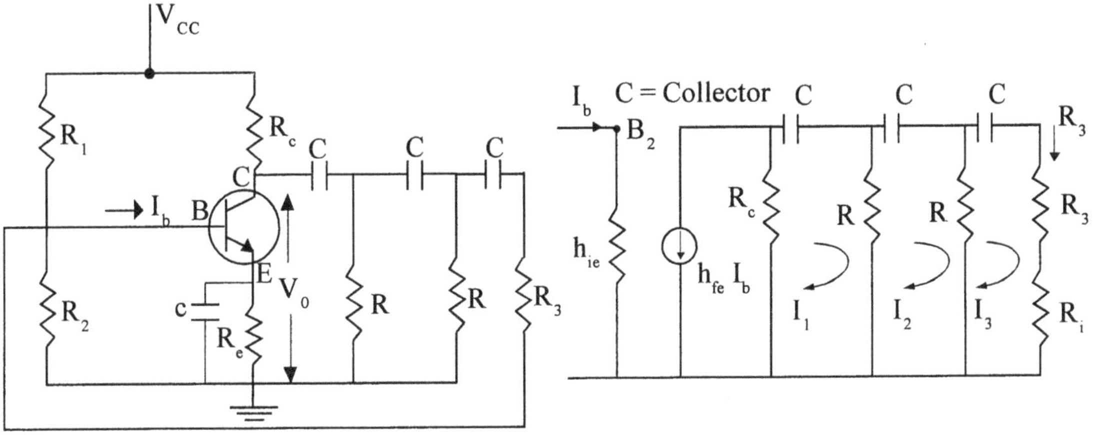

(a) *(b)*

Fig. 7.11 Transistor phase shift oscillator.

The value of $R_3 = R - R_i$, where $R_i \simeq h_{ie}$, the input resistance of transistor. ∴ The three RC sections of the phase-shifting network are identical R_1 R_2 and R_e are biasing resistors.

The feedback current $x_f = -I_3$ Input current $x_i = I_b$ (Negative sign is because it is negative feedback)

$$\therefore \text{ Loop current gain } = \frac{-x_f}{x_i} = \frac{I_3}{I_b}$$

By writing K V L for the three nodes, $\frac{I_3}{I_b}$ can be found out.

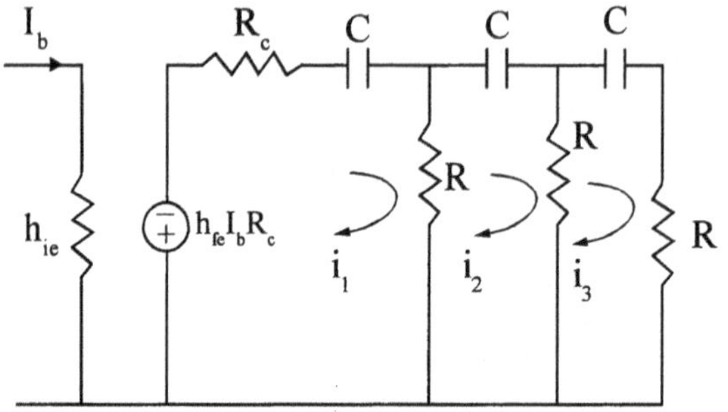

Fig. 7.12 R - C Equivalent circuit.

Loop 1 : $(R_C + R - \frac{j}{\omega c})\, I_1 - R\, I_2 = -\, h_{fe}\, I_b\, R_C$ .....(7.4)

Loop 2 : $-RI_1 + (2R - \frac{j}{\omega c})\, I_2 - RI_3 = 0$ .....(7.5)

Loop 3 : $-RI_2 + (2R - \frac{j}{\omega c})\, I_3 = 0$ .....(7.6)

By taking 'R' as common, the modified given by equations are

$$\left(\frac{R_C}{R} + 1 - \frac{j}{\omega_{RC}}\right) I_1 - I_2 = -h_{fe}\, I_b\, \frac{R_C}{R} \quad .....(7.7)$$

$$-I_1 + \left(2 - \frac{j}{\omega RC}\right) I_2 - I_3 = 0 \quad .....(7.8)$$

$$-I_2 + \left(2 - \frac{j}{\omega RC}\right) I_3 = 0 \quad .....(7.9)$$

Let $\frac{R_C}{R} = K$ and $\alpha = \frac{1}{\omega RC}$, the equations are

$$(1 + K - j\alpha)\, I_1 - I_2 = -\, h_{fe}\, I_b = 0 \quad .....(7.10)$$

$$-I_1 + (2 - j\alpha)\, I_2 - I_3 = 0 \quad .....(7.11)$$

$$-I_2 + (2 - j\alpha)\, I_3 = 0 \quad .....(7.12)$$

$$\begin{bmatrix} 1+k-j\alpha & -1 & 0 \\ -1 & 2-j\alpha & -1 \\ 0 & -1 & 2-j\alpha \end{bmatrix}\begin{bmatrix} I_1 \\ I_2 \\ I_3 \end{bmatrix} = \begin{bmatrix} -h_{fe}I_b k \\ 0 \\ 0 \end{bmatrix}$$

$$I_3 = \frac{1}{\Delta}\begin{vmatrix} 1-k-j\alpha & -1 & -h_{fe}I_b K \\ -1 & 2-j\alpha & 0 \\ 0 & -1 & 0 \end{vmatrix} = \frac{1}{\Delta}\left[-h_{fe}I_b K\right]$$

$$\Delta = [1 + K - j\alpha]\,[(2 - j\alpha)^2 - 1] + [-(2 - j\alpha)]$$
$$= [1 + K - j\alpha]\,[4 - \alpha^2 - j4\alpha - 1] - 2 + j\alpha$$
$$= - j\alpha^3 - j6\alpha - j4K\alpha - 5\alpha^2 - \alpha^2 K + 3K + 1$$
$$\Delta = -(5 + K)\,\alpha^2 + 3K + 1 + j\,[\alpha^3 - (6 + 4K)\,\alpha]$$

$$\therefore\ I_3 / I_b = \frac{-h_{fe}K}{-(5+K)\alpha^2 + 3K + 1 + j\,[\alpha^3 - (6+4K)\alpha]} \qquad .....(7.13)$$

The Barkhausen condition that the loop gain $\frac{I_3}{I_b}$ phase shift must equal zero. The phase shift equals zero provided imaginary part is zero.

Hence $\alpha^3 = (6 + 4K)\,\alpha$

$$\alpha^2 = 6 + 4K$$

$$\alpha = \sqrt{6+4K}$$

Since $\alpha = \frac{1}{\omega RC}$, $\omega = \frac{1}{RC\sqrt{6+4K}}$

$\therefore$ The frequency of oscillation f is given by

$$f = \frac{1}{2\pi RC\sqrt{6+4K}} \qquad .....(7.14)$$

For mantaining the oscillations at the above frequency, $|I_3 / I_b| > 1$

$$\left|\frac{I_3}{I_b}\right| = \left|\frac{-h_{fe}K}{-(5+K)(6+4K)+3K+1}\right| > 1$$

$$\left|-h_{fe}K\right| > \left|-(5+K)(6+4K)+3K+1\right|$$

$$\left|-h_{fe}K\right| > \left|-4K^2 - 23K - 29\right|$$

$$h_{fe}\,K > 4K^2 + 23K + 29$$

$$h_{fe} > 4K + 23 + \frac{29}{K} \qquad .....(7.15)$$

To determine the minimum value of h_{fe}, the optimum value of K should be determined by differentiating h_{fe} w.r.t K and equate it to zero.

$$\frac{dh_{fe}}{dk} > \frac{d}{dk}\left(4K + 23 + \frac{29}{K}\right)$$

$$0 > 4 - \frac{29}{K^2}$$

$$\therefore K^2 < \frac{29}{4}, \qquad \therefore \mathbf{K < 2.7} \qquad .....(7.16)$$

Substitute the optimum K = 2.7 in eq. (7.15)

$$h_{fe} > 4 \times 2.7 + 23 + \frac{29}{2.7}$$

$$\therefore h_{fe} > 44.5 \qquad .....(7.17)$$

The BJT with a small-signal common-emitter short-circuit gain h_{fe} lessthan 44.5 cannot be used in this phase shift oscillator.

In R–C phase shift oscillator circuit, each RC network produces 60° phase shift. Thus 3 sections produce the required 180^0 phase shift. If there are 4-sections, each sections must produce 45° phase shift. But more number of components are to be used. If only 2 sections are these, 90° phase shift must be produced, which is not possible for practical R-C network.

7.6 A GENERAL FORM OF LC OSCILLATOR CIRCUIT

Many Oscillator Circuits fall in to the general form as shown in Fig.7.13 (a) and (b).

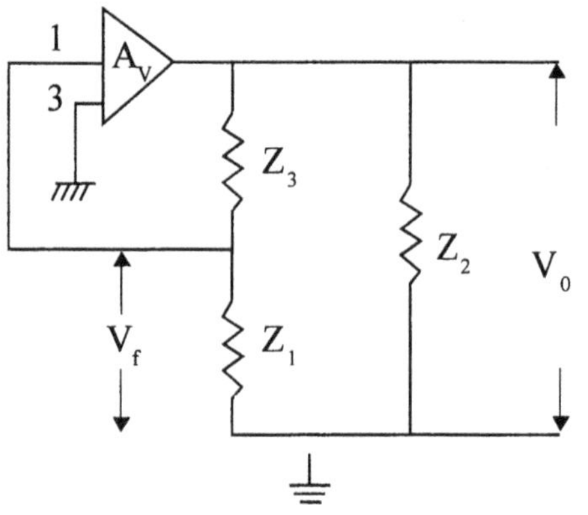

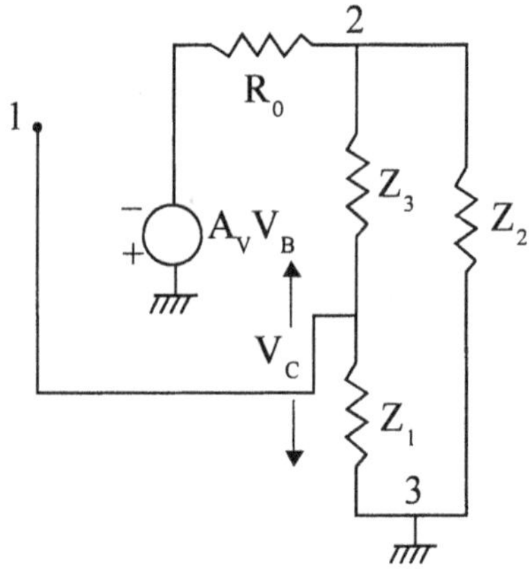

Fig. 7.13 General form of oscillator circuit

The active device can be FET, transistor or operational amplifier, Fig.7.13(b) shows the equivalent circuit using an amplifier with negative gain A_v and output resistance R_o. This is Voltage Series Feedback.

7.7 LOOP GAIN

$$\beta = -\frac{V_f}{V_o}$$

$$\therefore \qquad \beta = \frac{-V_f}{V_0} = -\frac{Z_1}{Z_1 + Z_3}.$$

The load impedance, Z_L = (Z_3 in series with Z_1) parallel with Z_L.

R_o is the output resistance.(See Fig. 7.14).

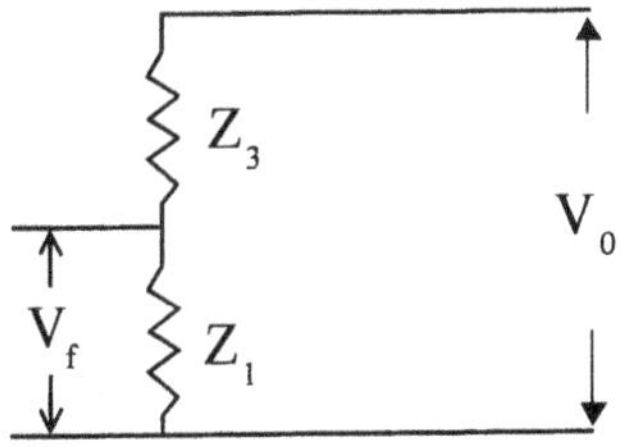

Fig. 7.14 Potential divider network.

Gain without feedback $A = \frac{-A_V.Z_L}{Z_L + R_O}$

$$\therefore \qquad -\beta A = \frac{-A_V.Z_1Z_L}{\left(Z_L + R_O\right)\left(Z_1 + Z_3\right)}; \; Z_L = \frac{Z_2(Z_1 + Z_3)}{Z_2 + Z_1 + Z_3}$$

$$= \frac{-A_V.Z_2(Z_1 + Z_3)Z_1}{(Z_1 + Z_2 + Z_3)\frac{(Z_2(Z_1 + Z_3)}{(Z_1 + Z_2 + Z_3)} + R_O)(Z_1 + Z_3)}$$

$$-\beta A = \frac{-A_v.Z_1 \; Z_2}{R_O(Z_1 + Z_2 + Z_3) + Z_2(Z_1 + Z_3)}$$

If the impedances are pure reactances, then $Z_1 = j\,x_1, \quad Z_2 = jx_2; \; Z_3 = jx_3$

$$\therefore \qquad -A\,\beta = \frac{A_v X_1 \; X_2 (j)^2}{jR_o(X_1 + X_2 + X_3) - X_2(X_1 + X_3)}$$

If $\beta A = 1$, or for zero phase-shift, imaginary part must be zero.

$$jR_o\,(X_1 + X_2 + X_3) = 0.$$

or $$X_1 + X_2 + X_3 = 0$$

$$-A\beta = \frac{A_v X_1 \; X_2}{-X_2(X_1 + X_3)} = -\frac{A_v \; X_1}{X_1 + X_3}$$

But $\quad X_1 + X_2 + X_3 = 0 \quad \therefore \; X_1 + X_3 = -X_2$

$$\therefore \qquad -A\beta = \frac{A_V X_1}{X_2}.$$

$\therefore$ $-A\beta$ must be positive, and at least unity in magnitude. Than X_1 and X_2 must have the same sign.

So if X_1 and X_2 are capacitive, X_3 should be inductive and vice versa.

If X_1 and X_2 are capacitors, the circuit is called ***Colpitts Oscillator (Fig. 7.15(a))***
If X_1 and X_2 are inductors, the circuit is called ***Hartley Oscillators (Fig. 7.15(b))***

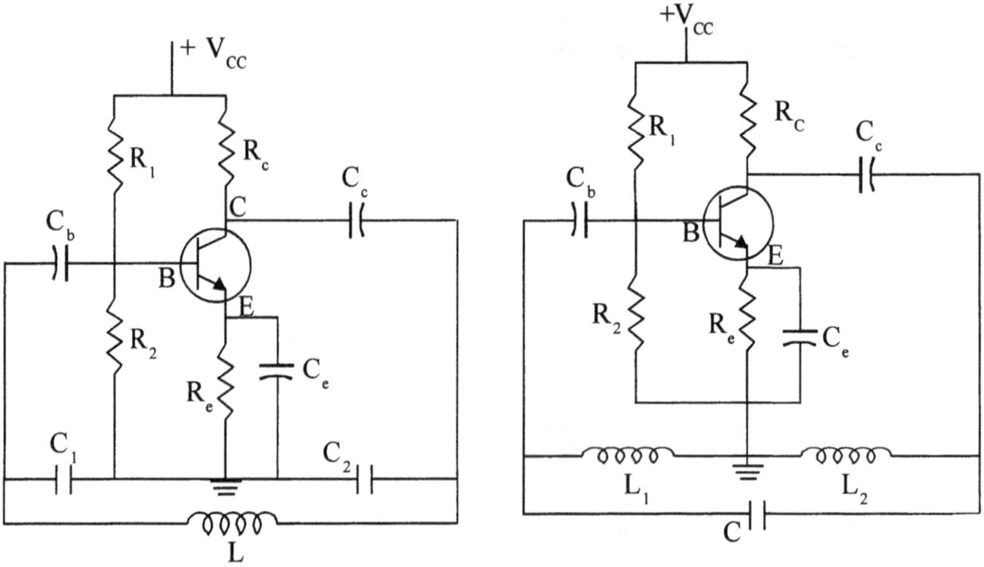

(a) Colpitts oscillator *(b) Hartley oscillator circuit*

Fig. 7.15

7.7.1 For Hartley Oscillator

Condition for oscillations, $X_1 + X_2 + X_3 = 0$.

where $$X_1 = j\omega L_1 \; X_2 = j\omega L_2; \; X_3 = +\frac{1}{j\omega C}$$

$\therefore$ $$j\omega L_1 + j\omega L_2 + \frac{1}{j\omega C} = 0 \quad \text{or} \quad \omega L_1 + \omega L_2 = \frac{1}{\omega C}$$

$$\omega^2(L_1 + L_2) = \frac{1}{C}; \quad \frac{1}{\sqrt{(L_1 + L_2)C}} = \omega$$

$$\boxed{f = \frac{1}{2\pi\sqrt{(L_1 + L_2)C_3}}}$$

7.7.2 For Colpitts Oscillator

$$X_1 = -\frac{j}{\omega C_1}; \quad X_2 = \frac{-j}{\omega C_2}; \quad X_3 = j\omega L.$$

$$X_1 + X_2 + X_3 = 0;$$

$$\frac{-j}{\omega C_1} - \frac{-j}{\omega C_2} + j\omega L = 0 \quad \text{or}$$

$$\omega L = \frac{1}{\omega C_1} + \frac{1}{\omega C_2}$$

$$\omega L = \frac{\omega C_2 + \omega C_1}{\omega C_1\, C_2} = \frac{C_2 + C_1}{\omega C_1\, C_2} \quad \text{or} \quad \omega = \left(\frac{1}{\sqrt{L_3}}\sqrt{\frac{C_2 + C_1}{C_1\, C_2}}\right)$$

$$\boxed{f = \frac{1}{2\pi\sqrt{L\,C_T}}}$$

where $$C_T = \frac{C_1 C_2}{C_1 + C_2}$$

7.8 WIEN BRIDGE OSCILLATOR

In this circuit, a balanced bridge is used as the feedback network. The active element is an operational amplifier. It employs lead-lag Network. Frequency f_0 can be varied in the ratio of 10 : 1 compared to 3 : 1 in other oscillator circuits.

External voltage V_0' is applied betweeen 3 and 4, as shown in Fig.7.16.

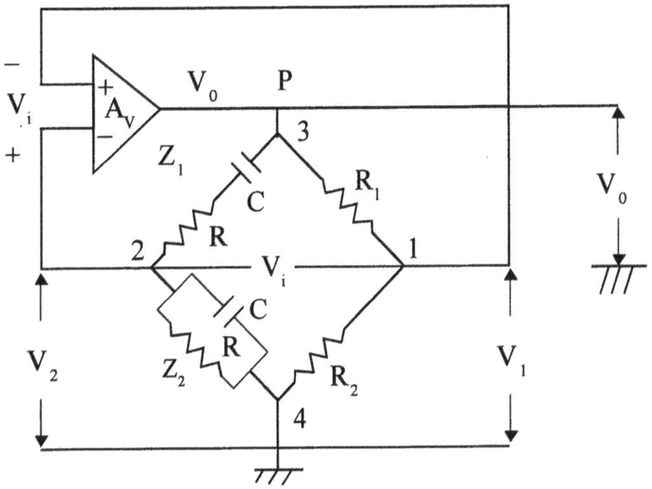

Fig. 7.16 Wien bridge oscillator circuit.

To find loop gain, – βA, (–sign because phase-shift feedback)

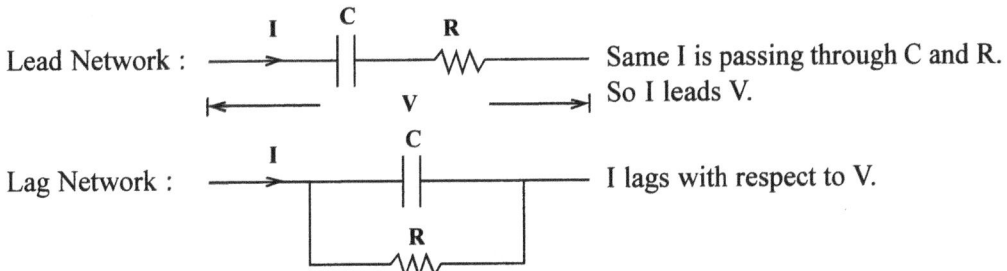

Lead Network :

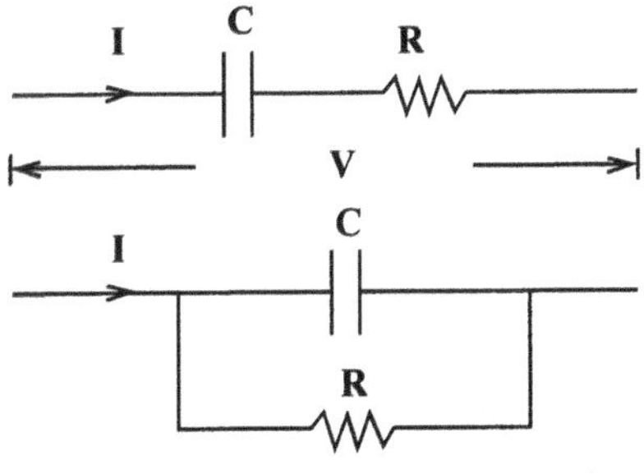

Same I is passing through C and R. So I leads V.

Lag Network :

I lags with respect to V.

Fig. 7.17

Frequency variation of 10 : 1 is possible in Wein Bridge compared to 3 : 1 in other oscillator circuits.

$$V_o = A_v V_i;\ V_i \text{ is } (V_2 - V_1);\ V_i = V_f.$$

Loop gain $$= \frac{V_o}{V_o'} = \frac{A_V . V_i}{V_O'} = -\beta A. \quad .....(7.18)$$

V_1 and V_2 are auxillary voltages. $V_i = V_2 - V_1$.

$$-\beta = \frac{V_i}{V_o'} \quad \because V_f = V_i \quad \therefore A = A_v$$

$$-\beta = \frac{V_i}{V_o'} = \frac{V_2 - V_1}{V_o'} = \left[\frac{Z_2}{Z_1 + Z_2} - \frac{R_L}{R_1 + R_2}\right]$$

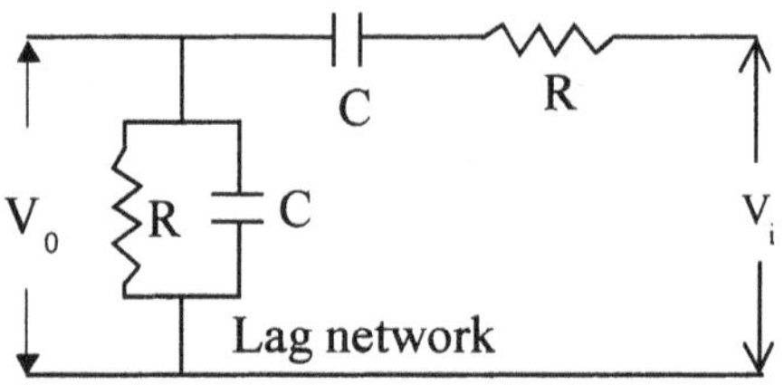

Fig. 7.18

7.9 EXPRESSION FOR *f*

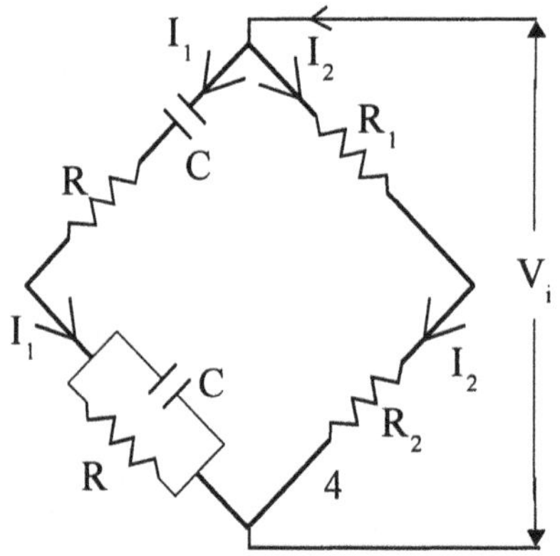

Fig. 7.19 Wien Bridge oscillator circuit.

$$I_1 = \frac{\left(R + \frac{1}{j\omega c}\right)}{I_1\left(\frac{R}{1 + j\omega cR}\right)} = \frac{I_2 R_1}{I_2 R_L}$$

$$R_1 = \frac{\left(R_2 R + \frac{R_2}{j\omega C}\right)(1 + j\omega CR)}{R};$$

$$R_1R = \left(R_2 R + \frac{R_2}{jwC}\right)(1 + jwCR)$$

$$R_2 . R + \frac{R_2}{j\omega C} = \left(\frac{R_1 R}{1 + j\omega CR}\right)$$

$$[(R R_2)(j\omega C) + R_2] R_2 j\omega CR - \omega^2 C^2 R^2 R_2 = R_1 R j\omega C$$

$$(R_1 R_2)(j\omega C) + R_2 + R_2 j\omega CR - \omega^2 c^2 R^2 R_2 = R_1 R j\omega C$$

Equating imaginary parts,

$$R_1 R_2 \omega C + R_2 \omega CR = R_1 R\omega C$$

Equating real parts,

$$A = 1 + \frac{R_1}{R_2} + \frac{C_1}{C_2}, \quad R_2 = \omega^2 C^2 R^2 R_2$$

$$\therefore \quad \omega^2 = \frac{1}{C^2R^2} \quad \text{or} \quad \boxed{f = \frac{1}{2\pi RC}}$$

$$R_1 = \frac{R_2.R}{R} + \frac{R_2}{j\omega CR} + j\omega CRR_2 + R_2$$

$$2R_2 = R_1 \quad \text{or} \quad \boxed{\frac{R_2}{R_1 + R_2} = \frac{1}{3}}$$ So the minimum gain of the amplifier must be 3.

$$\frac{-R_2}{\omega CR} = \omega CRR_2$$

$$\omega^2 = \frac{1}{(RC)^2}; \quad \text{or} \quad \boxed{f = \frac{1}{2\pi RC}}$$

This is the frequency at which the circuit oscillates. Continuous variation of frequency is accomplished by using the capacitors 'C'.

7.10 THERMISTOR

Conductivity of Germanium and Si increases with temperature. A semiconductor when used in this way, taking advantage of this property is called a ***Thermistor***. Ex : NiO, Mn_2O_3 etc. These are used for temperature compensation in oscillator circuits.

7.11 SENSISTOR

A heavily doped semiconductor can exhibit a positive temperature coefficient of resistance because under heavy doping, semiconductor acquires the properties of a metal. So R increases because mobility decreases. Such a device is called ***Sensistor***. These are also used for temperature compensation like thermistors.

7.12 AMPLITUDE STABILIZATION

The amplitude of oscillations can be stabilized by replacing R_2 with a senistor. If β is fixed, as A increases, amplitude of oscillations increases. If a senistor is introduced, as its 'R' changes with temperature, it changes β, so that βA is always constant, (when A changes).

By equating the imaginary part to zero we get $f = \frac{1}{2\pi RC\sqrt{6+4K}}$ where $K = \frac{R_C}{R}$

Forward Current Gain $\boxed{h_{fe} = 4k + 23 + \frac{29}{K}.}$

Practically RC phase-shift oscillators can be used from several hertz to several hundred kilohertzs. In the mega hertz range, tuned LC circuits are more advantageous. Frequency of oscillators can be changed by changing R and C. Amplitude of oscillations will not vary if any C is varied, because X_C varies, but the imaginary part will be zero. Phase-shift oscillator is operated in class A in order to keep distortion minimum.

7.13 APPLICATIONS

Elevation levelling systems, Burglar detection: frequency of oscillators change as a result of local disturbance. The change in frequency causes further electronic action or alarm signal.

7.14 RESONANT CIRCUIT OSCILLATORS

Initial transient due to switching will initiate electrical signals. These are feedback to the transistor as input. The input is amplified and obtained as output. Oscillations are sustained. Output is coupled to the base of transistor through L_1. LL_1 is a transformer. R represents the resistance in series with the winding in order to account for the losses in the transformer shown in Fig. 7.20. If 'r' is very small, then at $\omega = \frac{1}{\sqrt{LC}}$ the Z is purely resistive. Then voltage drop across inductor is 180^0 out of phase with the applied input voltage to the FET. If the direction of winding of the secondary connected to the gate is such that it introduces another 180^0 phase-shift, the total phase-shift is zero.

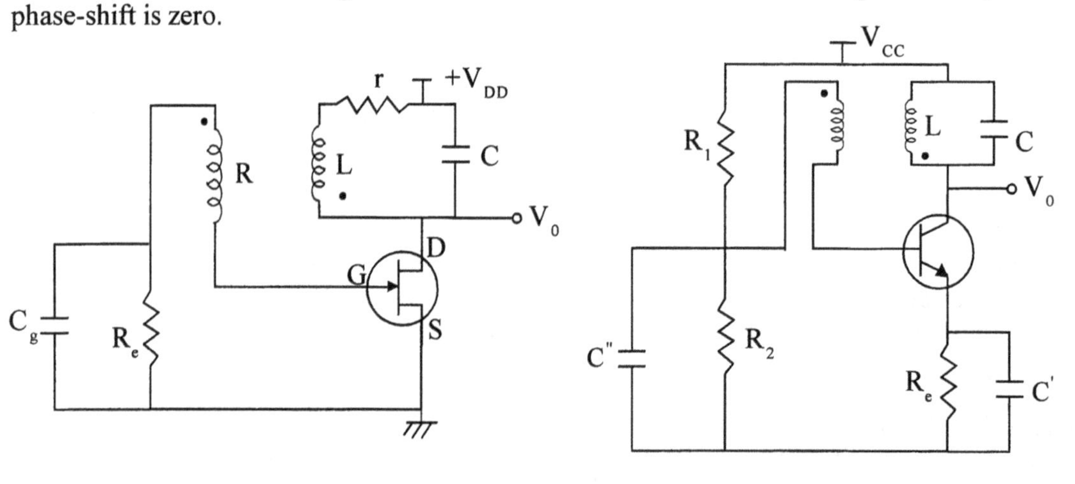

Fig. 7.20 Resonant circuit oscillator.

The ratio of the amplitude of secondary to the primary voltage is M/L where M is the inductance.

$$\text{i.e.,} \quad \frac{V_f}{V_o} = \frac{M}{L} = \beta$$

Voltage gain $A_v = -\mu$

$$\beta A_v = -1$$

$$\therefore \quad -\mu\beta = -1;$$

$$\mu = \frac{1}{\beta}$$

$$\therefore \quad \mu = \frac{L}{M}$$

$$\therefore \quad \mu = L/M$$

If we consider resistance 'r' also, $\omega^2 = \frac{1}{LC}(1 + \frac{r}{r_d})$

$$\boxed{g_m = \frac{\mu r C}{\mu M - L}.}$$

7.15 CRYSTAL OSCILLATORS

When certain solid materials are deformed, they generate within them, an electric charge. This effect is reversible in that, if a charge is applied, the material will mechanically deform in response. This is called ***Piezoelectric effect.***

Naturally available materials : 1. Quartz 2. Rochelle salt.

Synthetic materials : 1. Lithium sulphate 2. Ammonium-di-hydrogen phosphate, PZT (Lead Zirconate Titanate), $BaTiO_3$ (Barium Titanate).

If the crystal is properly mounted, deformations take place within the crystal, and an electro mechanical system is formed which will vibrate when properly excited. The resonant frequency and Q depend upon crystal dimensions etc. With these, frequencies from few KHz to MHz and Q in the range from 1000s to 100,000 can be obtained. Since Q is high, and for Quartz, the characteristics are extremely stable, with respect to time, temperature etc., very stable oscillators can be designed. The frequency stability will be $\pm$ 0.001%. It is same as $\pm$ 10 parts per million (10 ppm).

The electrical equivalent circuit of a crystal is as shown in Fig.7.21. L,C,R are analogous to mass, spring constant, and viscous damping factor of the mechanical system.

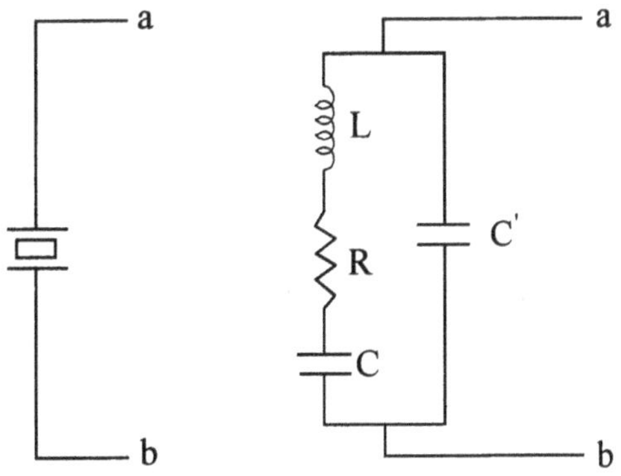

Crystal ***Electrical Model***

Fig. 7.21

Values for a 90 kHz crystal are L = 137 H, C = 0.023pF; R = 15kΩ corresponding to Q = 5,500. The dimension of a crystal will be 30 × 4 ×1.5 mm. C' is the electostatic capacitance between electrodes with the crystal as a dielectric. $C' = 3.5$ pF and is larger than C.

When the crystal slabs are cut in proper directions, with regard to the crystal axis, a potential difference exists between the faces of the crystal slab when pressure is brought to bear on them. And if the slab is placed in an electrostatic field, the slab undergoes deformation. (Fig.7.22). If the electric field is an alternating one, with a frequency which sets the slab into mechanical resonance, the slab will physically vibrate vigorously. Such a crystal can be employed to maintain an oscillation of great frequency stability. When the L.C. circuit in the plate circuit is tuned close to the crystal resonant frequency, steady oscillations will be established. These are maintained by C whose oscillations value is small. By placing crystal between the gate and source, of the FET amplifier, and feeding back a small A.C. voltage from the output, to keep crystal vibrating, the circuit becomes an oscillator with precise stability. Accuracy ≃ 0.01% '*f*' range is 0.1 to 20 MHz. By keeping crystal in an oven, accuracy can be improved to 0.001%.

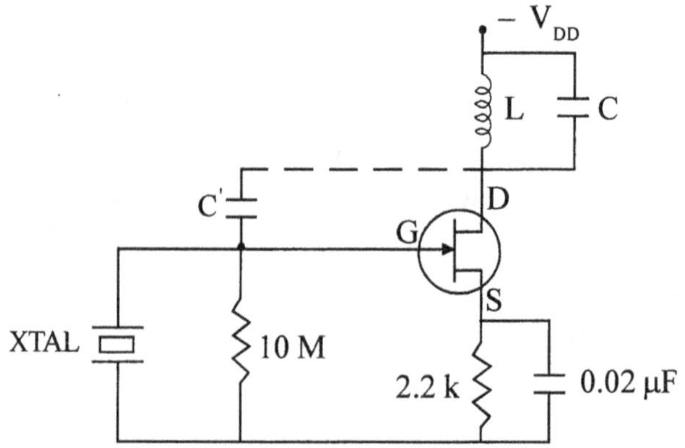

Fig. 7.22 Crystal oscillator circuit.

7.16 FREQUENCY STABILITY

It is a measure of the ability of the circuit to maintain exactly the same frequency for which it is designed over a long time interval. But actually in many circuits the '*f*' will not remain fixed but it drifts from the designed frequency continuously. This is because of variations of number of parameters, circuit components, transistor parameters, supply voltages temperature, stray capacitances etc. In order to increase the stability the factors which effect the '*f*' largely should be taken care of. If '*f*' depends only on R and C high precision R and C should be employed. Also temperature compensating elements are to be employed.

Effect of temperature on inductors and capacitors amounts to more than 10 parts per million per degree change.

Crystal will have mass, elasticity and damping. Crystal will have very high mass to elastic ratio $\left(\frac{L}{C}\right)$ and to a high ratio of mass to damping (high Q). Its value is of the order of 10,000 to 30,000.

The crystal is coupled with external C, and the LC circuit oscillates. In the oscillator circuits, instead of external L and C, crystal is connected.i.e. crystal L and C are being used. So 'f' is fixed by crystal itself. 'f' will not vary with temperature. To change 'f' another crystal is used.

7.17 FREQUENCY OF OSCILLATIONS FOR PARALLEL RESONANCE CIRCUIT

$$Z = \frac{(R - j\omega C)\times \frac{1}{j\omega C}}{(R + j\omega C) + \frac{1}{j\omega C}} = \frac{R + j\omega C}{(1-\omega^2 LC) + j\omega RC}$$

$$Y_1 = \frac{1}{R + j\omega C} = \frac{R - j\omega C}{R^2 + \omega^2 C^2} \quad Y_2 = j\omega C$$

Total admittance $Y = Y_1 + Y_2 = \frac{R}{R^2+\omega^2 C^2} - j\left(\frac{\omega C}{R^2+\omega^2 C^2} - \omega C\right)$

At resonance, imaginary part is zero $\therefore$ $R \to \infty$.

$$\therefore \quad = \frac{\omega_2}{R^2+\omega^2 C^2} = \omega C \text{ or } \frac{L}{C} = R^2 + \omega^2 C^2$$

$$\omega^2 C^2 = \frac{L}{C} - R^2 \text{ or } \omega^2 = \frac{1}{LC} - \frac{R^2}{L^2}$$

or $$f = \frac{1}{2\pi}\sqrt{\frac{1}{LC} - \frac{R^2}{L^2}}$$

This is the expression for frequency of oscillations.

Amplitude of oscillator will be very small (v is Less) for series resonance circuit. So parallel tuned circuits are employed.

7.18 1 MHz FET CRYSTAL OSCILLATOR CIRCUIT

In the basic configuration of oscillator circuit,

$$Z_1 + Z_2 + Z_3 = 0.$$

Z_1 is crystal Z_2 is L and C combination.

Z_3 is C_{gd}. The frequency of oscillator essentially depends up on crystal only as shown in Fig. 7.23. Z_2 and Z_3 values are insignificant compared to Z_1 of the crystal. Therefore the stability is high.

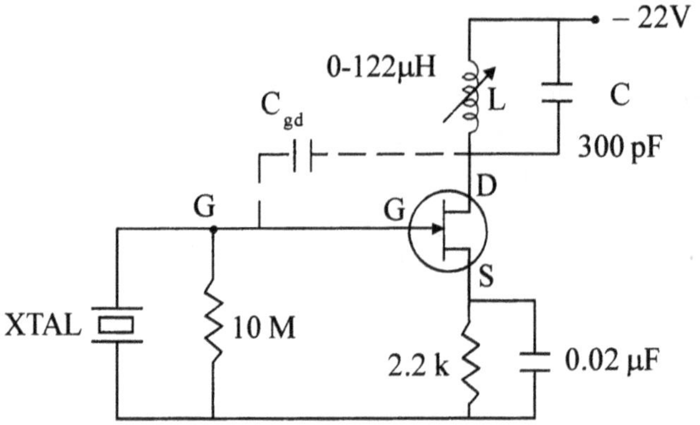

Fig. 7.23 Crystal oscillator circuit.

When electrical input is given, mechanical vibrations are set in the crystal, due to piezoelectric effect. But the crystal is being considered as an inductor capacitor combination only. So it acts like a $|Z|$ R_o L, C combination. Frequency of oscillations depend only on crystal. Other L and Cs are insignificant, as their values are less compared to L, C and R of the crystal.

SUMMARY

- Oscillators generate A.C. output without external A.C input by using Noise A.C signal generated in switching. D.C power from V_{CC} or V_{DD} is converted to A.C. power. The range of frequency signals generated by the circuit can be high and output power is small.
- For sustained oscillations, the conditions known as Barkhausen criterion to be satisfied are (i) $|\beta A| \geq 1$ (ii) Total loop phase shift must be 0^0 or 360^0 (or phase shift of feedback network must be 180^0 when the amplifying device produces another 180^0).
- In the case of JFET, R - C phase shift oscillator circuit, $f_o = \dfrac{1}{2\pi RC\sqrt{6}}$
- In the case BJT, R - C phase shift oscillator circuit, $f_o = \dfrac{1}{2\pi RC\sqrt{6+4k}}$ where $K = \dfrac{R_C}{R}$
- For Hartley oscillator circuit which employs two inductors and one capacitor in the feedback network is, $f_o = \dfrac{1}{2\pi\sqrt{(L_1+L_2)C_3}}$
- For Colpitts oscillator circuit with two capacitors and one inductor in the feedback network, $f_o = \dfrac{1}{2\pi}\dfrac{1}{\sqrt{L_3}}\sqrt{\dfrac{C_2+C_1}{C_1 C_2}}$
- For Wein Bridge oscillator circuit, the minimum gain of amplifier must be 3. It can produce variations in f_0 in the ratio of 10 : 1 compared to a variation of 3 : 1 in other types of oscillator circuits.
- Thermistors with (NTCR) and sensistor with positive temperature coefficient of resistance (PTCR) are used for frequency stability in oscillator circuits.
- The specification parameters of oscillator circuits are (i) Amplitude stability (ii) Frequency stability (iii) Frequency range (iv) Distortion in output waveform etc.
- Crystal oscillators produce highly stable output waveform in the high frequency range of MHz also.

OBJECTIVE TYPE QUESTIONS

1. The difference between an amplifier circuit and oscillator circuit is ______________.
2. A.C. output signal is generated by the oscillator circuits without external A.C input, by amplifying ______________ signal.
3. Oscillator circuits employ ______________ type of feedback.
4. One condition for sustained oscillations is total loop phase shift must be ______________ degrees.
5. As per Berkauhsen criteria for sustained oscillations, $|\beta A|$ must be ______________.

6. The range of frequencies over which R - C phase shift oscillator circuit is used is ________________.

7. In the Mega Hertzs frequency range, the type of oscillator circuit used is ________________.

8. The range of frequency over which Wein Bridge oscillator used is ________________.

9. In the feedback network if two inductors and one capacitor elements are used, the oscillator circuit is ________________.

10. The oscillator circuit which employs two capacitors and one inductor in the feedback network, is ________________ oscillator circuit.

11. Naturally occuring materials used for in crystal oscillator circuits, exhibiting piezoelectric effect are 1. ________________ 2. ________________.

12. Synthetic materials which exhibit piezoelectric effect are,

    1. ________________ 2. ________________

    3. ________________ 4. ________________

13. The ratio of frequency variation possible with Wein Bridge Oscillator circuit is ________________ is compared to the ratio of ________________ with others oscillator circuits.

14. Thermistors have temperature coefficient and the materials used are ________________.

15. Typical values of L, C, R and Q of a crystal used in oscillator circuits are L = ________________ C = ________________ R = ________________ Q = ________________.

16. Expression for frequency of Osillations in the case of Phase-shaft oscillator circuit is f_o = ________________ and the value of B must be at least ________________

17. Expression for frequency of oscillations in the case of Wien Bridge Oscillator is f_o = ________________.

18. In the case of Colpitts oscillator f_o = ________________

19. For Hartley oscillator, f_o = ________________

20. Frequency of oscillations, in the case of uJT oscillator circuit is $f_o \simeq$ ________________

ESSAY TYPE QUESTIONS

1. Explain the basic principle of generation of oscillations in LC tank circuits. What are the considerations to be made in the case of practical L.C. Oscillator Circuits ?

2. Deduce the Barkausen Criterion for the generation of sustained oscillations. How are the oscillations initiated?

3. Draw the circuit and explain the princple of operaiton of R.C.phase-shift oscillator circuit. What is the frequency range of generation of oscillations ? Derive the expression for the frequency of oscillations.

4. Derive the expression for the frequency of Hartely oscillators.
5. Derive the expression for the frequency of Colpitt Oscillators.
6. Derive the expression for the frequency of Wein Bridge Oscillators.
7. Derive the expression for the frequency of Crystal Oscillators.
8. Explain how better frequency stability is obtained in crystal oscillator ?
9. Draw the equivalent circuit for a crystal and explain how oscillations can be generated in electronic circuits, using crystals.
10. Why three identical R-C sections are used in R-C phase-shift oscillator circuits? Consider the other possible combinations and limitations.

MULTIPLE CHOICE QUESTIONS

1. Johnson Noise is due to .....

(a) Humidity (b) Atmospheric conditions
(c) Temperature (d) Interference

2. The noise that arises in electronic circuits due to variation in concentrations of carriers in semiconductor devices is

(a) shot noise (b) thermal noise (c) Johnson noise (d) None of these

3. Expression for frequency of oscillations for the R-C phase shift oscillator circuit using JFET is, f_0 =

(a) $\frac{1}{2\pi RC}$ (b) $\frac{1}{2\pi RC^2\sqrt{6}}$ (c) $\frac{1}{2\pi R^2C\sqrt{6}}$ (d) $\frac{1}{2\pi RC\sqrt{6}}$

4. In the case of JFET R-C phase shift oscillator circuit, in order that $|\beta A|$ is not less than unity, μ of JFET must be at least

(a) 16 (b) 92 (c) 29 (d) None of these

5. Oscillator circuit in which the reactances α_1, α_2 are capacitive and α_3 is inductive is ....

(a) Colpitts oscillator (b) Hartely oscillator
(c) Crystal oscillator (d) None of these

6. In wein bridge oscillator circuits the range over which frequency of oscillations can be varied is ...

(a) 3 : 1 (b) 10 : 1 (c) 6 : 1 (d) None of these

7. Example for naturally occurring piezoelectric crystal material is ....

(a) $BaTio_3$ (b) Rochelle salt (c) PZT (d) None of these

8. Typical values for 90 KHz crystal are :

(a) L = 137 H; C = 0.0235 μf; R = 15 KΩ; Q = 5,500

(b) L = 1 H; C = 0.02 f; R = 1Ω; Q = 10

(c) L = 137 H; C = 0.02 μf; R = 10 KΩ; Q = 1

(d) None of these

9. One characteristic feature of crystals is ...

(a) They have low mass to elastic ratio

(b) They have high mass to elastic ratio

(c) They have low Q

(d) None of the above is correct

10. For Wein Bridge oscillator circuits, the minimum gain of amplifier must be

(a) 1 (b) 3 (c) 10 (d) None of these

CONCEPTUAL QUESTIONS (Interview Questions)

1. How oscillator circuit gives output signal without input singnal ?

Ans : The circuit needs D-C input voltage supply. When the supply is switched on, the transient signal produced acts as the a.c input signal. It is amplified by the amplifier and fedback to the input. By the time the transient dies down, the feed back signal acts as the input signal. Output oscillations will be sustained for a particular frequency component for which Barkhausen criteria is satisfied. The wave shape is sinosoidal because sinosoidal function is the only function that satisfies the condition governing the exchange of energy between the energy storage elements in the circuit.

2. Why R - C Phase shift Oscillator Circuit must have three R - C networks ? Why not two or four RC networks?

Ans : If only two R - C networks are used, each network must produce 90° phase shift, to get a total phase shift of 180° in the feed back circuit. So we must have ideal capacitor, which is not practically possible. On the otherhand we can have 4 - RC networks each producing 45° phase shift. But we will be using more number of R-C components which is not economical.

3. Why the frequency of oscillations of a crystal oscillator circuit are stable ?

Ans : A crystal can be represented as a combination of R L and C components in its equivalent circuit form. When a crystal is connected in the circuit, it is as if R, L and C are connected in the circuit. For a crystal, these equivalent parameters of R, L and C do not change much with temperature. So frequency of oscillations of crystal oscillator circuit do not change with temperature. Thus frequency stability is high. For other oscillator circuits the R, L, C component values change significantly with temperature. So frequency stability is less.

Additional Objective Type Questions (Chapter 1-7)

1. In a BJT the arrow on the emitter lead specifies the direction of _______________ where the emitter-base junction is _______________ biased.

2. The emitter efficiency of a BJT is defined as the ratio of current of injected carriers at _______________ junction to total _______________ current.

3. _______________ type of npn junction transistor is made by drawing a single crystal from a melt of silicon whose _______________ concentration is changed during the crystal drawing operation by adding n or p type atoms as requires.

4. In an _______________ type of pnp transistor two small dots of indium are attached to opposite sides of a semiconductor.

5. In the active region of common base output characteristics, the collector junction is _______________ biased and the emitter junction is _______________ biased.

6. If both emitter and collector junctions are (a) forward biased, the transistor is said to be operating the _______________ region. (b) reverse biased, the transistor is said to be operating in the _______________ region.

7. The operation of a FET depends upon the flow of _______________ carriers only. It is therefore a _______________ device.

8. The maximum voltage that can be applied between any two terminals of the FET is the _______________ voltage that will cause avalanche breakdown across the _______________.

9. A MOSFET of the depletion type may also be operated in an _______________ mode. Some times the symbol for the JFET is also is used for the MOSFET with the understanding that _______________ is internally connected to source.

10. The emitter diode of a UJT drives the junction of R_{B1} and R_{B2} and $R_{BB} = R_{B1} + R_{B2}$. The intrinsic stand off ratio is defined as η = _______________ and its value usually lies, between _______________.

Answers to Additional Objective Type Questions

1. Holes or conventional current, forward
2. Emitter base, emitter
3. Grown junction, doping
4. Alloy junction
5. Reverse, Forward
6. Saturation, Cut-off
7. Majority, Unipolar
8. Breakdown, drain-source junction
9. Enhancement, the gate
10. $\frac{R_{B_1} + R_{B_2}}{R_{BB}}$, 0.5 & 0.82

Appendices

APPENDIX - 1

Colour Codes for Electronic Components

RESISTOR COLOUR CODE :

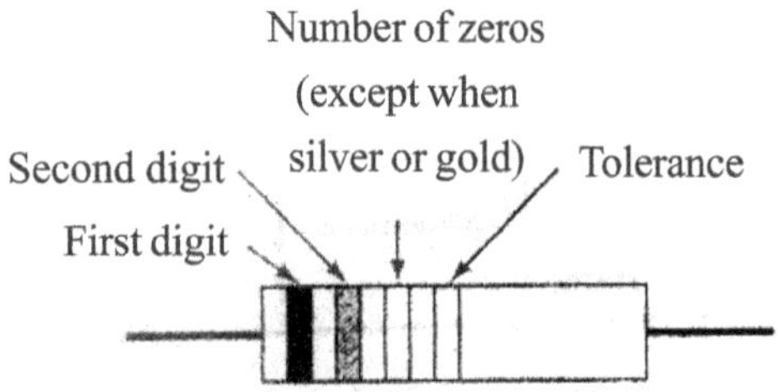

First Three Bands				Fourth Band	
Black	– 0	Blue	– 6	Gold	+ 5%
Brown	– 1	Violet	– 7	Silver	+ 10%
Red	– 2	Grey	– 8	None	+ 20%
Orange	– 3	White	– 9		
Yellow	– 4	Silver	0.01		
Green	– 5	Gold	0.1		

CAPACITOR COLOUR CODE :

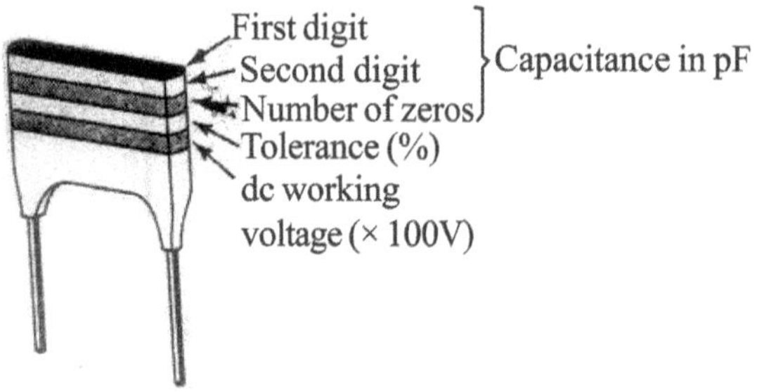

Colour	Figure Significant	Tolerance (%)
Black	0	20
Brown	1	1
Red	2	2
Orange	3	3
Yellow	4	4
Green	5	5
Blue	6	6
Violet	7	7
Grey	8	8
White	9	9
Silver	0.01	10
Gold	0.1	5
No Band		20

INDUCTOR COLOUR CODE :

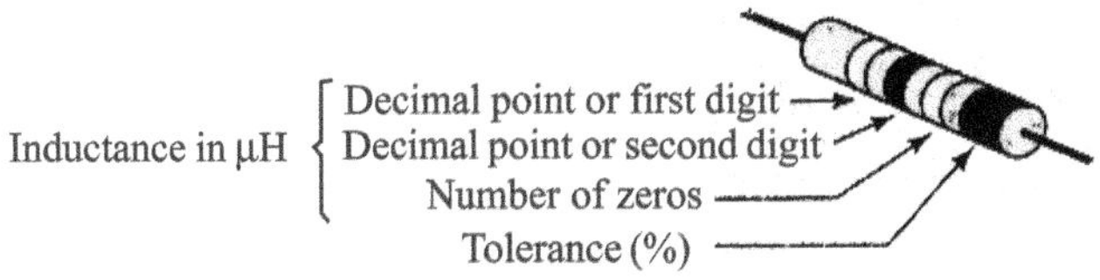

Color	Significant Figure	Tolerance (%)
Black	0	
Brown	1	
Red	2	
Orange	3	
Yellow	4	
Green	5	
Blue	6	
Violet	7	
Grey	8	
White	9	
Silver		10
Gold	Decimal point	5
No Band		20

COLOUR CODE MEMORY AID : W^G VIBGYOR BB (W^G Vibgyor BB)

Memory aid	Color	Number	
Black	Black	0	
Bruins	Brown	1	
Relish	Red	2	
Ornery	Orange	3	
Young	Yellow	4	
Greenhorns	Green	5	
Blue	Blue	6	
Violets	Violet	7	
Growing	Grey	8	
Wild	White	9	
Smell	Silver	0.01	10%
Good	Gold	0.1	5%

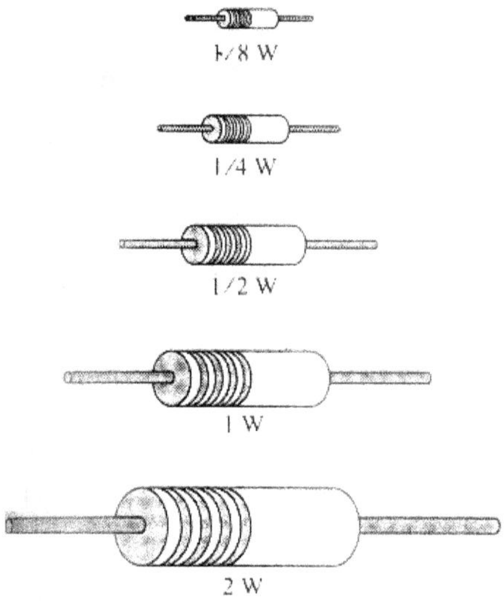

Fig. A-1.1 Relative size of carbon composition resistors with various power ratings

Specifications of Power Transistors

Device	Type	P_D(W)	I_C(A)	V_{CEO}(V)	V_{CBO}(V)	h_{FE} MIN	Max	f_T M.HZs
2N 6688	NPN	200	20	200	300	20	80	20
2N 3442	NPN	117	10	140	160	20	70	0.08
BUX 39	NPN	120	30	90	120	15	45	8
ECP 149	PNP	30	4	40	50	30	–	2.5

Darlington Pair

2N 6052	PNP	150	12	100	100	750	–	4
2N 6059	NPN	150	12	100	100	750	–	4

APPENDIX - 2

Resistor and Capacitor Values

Typical Standard Resistor Values (± 10% Tolerance)							
Ω	Ω	Ω	kΩ	kΩ	kΩ	MΩ	MΩ
–	10	100	1	10	100	1	10
–	12	120	1.2	12	120	1.2	–
–	15	150	1.5	15	150	1.5	15
–	18	180	1.8	18	180	1.8	–
–	22	220	2.2	22	220	2.2	22
2.7	27	270	2.7	27	270	2.7	–
3.3	33	330	3.3	33	330	3.3	–
3.9	39	390	3.9	39	390	3.9	–
4.7	47	470	4.7	47	470	4.7	–
5.6	56	560	5.6	56	560	5.6	–
6.8	68	680	6.8	68	680	6.8	–
–	82	820	8.2	82	820	–	–

Typical Standard Capacitor Values (± 10% Tolerance)

pF	pF	pF	pF	μF	μF	μF	μF	μF	μF	μF
5	50	500	5000		0.05	0.5	5	50	50	5000
–	51	510	5100		–	–	–	–	–	–
–	56	560	5600		0.056	0.56	5.6	56	–	5600
–	–	–	6000		0.06	–	6	–	–	6000
–	62	620	6200		–	–	–	–	–	–
–	68	680	6800		0.068	0.68	6.8	–	–	–
–	75	750	7500		–	–	–	75	–	–
–	–	–	8000		–	–	8	80	–	–
–	82	820	8200		0.082	0.82	8.2	82	–	–
–	91	910	9100		–	–	–	–	–	–
10	100	1000		0.01	0.1	1	10	100	1000	10,000
–	110	1100		–	–	–	–	–	–	
12	120	1200		0.012	0.12	1.2	–	–	–	
–	130	1300		–	–	–	–	–	–	
15	150	1500		0.015	0.15	1.5	15	150	1500	
–	160	1600		–	–	–	–	–	–	
18	180	1800		0.018	0.18	1.8	18	180	–	
20	200	2000		0.02	0.2	2	20	200	2000	
24	240	2400		–	–	–	–	240	–	
–	250	2500		–	0.25	–	25	250	2500	
27	270	2700		0.027	0.27	2.7	27	270	–	
30	300	3000		0.03	0.3	3	30	300	3000	
33	330	3300		0.033	0.33	3.3	33	330	3300	
36	360	3600		–	–	–	–	–	–	
39	390	3900		0.039	0.39	3.9	39	–	–	
–	–	4000		0.04	–	4	–	400	–	
43	430	4300		–	–	–	–	–	–	
47	470	4700		0.047	0.47	4.7	47	–	–	

Physical constants

Charge of an electron	: e :	1.60×10^{-19} coulombs
Mass of an electron	: m :	9.09×10^{-31} Kg
e/m ratio of an electron	: e/m :	1.759×10^{11} C/Kg
Plank's constant	: h :	6.626×10^{-34} J-sec
Boltzman's constant	: $\overline{K}$:	1.381×10^{-23} J/°K
	: K :	8.62×10^{-5} ev/°K
Avogadro's number	: N_A :	6.023×10^{23} molecules/mole
Velocity of light	: c :	3×10^{8} m/sec
Permeability of free space	: m_o :	1.257×10^{-6} H/m
Permittivity of free space	: $\hat{I}_o$:	8.85×10^{-12} F/m
Intrinsic concentration in silicon at 300 °K	: n_i	$= 1.5 \times 10^{10}$ /cm^3
Intrinsic resistivity in silicon at 300 °K	: r_i	= 230,000 W–cm
Mobility of electronics in silicon	: m_n	= 1300 cm^2/ V–sec
Mobility of holes in silicon	: m_p	= 500 cm^2/ V–sec
Energy gap at in silicon at 300 °K	:	= 1.1 ev.

APPENDIX - 3

Capacitors

Capacitance

The farad (F) *is the SI unit of capacitance.*

The farad is the capacitance of a capacitor that contains a charge of 1 coulomb when the potential difference between its terminals is 1 volt.

Leakage Current

Despite the fact that the dielectric is an insulator, small leakage currents flow between the plates of a capacitor. The actual level of leakage current depends on the insulation resistance of the dielectric. Plastic film capacitors, for example, may have insulation resistances higher than 100 000 MW. At the other extreme, an electrolytic capacitor may have a microamp (or more) of leakage current, with only 10 V applied to its terminals.

Polarization

Electrolytic capacitors normally have one terminal identified as the most positive connection. Thus, they are said to be polarized. This usually limits their application to situations where the polarity of the applied voltage will not change. This is further discussed for electrolytic capacitors.

Capacitor Equivalent Circuit

An ideal capacitor has a dielectric that has an infinite resistance and plates that have zero resistance. However, an ideal capacitor does not exist, as all dielectrics have some leakage current and all capacitor plates have some resistance. The complete equivalent circuit for a capacitor [shown in Fig. A-3.1(a)] consists of an ideal capacitor C in series with a resistance R_D representing the resistance of the plates, and in parallel with a resistance R_L representing the leakage resistance of the dielectric. Usually, the plate resistance can be completely neglected, and the equivalent circuit becomes that shown in Fig. A-3.1(b). With capacitors that have a very high leakage resistance (e.g., mica and plastic film capacitors), the parallel resistor is frequently omitted in the equivalent circuit, and the capacitor is then treated as an ideal capacitor. This cannot normally be done for electrolytic capacitors, for example, which have relatively low leakage resistances. The parallel R_C circuit in,

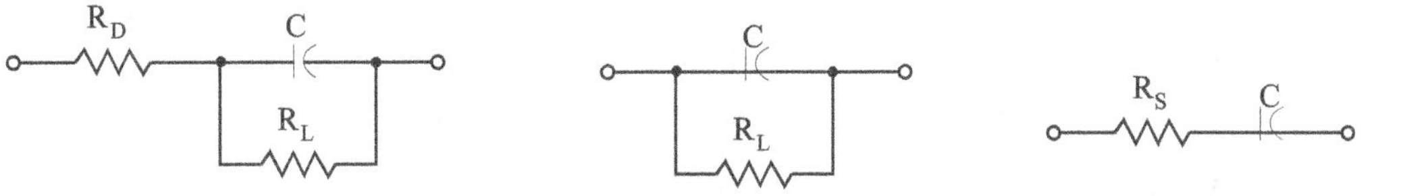

(a) Complete equivalent circuit ***(b) Parallel equivalent circuit*** ***(c) Series equivalent circuit***

Fig. A. 3.1

A capacitor equivalent circuit consists of the capacitance C, the leakage resistance R_L *in parallel with* C, *and the plate resistance* R_D *in series with C and* R_L.

Fig. A. 3.1 (b) can be shown to have an equivalent series *RC* circuit, as in Fig. A. 3.1(c). This is treated in Section 20-6.

A variable air capacitor is made up of a set of movable plates and a set of fixed plates separated by air.

Because a capacitor's dielectric is largely responsible for determining its most important characteristics, capacitors are usually identified by the type of dielectric used.

Air Capacitors

A typical capacitor using air as a dielectric is illustrated in Fig. A.3.2. The capacitance is variable, as is the case with virtually all air capacitors. There are two sets of metal plates, one set fixed and one movable. The movable plates can be adjusted into or out of the spaces between the fixed plates by means of the rotatable shaft. Thus, the area of the plates opposite each other is increased or decreased, and the capacitance value if altered.

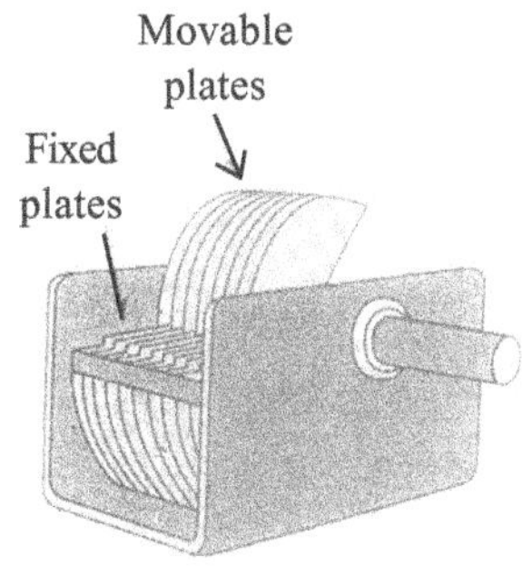

Fig. A. 3.2 A variable air capacitor is made up of a set of movable plates and a set of fixed plates separated by air

Paper Capacitors

In its simplest form, a paper capacitor consists of a layer of paper between two layers of metal foil. The metal foil and paper are rolled up, as illustrated in Fig. A.3.3 (a); external connections are brought out from the foil layers, and the complete assembly is dipped in wax or plastic. A variation of this is the metalized paper construction, in which the foil is replaced by thin films of metal deposited on the surface of the paper. One end of the capacitor sometimes has a band around it [see Fig. A.3.3 (b)]. This does not mean that the device is polarized but simply identifies the terminal that connects to the outside metal film, so that it can be grounded to avoid pickup of unwanted signals.

Paper capacitors are available in values ranging from about 500 pF to 50μF, and in dc working voltages up to about 600 V. They are among the lower-cost capacitors for a given capacitance value but are physically larger than several other types having the same capacitance value.

Plastic Film Capacitors

The construction of plastic film capacitors is similar to that of paper capacitors, except that the paper is replaced by a thin film that is typically polystyrene or Mylar. This type of dielectric gives insulation resistances greater than 100 000 MΩ. Working voltages are as high as 600 V, with the capacitor surviving 1500 V surges for a brief period. Capacitance tolerances of + 2.5% are typical, as are temperature coefficients of 60 to 150 ppm/°C.

Plastic film capacitors are physically smaller but more expensive than paper capacitors. They are typically available in values ranging from 5 pF to 0.47 μF.

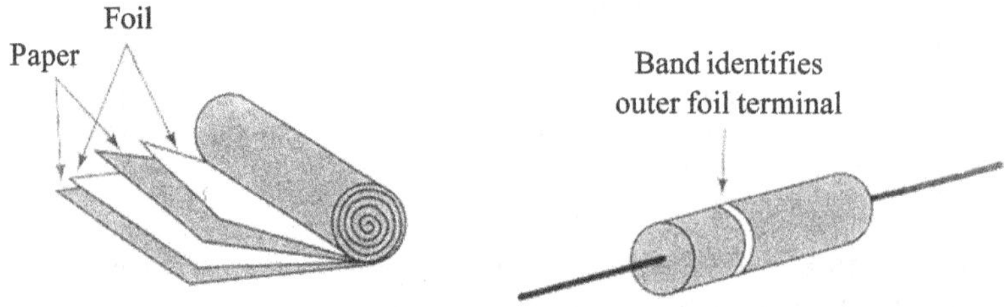

(a) Construction of a paper capacitor (b) Appearance of a paper capacitor

Fig. A. 3.3 In a paper capacitor, two sheets of metal foil separated by a sheet of paper are rolled up together. External connections are made to the foil sheets.

Mica Capacitors

As illustrated in Fig. A. 3.4(a), mica capacitors consist of layers of mica alternated with layers of metal foil. Connections are made to the metal foil for capacitor leads, and the entire assembly is dipped in plastic or encapsulated in a molded plastic jacket. Typical capacitance values range from 1pF to 0.1μF, and voltage ratings as high as 35 000 V are possible. Precise capacitance values and wide operating temperatures are obtainable with mica capacitors. In a variation of the process, silvered mica capacitors use films of silver deposited on the mica layers instead of metal foil.

Ceramic Capacitors

The construction of a typical ceramic capacitor is illustrated in Fig. A. 3.4(b). Films of metal are deposited on each side of a thin ceramic disc, and copper wire terminals are connected to the metal. The entire units is then encapsulated in a protective coating of plastic. Two different types of ceramic are used, one of which has extremely high relative permitivity. This gives capacitors that are much smaller than paper or mica capacitors having the same capacitance value. One disadvantage of this particular ceramic dielectric is that its leakage resistance is not as high as with other types. Another type of ceramic gives leakage resistances on the order of 7500 MW. Because of its lower permitivity, this ceramic produces capacitors that are relatively large for a given value of capacitance.

The range of capacitance values available with ceramic capacitors is typically 1 pF to 0.1 μF, with dc working voltages up to 1000 V.

(a) Construction of mica capacitor

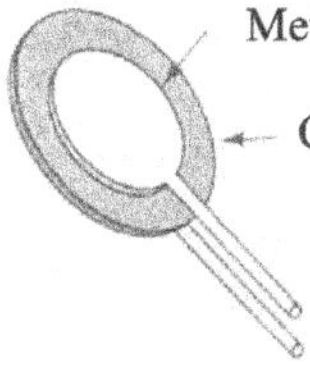

(b) Ceramic capacitor

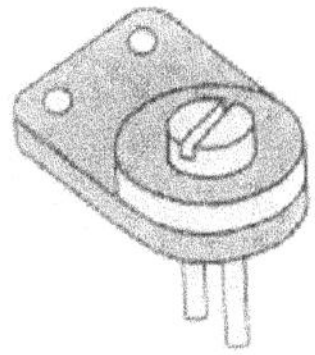

(c) Ceramic trimmer

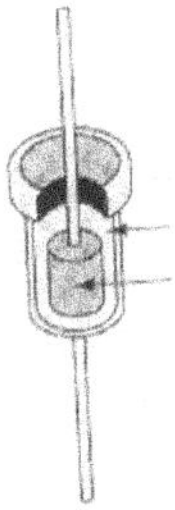

(d) Construction of tantalum capacitor

Fig. A. 3.4 Mica capacitors consist of sheets of mica interleaved with foil. A ceramic disc silvered on each side makes a ceramic capacitor; in a ceramic trimmer, the plates area is screwdriver adjustable. A tantalum capacitor has a relatively large capacitance in a small volume.

Fig. A. 3.4(c) shows a variable ceramic capacitor known as a *trimmer*. By means of a screwdriver, the area of plate on each side of a dielectric can be adjusted to alter the capacitance value. Typical ranges of adjustment available are 1.5 pF to 3 pF and 7 pF to 45 pF.

Electrolytic Capacitors

The most important feature of electrolytic capacitors is that they can have a very large capacitance in a physically small container. For example, a capacitance of 5000 μF can be obtained in a cylindrical package approximately 5 cm long by 2 cm in diameter. In this case the dc working voltage is only voltage is only 10V. Similarly, a 1 F capacitor is available in a 22 cm by 7.5 cm cylinder, with a working voltage of only 3 V. Typical values for electrolytic capacitors range from 1 μF through 100 000 μF.

The construction of an electrolytic capacitor is similar to that of a paper capacitor (Fig. A.3.5(a)). Two sheets of aluminium foil separated by a fine gauze soaked in electrolyte are rolled up and encased in an aluminium cylinder for protection. When assembled, a direct voltage is applied to the capacitor terminals, and this causes a thin layer of aluminium oxide to form on the surface of the positive plate next to the electrolyte (Fig. A.3.5(b)). The aluminium oxide is the dielectric, and the electrolyte and positive sheet of foil are the capacitor plates. The extremely thin oxide dielectric gives the very large value of capacitance.

It is very important that electrolytic capacitors be connected with the correct polarity. When incorrectly connected, gas forms within the electrolyte and the capacitor may **explode**! Such an explosion blows the capacitor apart and spreads its contents around. This could have **tragic consequences** for the eyes of an experimenter who happens to be closely examining the circuit when the explosion occurs. The terminal designated as positive must be connected to the most positive of

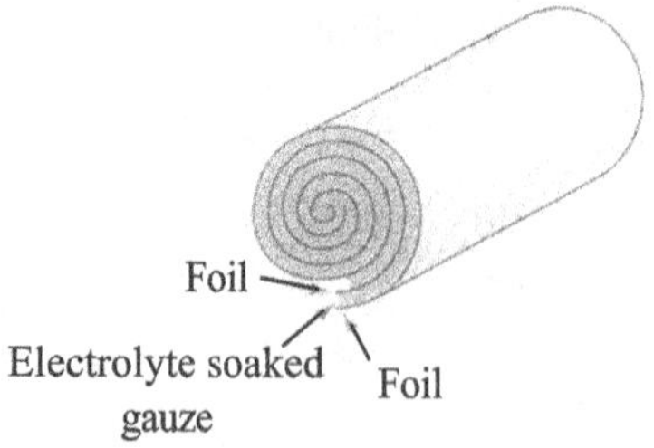

(a) Rolled-up foil sheets and electrolyte-soaked gauze

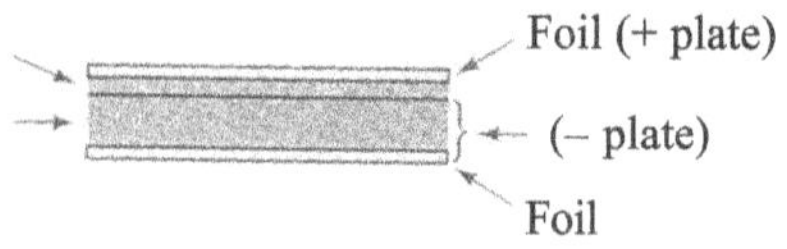

(b) The dielectric is a thin layer of aluminium oxide

Fig. A. 3.5 An electrolyte capacitor is constructed of rolled-up foil sheets separated by electrolyte-soaked gauze, the dielectric is a layer of aluminium oxide at the positive plate.

the two points in the circuit where the capacitor is to be installed. Fig. A. 3.6 illustrates some circuit situations where the capacitor must be correctly connected. Nonpolarized electrolytic capacitors can be obtained. They consist essentially of two capacitors in one package connected *back to back*, so one of the oxide films is always correctly biased.

Electrolytic capacitors are available with dc working voltages greater than 400 V, but in this case capacitance values do not exceed 100 mF. In addition to their low working voltage and polarized operation. Another disadvantage of electrolytic capacitors is their relatively high leakage current.

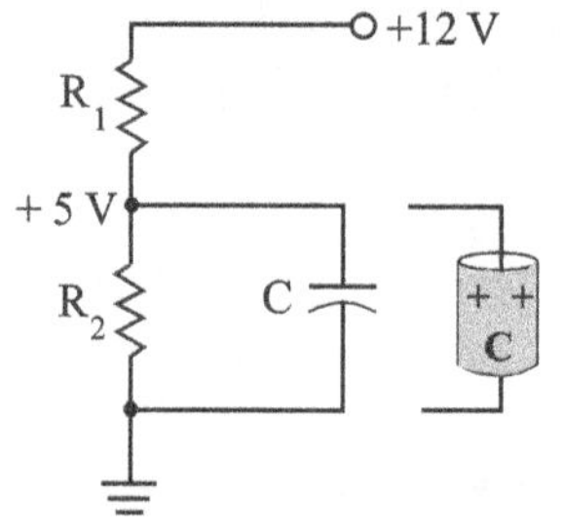

(a) Capacitor connected between +5 V and ground

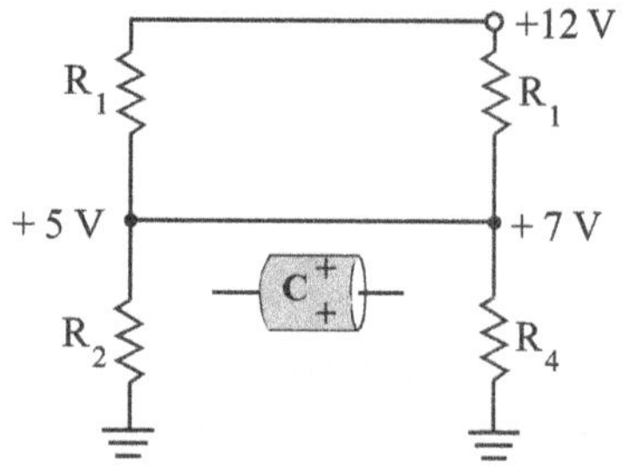

(b) Connected between +7 V and +5 V

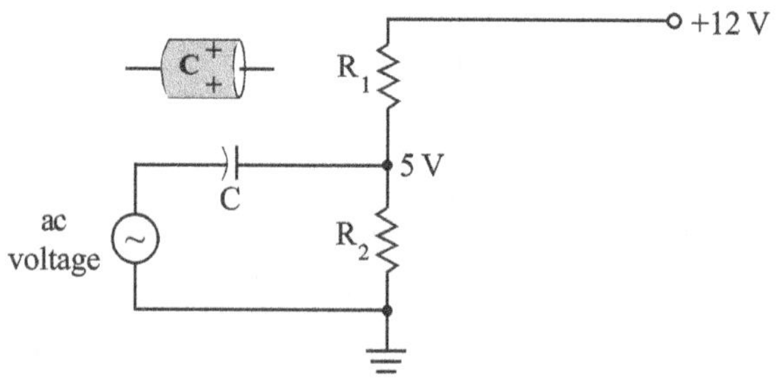

(c) Connected between + 5.7 V and a grounded ac voltage source

Fig. A. 3.6 It is very important that polarized capacitors be correctly connected. The capacitor positive terminal voltage must be more positive than the voltage at the negative terminal.

Tantalum Capacitors

This is another type of electrolytic capacitor. Powdered tantalum is sintered (or baked), typically into a cylindrical shape. The resulting solid is quite porous, so that when immersed in a container of electrolyte, the electrolyte is absorbed into the tantalum. The tantalum then has a large surface area in contact with the electrolyte (Fig. A. 3.5). When a dc *forming voltage* is applied, a thin oxide film is formed throughout the electrolyte-tantalum contact area. The result, again, is a large capacitance value in a small volume.

Capacitor Color Codes

Physically large capacitors usually have their capacitance value, tolerance and dc working voltage printed on the side of the case. Small capacitors (like small resistors) use a code of colored bands (or sometimes colored dots) to indicate the component parameters.

There are several capacitor color codes in current use. Here is one of the most common.

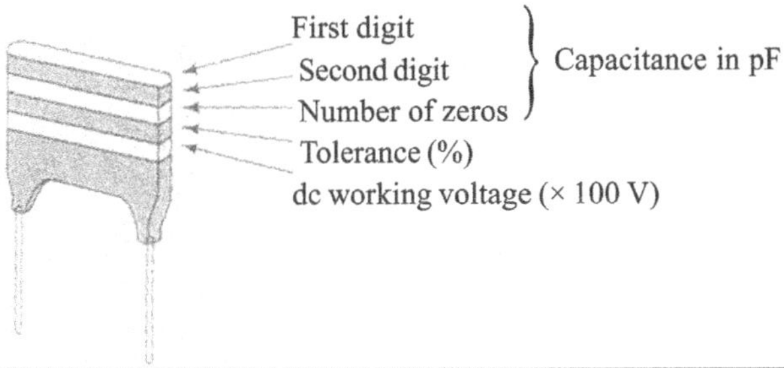

Color	Significant Figure	Tolerance (%)
Black	0	20
Brown	1	1
Red	2	2
Orange	3	3
Yellow	4	4
Green	5	5
Blue	6	6
Violet	7	7
Grey	8	8
White	9	9
Silver	0.01	10
Gold	0.1	5
No band		20

A typical tantalum capacitor in a cylindrical shape 2 cm by 1 cm might have a capacitance of 100 mF and a dc working voltage of 20 V. Other types are available with a working voltage up to 630 V, but with capacitance values on the order of 3.5 mF. Like aluminium-foil electrolytic capacitors, tantalum capacitors must be connected with the correct polarity size of the inductor, the maximum current can be anything from about 50 mA to 1 A. The core in such an inductor may be made adjustable so that it can be screwed into or partially out of the coil. Thus, the coil inductance is variable. Note the graphic symbol for an inductor with an adjustable core [Fig. A. 3.6(b)].

APPENDIX - 4

Inductors

Magnetic Flux and Flux Density

*The weber** (Wb) *is the SI unit of magnetic flux.*

The weber is defined as the magnetic flux which, linking a single-turn coil, produces an emf of 1 V when the flux is reduced to zero at a constant rate in 1 s.

*The tesla**** (T) *is the SI unit of magnetic flux density.*

The tesla is the flux density in a magnetic field when 1 Wb of flux occurs in a plane of 1 m^2*; that is, the tesla can be described as 1* Wb/m^2.

Inductance

The SI unit of inductance is the henry (H).

The inductance of a circuit is 1 henry (1 H) when an emf of 1 V is induced by the current changing at the rate of 1 A/s.

Molded Inductors

A small molded inductor is shown in Fig. A. 4.2(c). Typical available values for this type range from 1.2 μH to 10 mH, maximum currents of about 70 mA. The values of molded inductors are identified by a color code, similar to molded resistors. Fig. A. 4.2(d) shows a tiny-film inductor used in certain types of electronic circuits. In this case the inductor is simply a thin metal film deposited in the form of a spiral on ceramic base.

Laboratory Inductors

Laboratory-type variable inductors can be constructed in decade box format, in which precision inductors are switched into or out of a circuit by means of rotary switches. Alternatively, two coupled coils can be employed as a variable inductor. The coils may be connected in series or in parallel, and the total inductance is controlled by adjusting the position of one coil relative to the other.

Color Code For Small Inductors

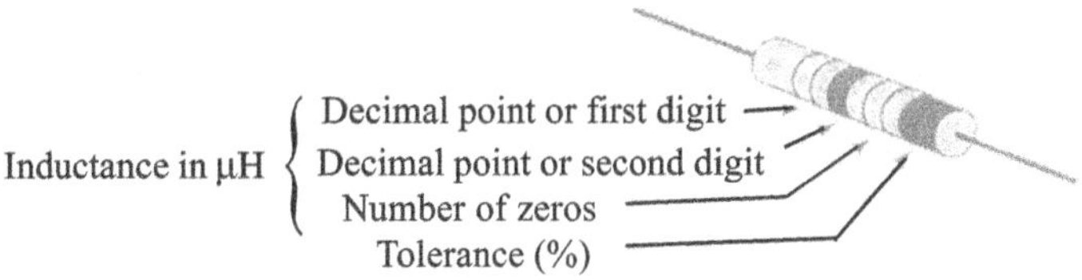

Color	Significant Figure	Tolerance (%)
Black	0	
Brown	1	
Red	2	
Orange	3	
Yellow	4	
Green	5	
Blue	6	
Violet	7	
Grey	8	
White	9	
Silver		10
Gold	decimal point	5
No band		20

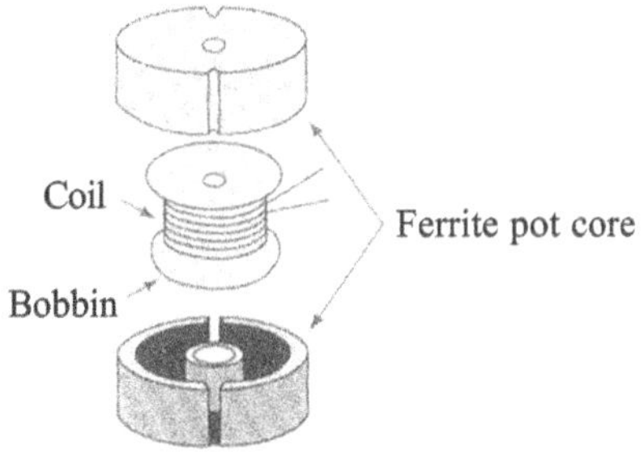

Fig. A. 4.1 Some low-current, high-frequency inductors are wound on bobbins contained in a ferrite pot core. The ferrite core increases the winding inductance and screens the inductor.

Coil to protect adjacent components against flux leakage and to protect the coil from external magnetic fields. The coil is wound on a bobbin, so its number of turns is easily modified.

Three different types of low-current inductors are illustrated in Fig. A. 4.2. Fig. A 4.2(a) shows a type that is available either as an air-cored inductor or with a ferromagnetic core. With an air core, the inductance values up to about 10 mH can be obtained. Depending on the thickness of wire used and the physical.

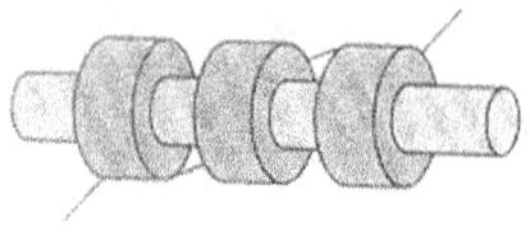

(a) Inductor with air core or ferromagnetic core

(b) Circuit symbol for an inductor with an adjustable ferromagnetic core

(c) Molded inductor

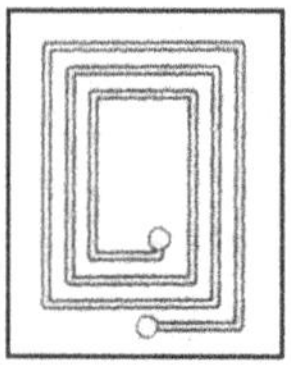

(d) Thin-film inductor

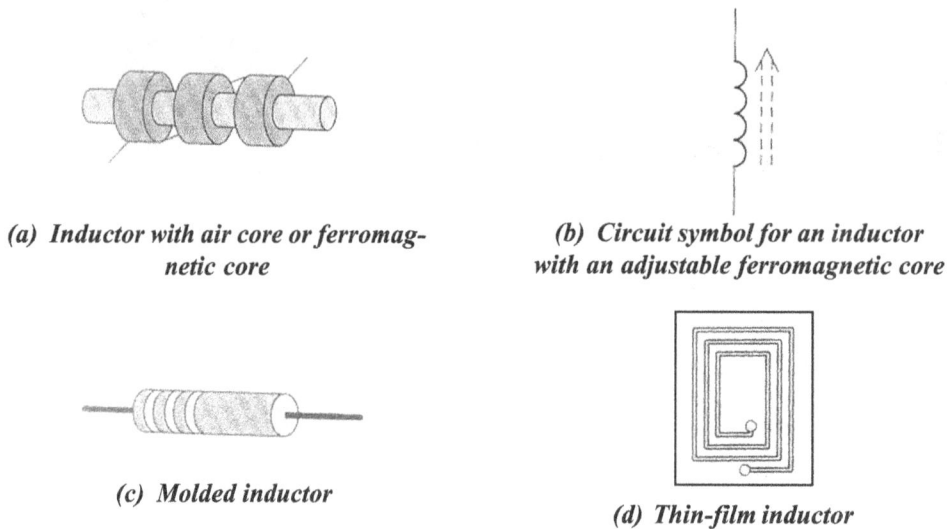

Fig. A. 4.2 Small inductors may be wound on an insulating tube with an adjustable ferrite core, molded like small resistors, or deposited as a conducting film on an insulating material.

If the mutual inductance between two adjacent coils is not known, it can be determined by measuring the total inductance of the coils in series-aiding and series-opposing connections. Then,

$$L_a = L_1 + L_2 + 2M \quad \text{for series-aiding}$$

and $$L_b = L_1 + L_2 - 2M \quad \text{for series-opposing}$$

Subtracting, $$L_a - L_b = 4M$$

Therefore, $$\boxed{M = \frac{L_a - L_b}{4}}$$

$$M = k\sqrt{L_1 L_2}$$

From these two equations, the coefficient of coupling of the two coils can be determined.

Stray Inductance

Inductance is (change in flux linkages) / (change in current). So every current-carrying conductor has some self-inductance, and every pair of conductors has inductance. These *stray inductance* are usually unwanted, although they are sometimes used as components in a circuit design. In dc applications, stray inductance is normally unimportant, but in radio frequency ac circuits it can be considerable nuisance. Stray inductance is normally minimized by keeping connecting wires as short as possible.

Summary of Formulae

Induced emf	:	$e_L = \dfrac{\Delta\Phi}{\Delta t}$
Induced emf	:	$e_L = \dfrac{\Delta\Phi}{\Delta t}$
Inductance	:	$L = \dfrac{e_L}{\Delta i / \Delta t}$
Inductance	:	$L = \dfrac{\Delta\Phi N}{\Delta i}$
Flux change	:	$\Delta\Phi = \mu, \mu_0 \, \Delta i \, N \, \dfrac{A}{t}$
Self - inductance	:	$L = \mu, \mu_0 \, \Delta i \, N^2 \, \dfrac{A}{t}$
Mutual inductance	:	$M = \dfrac{e_L}{\Delta i / \Delta t}$
Induced emf	:	$e_L = \dfrac{\Delta\Phi N_s}{\Delta t}$
Mutual inductance	:	$M = \dfrac{\Delta\Phi N_s}{\Delta i}$
Mutual inductance	:	$M = k \, \dfrac{\Delta\Phi N_s}{\Delta i}$
Mutual inductance	:	$M = k\sqrt{L_1 L_2}$
Energy stored	:	$W = \dfrac{1}{2} LI^2$
Energy stored	:	$W = \dfrac{B^2 Al}{2\mu_0}$
Inductances in series	:	$L_s = L_1 + L_2 + L_3 + \ldots\ldots$
Inductances in parallel	:	$\dfrac{1}{L_P} = \dfrac{1}{L_1} + \dfrac{1}{L_2} + \dfrac{1}{L_3} + \ldots\ldots$
Total inductance (series-aiding)	:	$L = L_1 + L_2 + 2M$
Total inductance (series-opposing)	:	$L = L_1 + L_2 - 2M$
Mutual inductance	:	$M = \dfrac{L_o - L_b}{4}$

APPENDIX - 5

Miscellaneous

Ionic Bonding

In some insulating materials, notably rubber and plastics, the bending process is also covalent. The valence electrons in these bonds are very strongly attached to their atoms, so the possibility of current flow is virtually zero. In other types of insulating materials, some atoms have parted with outer-shell electrons, but these have been accepted into the orbit of other atoms. Thus, the atoms are *ionized*; those which gave up electrons have become *positive ions*, and those which accepted the electrons become negative ions. This creates an electrostatic bonding force between the atoms, termed ionic bonding. Ionic bonding is found in such materials as glass and porcelain. Because there are virtually no free electrons, no current can flow, and the material is an insulator.

Insulators

Fig. A. 5.1 shows some typical arrangements of conductors and insulators. Electric cable usually consists of conducting copper wire surrounded by an insulating sheath of rubber or plastic. Sometimes there is more than one conductor, and these are, of course, individually insulated.

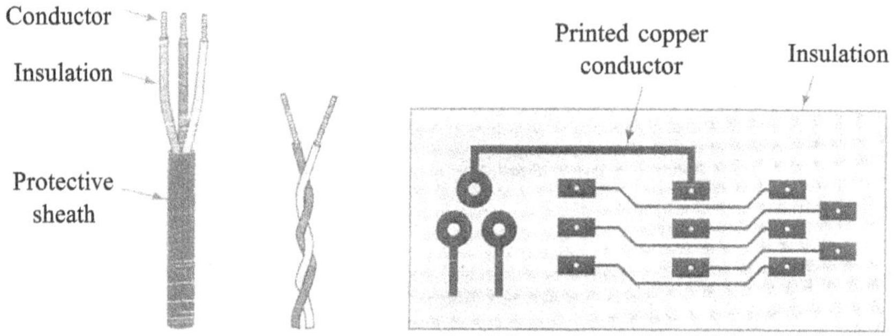

Fig. A. 5.1 Conductors employed for industrial and domestic purposes normally have stranded copper wires with rubber or plastic insulation. In electronics equipment, flat cables of fine wires and thin printed circuit conductors are widely used.

Conductors

The function of a conductor is to conduct current form one point to another in an electric circuit. As discussed, electric cables usually consist of copper conductors sheathed with rubber or plastic

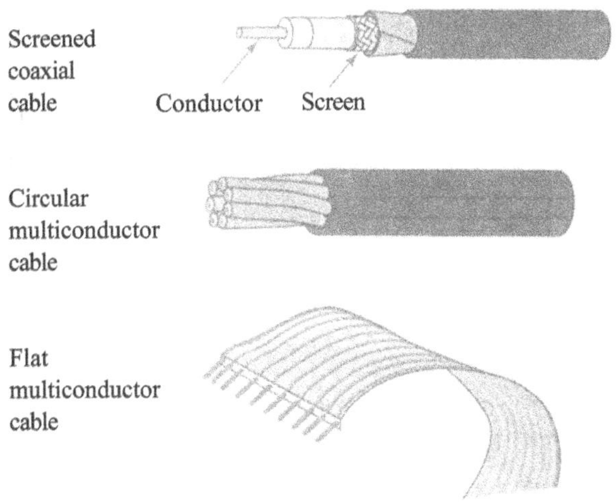

Fig. A. 5.2 Many different types of cables are used with electronics equipment.

insulating material. Cables that have to carry large currents must have relatively thick conductors. Where very small currents are involved, the conductor may be a thin strip of copper or even an aluminium film. Between these two extremes, a wide range of conductors exist for various applications. Three different types of cables used in electronics equipment are illustrated in Fig. A. 5.2 conductor and a circular plated conducting screen, as well as an outer insulating sheath. The other two are multiconductor cables, one circular, and one flat.

Because each conductor has a finite resistance, a current passing through it causes a voltage drop from one end of the conductor to the other (Fig. A. 5.3). When conductors are long and/or carry large currents, the conductor voltage drop may cause unsatisfactory performance of the equipment supplied. Power (I^2 R) is also dissipated in every current-carrying conductor, and this is, ofcourse, wasted power.

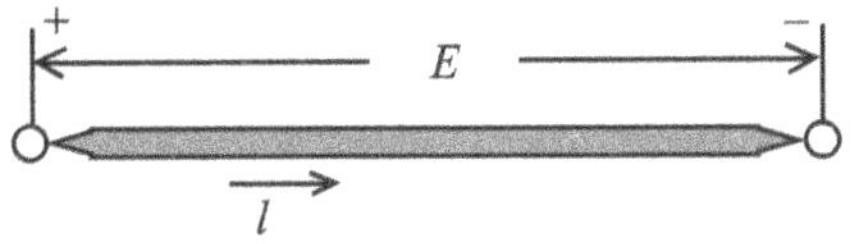

(a) Current flow through a conductor produces a voltage drop along the conductor

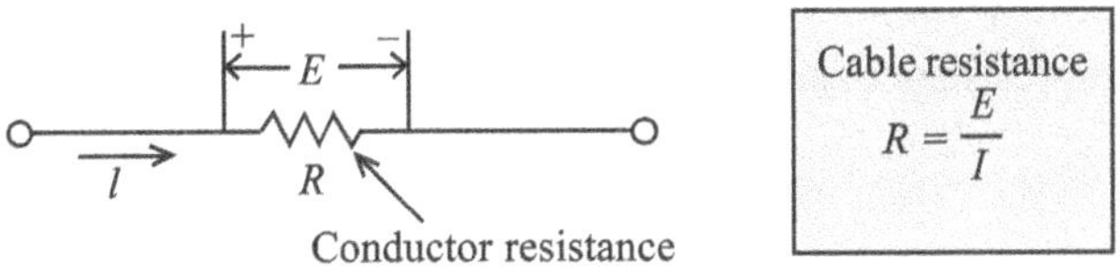

(b) Conductor resistance causes voltage drop when a current flows

Fig. A. 5.3 Conductor resistance (R) is determined by applying the voltage drop and current level to Ohm's law. The resistance per unit length (R/l) is then used to select a suitable wire gauge.

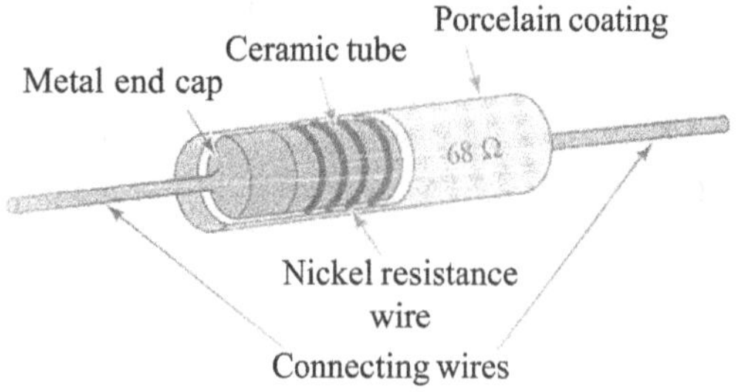

(a) Wire-wound resistor

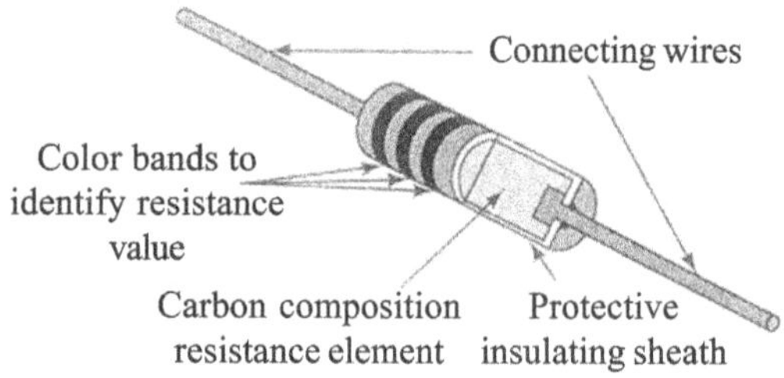

(b) Carbon composition resistor

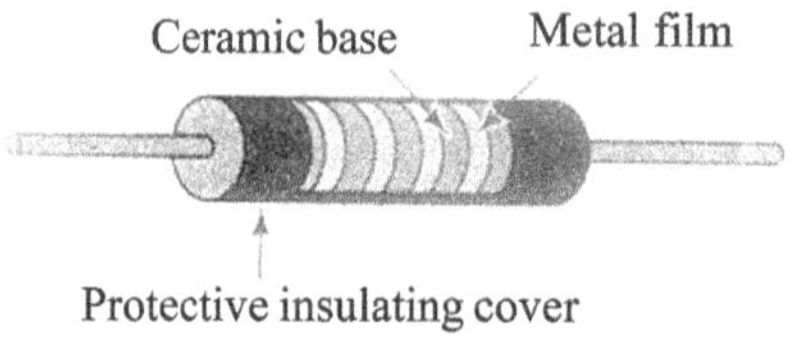

(c) Metal film resistor

Fig. A. 5.4 Individual resistors are typically wire-wound or carbon composition construction. Wirewound resistors are used where high power dissipation is required. Carbon composition type is the least expensive. Metal film resistance values can be more accurate than carbon composition type.

The illustration in Fig. A. 5.5(a) shows a coil of closely wound insulated resistance wire formed into partial circle. The coil has a low-resistance terminal at each end, and a third terminal is connected to a movable contact with a shaft adjustment facility. The movable contact can be set to any point on a connecting track that extends over one (unisulated) edge of the coil. Using the adjustable contact, the resistance from either end terminal to the center terminal may be adjusted from zero to the maximum coil resistance.

Another type of variable resistor, known as a decade resistance box, is shown in Fig. A. 5.5(c). This is a laboratory component that contains precise values of switched series-connected resistors. As illustrated, the first switch (from the right) controls resistance values in 1Ω steps from 0Ω to 9Ω and the second switches values of 10Ω, 20Ω, 30Ω, and so on. The decade box shown can be set to within + 1Ω of any value from 0Ω to 9999Ω. Other decade boxes are available with different resistance ranges.

Resistor Tolerance

Standard (fixed-value) resistors normally range from 2.7Ω to 22MΩ. The resistance tolerances on these standard values are typically + 20%, + 10%, + 5% or + 1%. A tolerance of + 10% on a 100Ω resistor means that the actual resistance may be as high as 100Ω + 10% (i.e., 110Ω) or as low as 100Ω – 10% (i.e., 90Ω). Obviously, the resistors with the smallest tolerance are the most accurate and the most expensive.

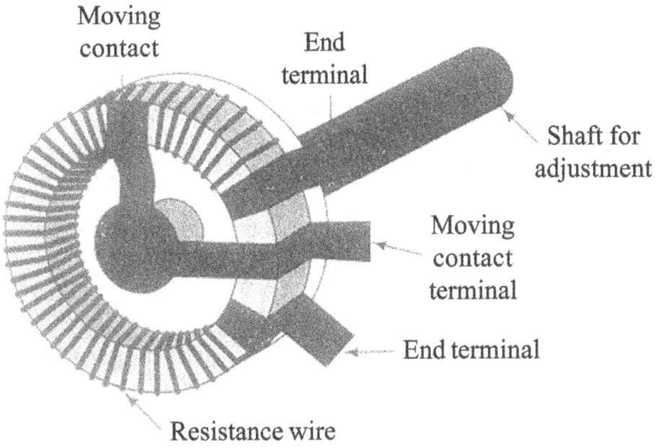

(a) Typical construction of a resistor variable resistor (and potentiometer)

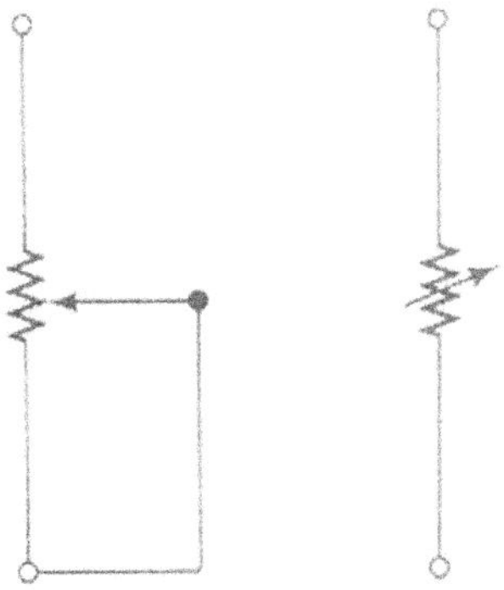

(b) Circuit symbols for a variable resistor

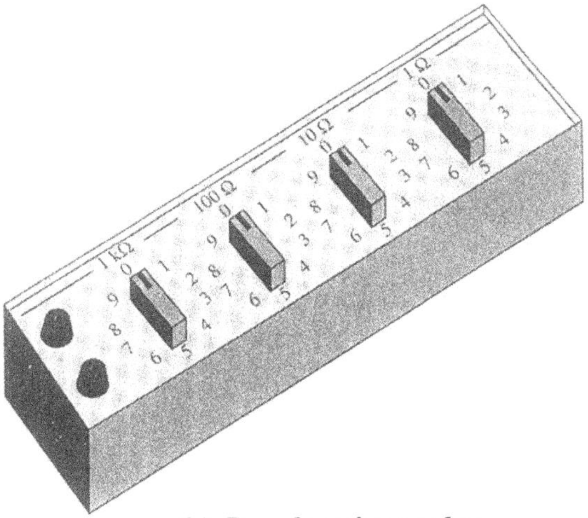

(c) Decade resistance box

Fig. A. 5.5 Small variable resistors are used in electronic circuit construction. Large decade resistance boxes are employed in electronics laboratories.

More Resistors

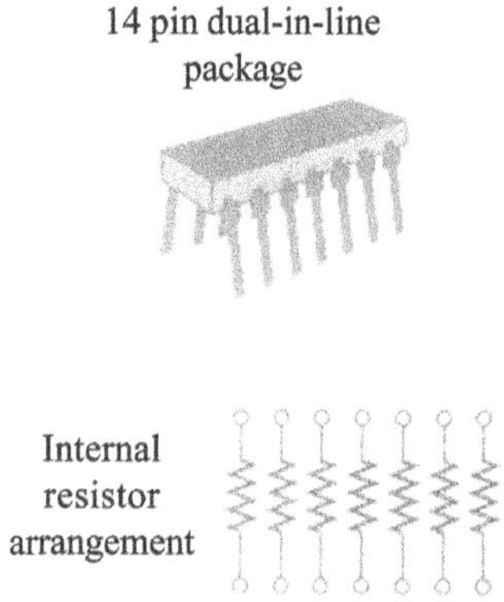

Resistor Networks

Resistor networks are available in *integrated circuits* type *dual-in-lin* package. One construction method uses a *thick film* technique in which conducting solutions are deposited in the required form.

Photoconductive Cell

This is simply a resistor constructed of photoconductive material (cadmium selenide or cadmium sulfide). When dark, the cell resistance is very high. When illuminated, the resistance decreases in proportion to the level of illumination.

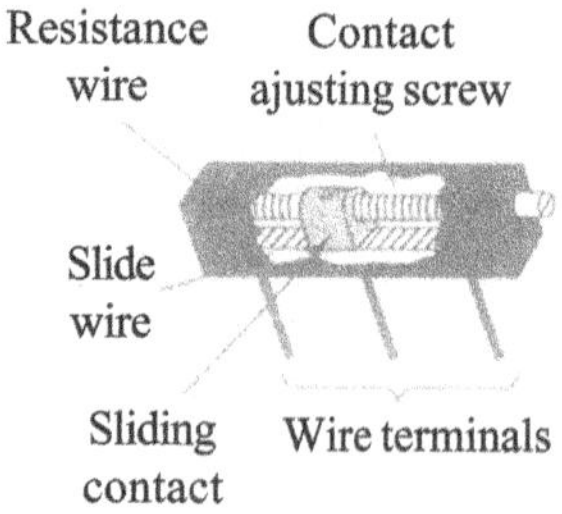

Low Power Variable Resistor

A small variable resistor suitable for mounting directly on a circuit board. A threaded shaft, which is adjustable by a screwdriver, sets the position of the moving contact on a resistance wire.

High Power Resistor

High power resistors are usually wire-wound on the surface of a ceramic tube. Air flow through the tube helps to keep the resistor from overheating.

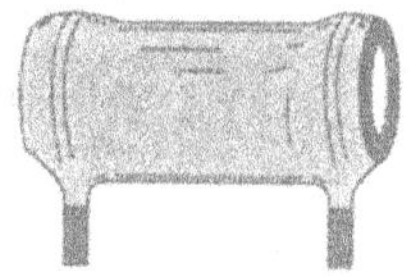

Two memory aids for determining the direction of the magnetic flux around a current-carrying conductor are shown in Fig. A. 5.6. The right-hand-screw rule as illustrated in Fig. A. 5.6(a) shows a wood screw being turned clockwise and progressing into a piece of wood. The horizontal direction of the screw is analogous to the direction of current in a conductor, and the circular motion of the screw shows the direction of magnetic flux around the conductor. In the right-hand rule, illustrated in Fig. A. 5.6(b), a right hand is closed around a conductor with the thumb p.ointing in the (conventional) direction of current flow. The fingers point in the direction of the magnetic lines of force around the conductor.

Because a current-carrying conductor has a magnetic field around it, when two current-carrying conductors are brought close together there will be interaction between the fields. Fig. A. 5.7(a) shows the effect on the fields when two conductors carrying in opposite direction are adjacent. The directions of the magnetic passes through the center of the coil. Therefore, the one-turn coil acts like a little magnet and has a magnetic field with an identifiable N pole and S pole. Instead of a single turn, the coil may have many turns, as illustrated in Fig. A. 5.7(c). In this case the flux generated by each of the individual current-carrying turns tends to link up and pass out of one end of the coil and back into the other end. This type of coil, known as a solenoid, obviously has a magnetic field pattern very similar to that of a bar magnet.

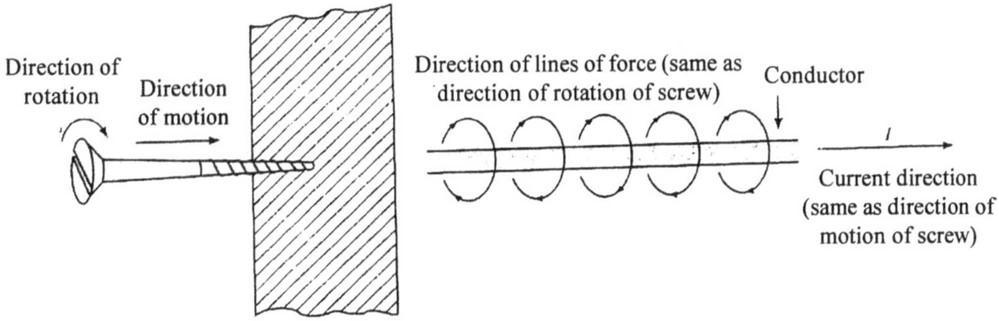

(a) Right-hand screw rule

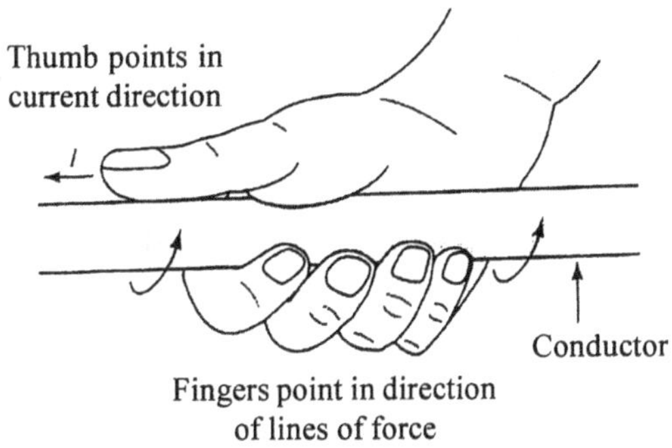

(b) Right-hand rule

Fig. A. 5.6 The right-hand-screw rule and the right-hand rule can be used for determining the direction of the magnetic lines of force around a current-carrying conductor.

The right-hand rule for determining the direction of flux from a solenoid is illustrated in Fig. A. 5.7(d). When the solenoid is gripped with the right hand so that the fingers are pointing in the direction of current flow in the coils, the thumb points in the direction of the flux (i.e., toward the N-pole end of the solenoid).

Electromagnetic Induction

It has been demonstrated that a magnetic flux is generated by an electric current flowing in a conductor. The converse is also possible; that is, a magnetic flux can produce a current flow in a conductor.

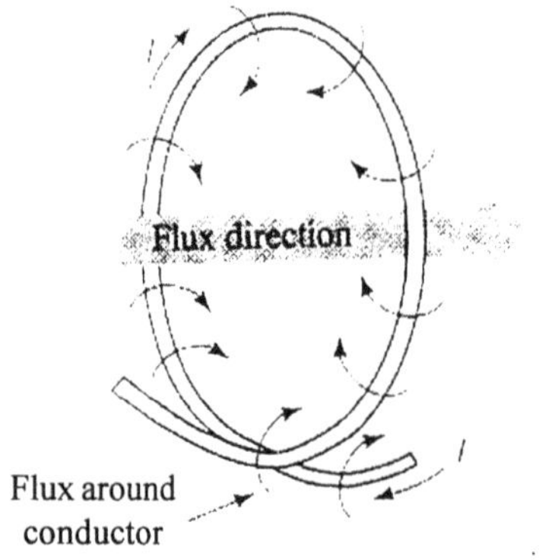

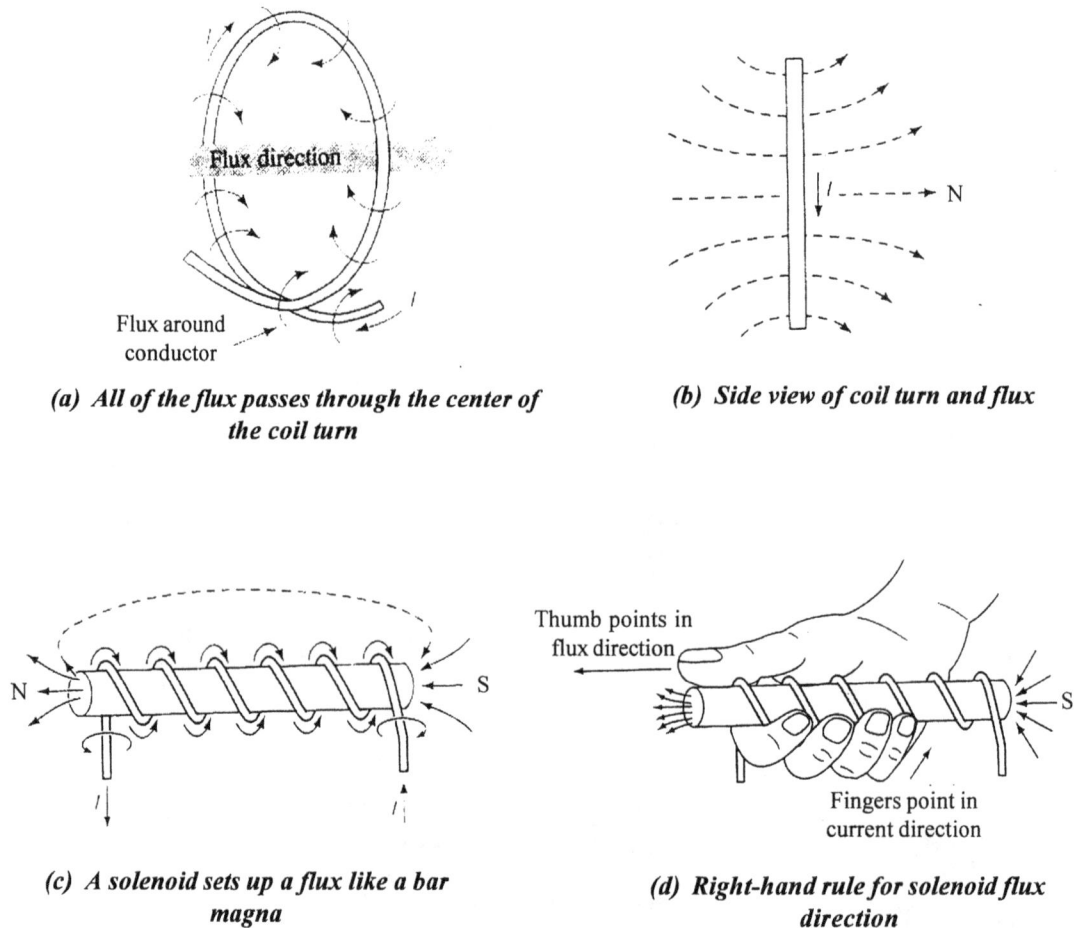

(a) All of the flux passes through the center of the coil turn

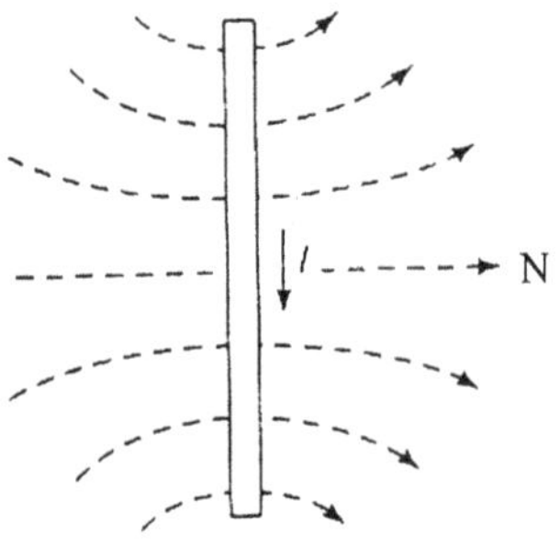

(b) Side view of coil turn and flux

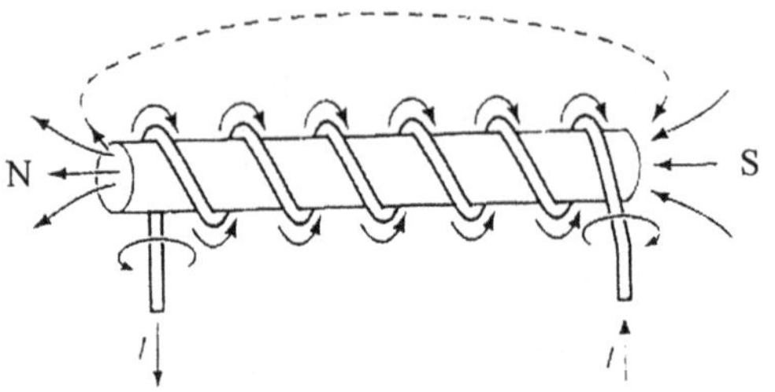

(c) A solenoid sets up a flux like a bar magna

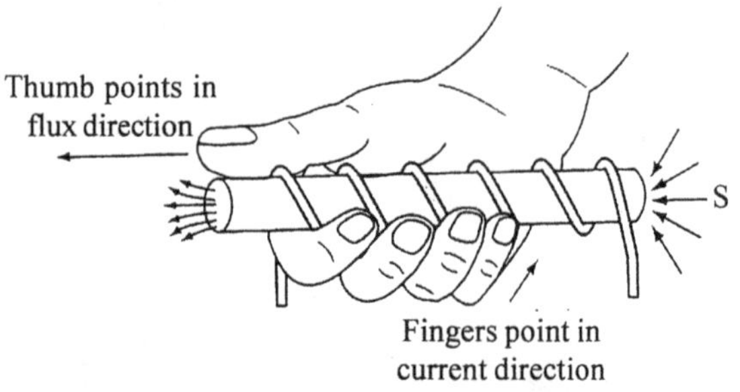

(d) Right-hand rule for solenoid flux direction

Fig. A. 5.7 In current-carrying coils, the magnetic lines of force around the conductors all pass through the center of the coil.

Consider Fig. A. 5.8(a), in which a handled bar magnet is shown being brought close to a coil of wire. As the bar magnet approaches the coil, the flux from the magnet *brushes across* the coil conductors or cuts the conductors. This produces a current flow in the conductors proportional to the total flux that cuts the coil. If the coil circuit is closed by a resistor (as shown broken in the figure), a current flows. Whether or not the circuit is closed, an *electromotive force* (emf) can be measured at the coil terminals. This effect is known as *electromagnetic* induction.

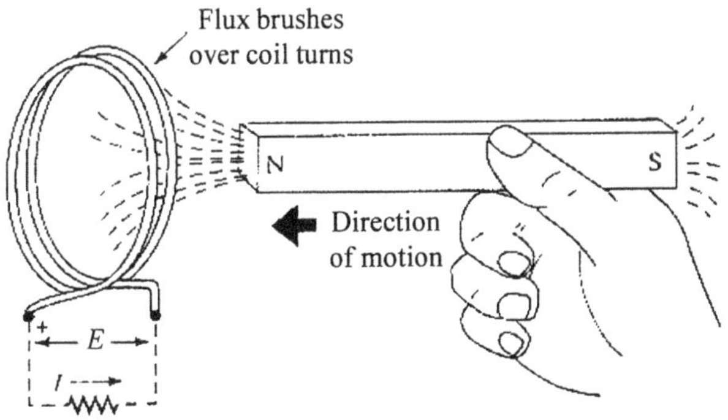

(a) emf induced in a coil by the motion of the flux from the bar magnet

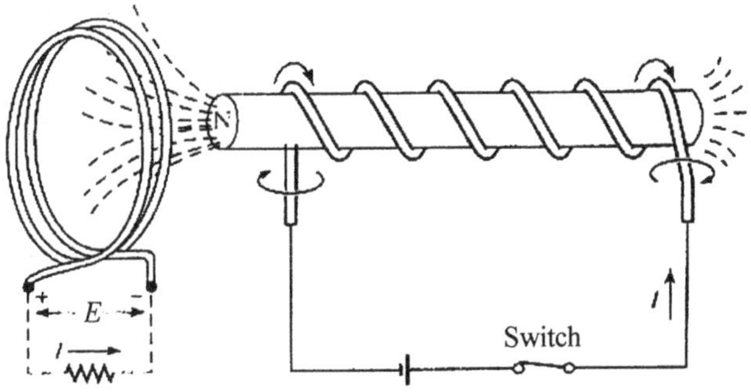

(b) emf induced in a coil by the motion of the flux from the solenoid when the current is switched on or off

Fig. A. 5.8 An electromotive force (emf) is induced in a coil when the coil is brushed by a magnetic field. The magnetic field may be from a bar magnet or from a current-carrying coil.

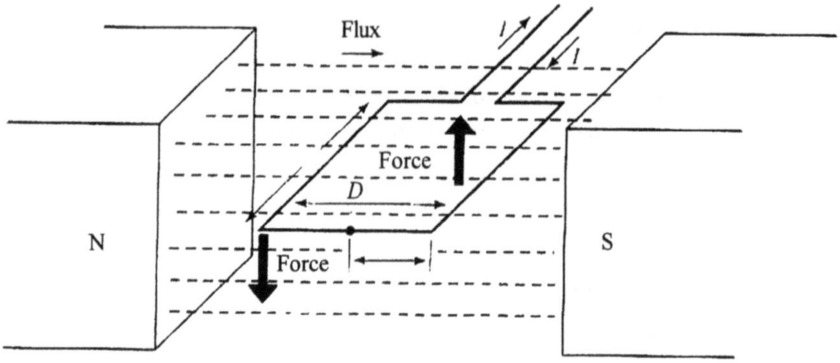

(a) Single-turn coil pivoted in a magnetic field

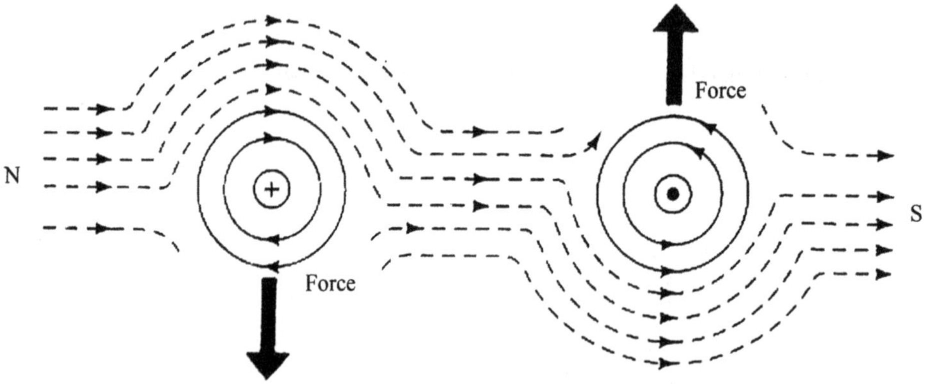

(b) Showing the force on each side of a single-turn pivoted in a magnetic field

Fig. A. 5.9 ***A force is exerted on each side of a current-carrying coil pivoted in a magnetic field. This force tends to cause the coil to rotate.***

Fibre Optic Cables :

Light and Other Rays

Light rays or waves are a type of energy called electromagnetic energy. They shine out, or radiate, from their source, so they are called electromagnetic radiation. Light rays are just a tiny part of a huge range of rays and waves called the electromagnetic spectrum (EMS)

The parts of the electromagnetic spectrum have different wavelengths and different names.

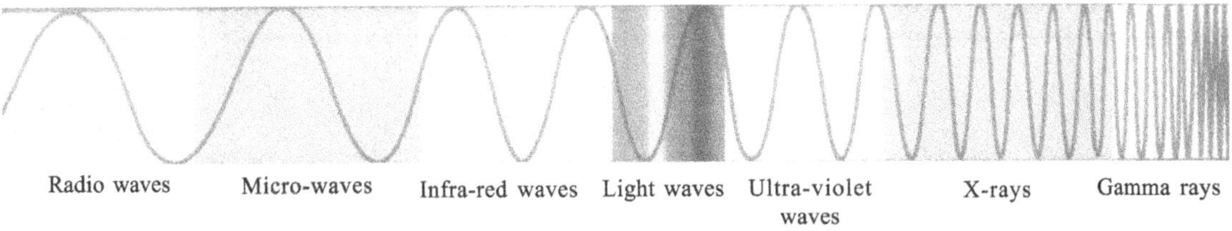
Radio waves
Micro-waves
Infra-red waves
Light waves
Ultra-violet waves
X-rays
Gamma rays

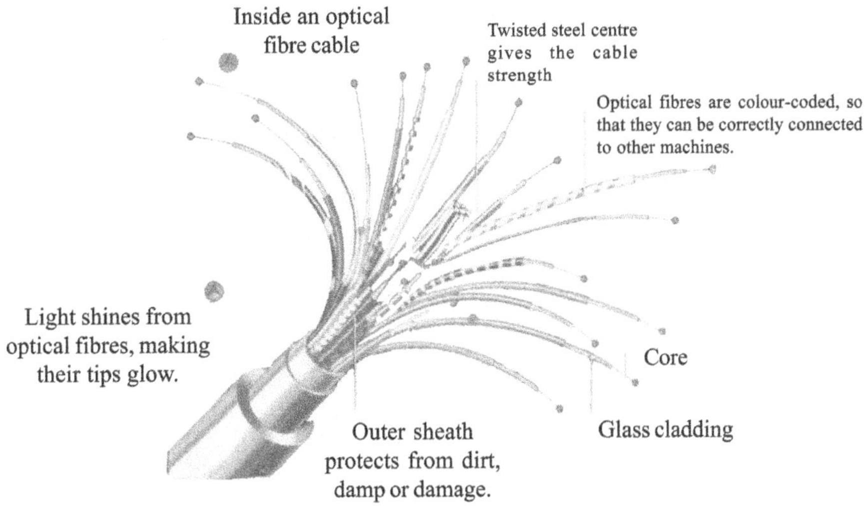
Inside an optical fibre cable
Twisted steel centre gives the cable strength
Optical fibres are colour-coded, so that they can be correctly connected to other machines.
Light shines from optical fibres, making their tips glow.
Core
Outer sheath protects from dirt, damp or damage.
Glass cladding

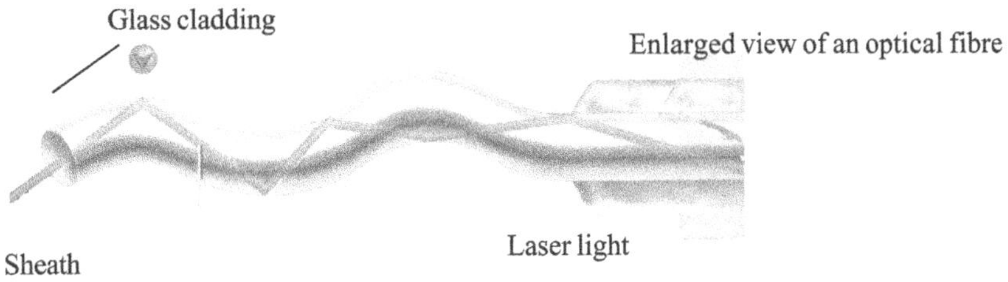
Glass cladding
Enlarged view of an optical fibre
Laser light
Sheath

Circuit Symbols

dc voltage source
or single-cell
battery

Multicell
battery

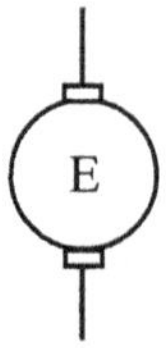

Generator

Current
source

ac voltage
source

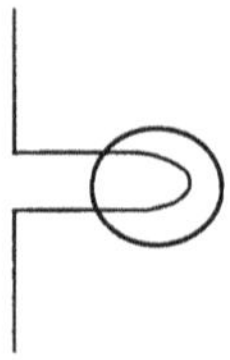

Lamp

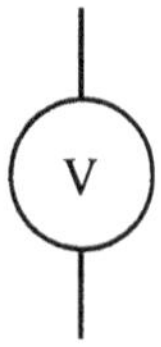

Voltmeter

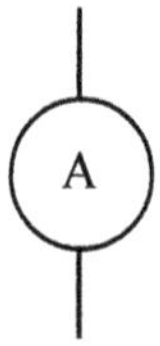

Ammeter

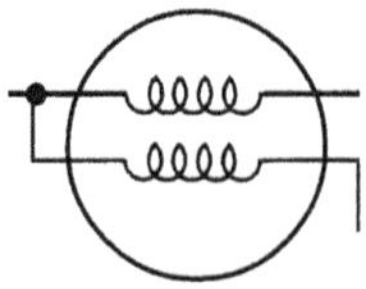

Wattmeter

Chassis

Ground

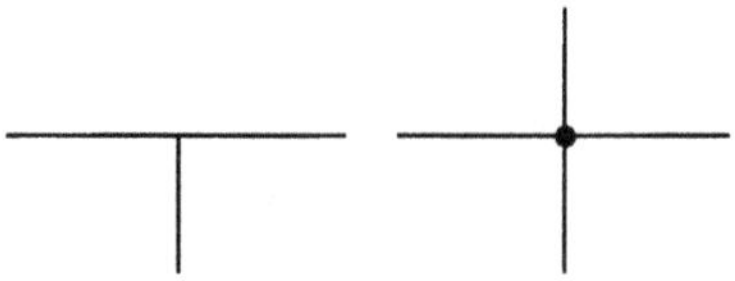

Conductor connection
or junction

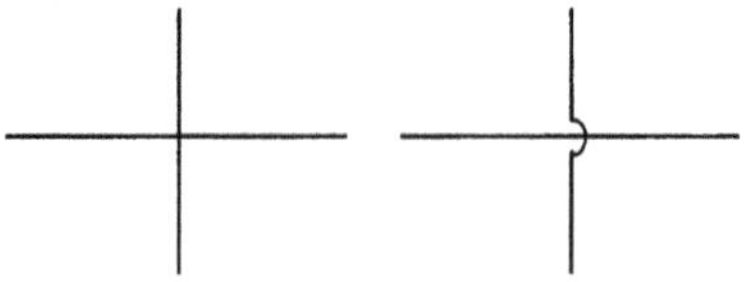

Unconnected crossing
conductors

Fuse

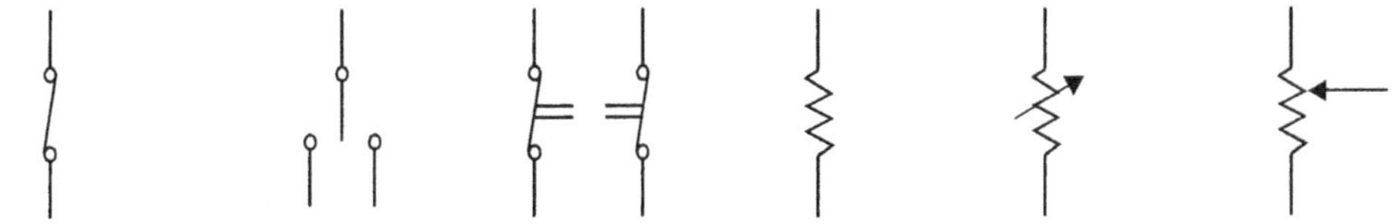

dc voltage source or single-cell battery | Two-way switch | Double-pole switch | Resistor | Variable resistor | Potentiometer

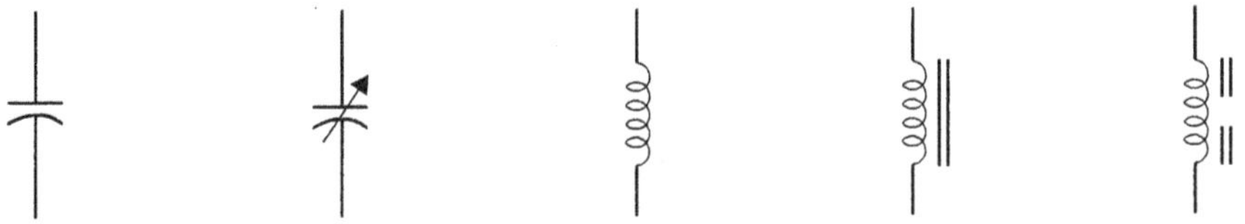

Capacitor | Variable capacitor | Air-cored inductor | Iron-cored inductor | Variable inductor

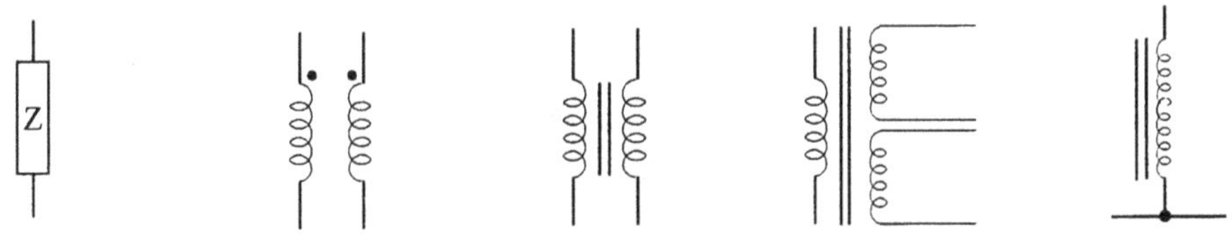

Capacitor | Inductor | Air-cored inductor | Iron-cored inductor | Variable inductor

APPENDIX - 7

Unit Conversion Factors

The following factors may be used for conversion between non-SI units and SI units.

To Convert	To	Multiply By
Area Units		
acres	square meters (m^2)	4047
acres	hectares (ha)	0.4047
circular mils	square meters (m^2)	5.067×10^{-10}
square feet	square meters (m^2)	0.0909
square inches	square centimeters (cm^2)	6.452
square miles	hectares (ha)	259
square miles	square kilometers (km^2)	2.59
square yards	square meters (m^2)	0.8361
Electric and Magnetic Units		
amperes/inch	amperes/meter (A/m)	39.37
gauses	teslas (T)	10^{-4}
gilberts	ampere (turns) (A)	0.7958
lines/sq. inch	teslas (T)	1.55×10^{-5}
Maxwells	webers (Wb)	10^{-8}
mhos	Siemens (S)	1
Oersteds	amperes/meter	79.577
Energy and Work Units		
Btu	joules (J)	1054.8
Btu	kilowatt-hours (kWh)	2.928×10^{-4}
ergs	joules (J)	10^{-7}
ergs	kilowatt-hours (kWh)	0.2778×10^{-13}
foot-pounds	joules (J)	1.356
foot-pounds	kilogram meters (kgm)	0.1383

Force Units

dynes	grams (g)	1.02×10^{-3}
dynes	newtons (N)	10^{-5}
pounds	newtons (N)	4.448
poundals	newtons (N)	0.1383
grams	newtons (N)	9.807×10^{-3}

Illumination Units

foot-candles	lumens/cm^2	10.764

To Convert	To	Multiply By
Linear Units		
angstroms	meters (m)	1×10^{-10}
feet	meters (m)	0.3048
fathoms	meters (m)	1.8288
inches	centimeters (cm)	2.54
microns	meters (m)	10^{-6}
miles (nautical)	kilometers (km)	1.853
miles (statute)	kilometers (km)	1.609
mils	centimeters (cm)	2.54×10^{-3}
yards	meters (m)	0.9144
Power Units		
horsepower	watts (W)	745.7
Pressure Units		
atmospheres	kilograms/sq. meter (kg/m^2)	10 332
atmospheres	kilopascals (kPa)	101.325
bars	kilopascals (kPa)	100
bars	kilograms/sq.meter (kg/m^2)	1.02 × 10–4
pounds/sq. foot	kilograms/sq.meter (kg/m^2)	4.882
pounds/sq. inch	kilograms/sq.meter (kg/m^2)	703
Temperature Units		
degrees Fahrenheit (°F)	degrees celsius (°C)	(°F – 32)/1.8
degrees Fahrenheit (°F)	degrees kelvin (K)	273.15 + (°F – 32)/1.8

Velocity Units		
miles/hour (mph)	kilometers/hour (km/h)	1.609
knots	kilometers/hour (km/h)	1.853
Volume Units		
bushels	cubic meters (m^3)	0.035 24
cubic feet	cubic meters (m^3)	0.028 32
cubic inches	cubic centimeters (cm^3)	16.387
cubic inches	liters (1)	0.016 39
cubic yards	cubic meters (m^3)	0.7646
gallons (U.S.)	cubic meters (m^3)	3.7853×10^{-3}
gallons (imperial)	cubic meters (m^3)	4.546×10^{-3}
gallons (U.S.)	liters (1)	3.7853
gallons (imperial)	liters (1)	4.546
gills	liters (1)	0.1183
pints (U.S.)	liters (1)	0.4732
pints (imperial)	liters (1)	0.5683

Gauge	Diameter (mm)	Copper Wire Resistance (Ω/km)	Diameter (mil)	Copper Wire Resistance (Ω/km)
36	0.127	1360	5	415
37	0.113	1715	4.5	523
38	0.101	2147	4	655
39	0.090	2704	3.5	832
40	0.080	3422	3.1	1044

To Convert	To	Multiply By
quarts (U.S.)	liters (1)	0.9463
quarts (imperial)	liters (1)	1.137
Weight Units		
ounces	grams (g)	28.35
pounds	kilograms (kg)	0.453 59
tons (long)	kilograms (kg)	1016
tons (short)	kilograms (kg)	907.18

The siemens* is the unit of conductance.

$$\text{conductance} = \frac{1}{\text{resis tan ce}}$$

American Wire Gauge Sizes and Metric Equivalents

Gauge	Diameter (mm)	Copper wire Resistance (Ω/km)	Diameter (mill)	Copper wire Resistance (Ω/1000 ft)
0000	11.68	0.160	460	0.049
000	10.40	0.203	409.6	0.062
00	9.266	0.255	364.8	0.078
0	8.252	0.316	324.9	0.098
1	7.348	0.406	289.3	0.124
2	6.543	0.511	257.6	0.156
3	5.827	0.645	229.4	0.197
4	5.189	0.813	204.3	0.248
5	4.620	1.026	181.9	0.313
6	4.115	1.29	162	0.395
7	3.665	1.63	144.3	0.498
8	3.264	2.06	128.5	0.628
9	2.906	2.59	114.4	0.792
10	2.588	3.27	101.9	0.999
11	2.30	4.10	90.7	1.26
12	2.05	5.20	80.8	1.59
13	1.83	6.55	72	2
14	1.63	8.26	64.1	2.52
15	1.45	10.4	57.1	3.18
16	1.29	13.1	50.8	4.02
17	1.15	16.6	45.3	5.06
18	1.02	21.0	40.3	6.39

Contd...

19	0.912	26.3	35.9	8.05
20	0.813	33.2	32	10.1
21	0.723	41.9	28.5	12.8
22	0.644	52.8	25.3	16.1
23	0.573	66.7	22.6	20.3
24	0.511	83.9	20.1	25.7
25	0.455	106	17.9	32.4
26	0.405	134	15.9	41
27	0.361	168	14.2	51.4
28	0.321	213	12.6	64.9
29	0.286	267	11.3	81.4
30	0.255	337	10	103
31	0.227	425	8.9	130
32	0.202	537	8	164
33	0.180	676	7.1	206
34	0.160	855	6.3	261
35	0.143	1071	5.6	329

Answers to Objective Type and Multiple Choice Questions

Chapter - 1

Answers to Objective Type Questions

1. $< 10^{28}/m^3$ and $> 10^7$ electrons/m^3
2. 10^9 Ω-cm.
3. $J = (ne\mu_n + pe\mu_p).E.e$
4. $n \times p = n_i^2$
5. $n_i \propto T^{3/2}$
6. Decreases as V_e increases
7. Valance Band
8. Con`duction Band
9. It gets doubled for every 10^0 rise in temperature
10. $V_T = 0.026V$
11. $\frac{D_p}{\mu_p} = \frac{D_n}{\mu_n} = V_T$
12. Insulator
13. Lower than, Conduction
14. Impurities added in, n_i^2
15. Immobile, Space
16. $E_o = KT \ln\left(\frac{N_C N_A}{n_i^2}\right)$
17. $E_o = 0.5eV$
18. $p_n(0) = p_{no} e^{V/V_T}$
19. $I = I_o(e^{\frac{V}{\eta V_T}} - 1)$
20. Germanium = 0.1V and Silicon = 0.5V
21. $V_B = \frac{eN_A}{2\in} W^2$
22. $I_o = KT^m e^{-\frac{V_{G_0}}{\eta V_T}}$
23. High
24. AC to unidirectional flow
25. Unidirectional flow to DC
26. Ripple Factor = 1.21
27. Inductance varies with current in permissable limits.
28. Good regulation and conduction angle of 180^0 for the diode

Answers to Multiple Choice Questions

1.	a	**2.**	b	**3.**	d	**4.**	a	**5.**	d	**6.**	a	**7.**	c
8.	b	**9.**	b	**10.**	c	**11.**	b	**12.**	c	**13.**	a	**14.**	d
15.	a	**16.**	c	**17.**	d	**18.**	c						

Chapter - 2

Answers to Objective Type Questions

1. NPN
2. the direction of conventional current when the E-B Junction is forward biased.
3. $\dfrac{I_{PE}}{I_E}$
4. $\dfrac{I_{pC}}{I_{pE}}$
5. $\alpha = \beta^* \gamma$
6. $\alpha = \dfrac{\beta}{(1+\beta)}$
7. $\beta = \dfrac{\alpha}{(1-\alpha)}$
8. $I_C = \beta I_B + (\beta + 1)I_{CBO}$
9. $I_{CEO} = \dfrac{I_{CBO}}{(1-\alpha)}$
10. $\beta' = \left.\dfrac{\partial I_C}{\partial I_B}\right|_{V_{CE}=K}$
11. Small Signal Common Emitter Forward Current Gain
12. $\beta' = \dfrac{\beta}{\left[1-(I_{CBO}+I_B)\dfrac{\partial h_{FE}}{\partial I_C}\right]}$
13. h_{fe} and h_{FE}
14. Large Signal Current Gain
15. Small Signal Current Gain
16. Change in base width with the change in the base voltage V_{BE}
17. MOSFET
18. UNIPOLAR
19. gain band width product is less.
20. C, B, E
21. Pinch off Voltage
22. Pinch off Voltage
23. few tens of ohms to few hundred ohms.
24. $\dfrac{1}{\sqrt{E_x}}$
25. $V_{GS} = \left(1-\dfrac{b}{a}\right)^2 . V_P$
26. $I_{DS} = I_{DSS}\left(1-\dfrac{V_{GS}}{V_P}\right)^2$
27. the saturation value of the drain current when gate is shorted to source. i.e., $V_{GS} = 0$
28. voltage variable
29. $\mu = r_d * g_m$
30. JFET
31. Depletion MOSFET (DMOSFET)

Answers to Multiple Choice Questions

1.	b	**2.**	a	**3.**	c	**4.**	d	**5.**	c	**6.**	a	**7.**	c
8.	d	**9.**	a	**10.**	b	**11.**	c	**12.**	d	**13.**	b	**14.**	a
15.	c	**16.**	d	**17.**	a								

Chapter - 3

Answers to Objective Type Questions

1. Quiescent point, Q - point, Biasing point
2. I_{CO}, V_{BE}, β
3. The centre of the active region on the load line.
4. $V_{CE} = 0.5\ V_{CC}$.
5. Base bias circuit.
6. Emitter bias, semiversal bias or Voltage divider bias circuit.
7. $S = \left.\dfrac{\partial I_C}{\partial I_{CO}}\right|_{\beta = K,\ V_{BE} = K}$
8. $S' = \left.\dfrac{\partial I_C}{\partial V_{BE}}\right|_{I_{CO} = K, \beta = K}$
9. $S'' = \left.\dfrac{\partial I_C}{\partial \beta}\right|_{V_{BE} = K, I_{CO} = K}$
10. $S = \dfrac{1+\beta}{1-\beta\left(\dfrac{\partial I_B}{\partial I_C}\right)}$
11. $S = \dfrac{1+\beta}{1+\beta\left(\dfrac{R_E}{R_E + R_B}\right)}$
12. 2.5mv/^{o}C
13. Sensistors
14. Oxides of Ni, Mn, Co.
15. base
16. smaller
17. Temperature, decreases
18. 25 to 75 (or about 50%)

Answers to Multiple Choice Questions

1.	c	**2.**	a	**3.**	b	**4.**	d	**5.**	a	**6.**	a	**7.**	b
8.	b												

Chapter - 4

Answers to Objective Type Questions

1. Ω, mhos and constants
2. the units of different parameters are not the same
3. $h_{11}I_1 + h_{12}V_2$
   $h_{21}I_1 + h_{22}V_2$
4. $\left.\frac{\partial V_B}{\partial V_C}\right|_{I_B=K}$
5. Audio
6. $h_{ie} = 1K\Omega$, $h_{re} = 1.5 \times 10^{-4}$, $h_{oe} = 6\mu mhos$, $h_{fe} = 200$
7. No units (constant)
8. $\frac{\text{AC Signal Power Delivered to the Load}}{\text{DC Input Power}} \times 100$
9. $\frac{h_{fe}}{1 + h_{oe}R_L}$
10. 1
11. Decreases
12. Low
13. Large
14. < 1
15. Common Base configuration
16. Common Collector Configuration
17. Voltatge follower/buffer

Answers to Multiple Choice Questions

1.	b	**2.**	d	**3.**	c	**4.**	b	**5.**	a	**6.**	c	**7.**	a
8.	b	**9.**	d	**10.**	c	**11.**	a	**12.**	a	**13.**	b	**14.**	c
15.	d	**16.**	c	**17.**	b								

Chapter - 5

Answers to Multiple Choice Questions

1. b **2.** b **3.** a

Chapter - 6

Answers to Objective Type Questions

1. Gain is reduced
2. $S = \frac{1}{1+\beta A}$
3. $D = (1 + \beta A)$
4. $(B.W)_f = (1 + \beta A)\ B.W$
5. $f_2' = f_2 (1 + \beta A)$
6. Degenerative feedback
7. $R_i = \infty$; $R_0 = \infty$
8. Low, Low
9. Node sampling
10. $A_V = \infty$ $R_i = \infty$ $R_0 = 0$
11. High current gain, Low R_i, High R_0
12. Return ratio
13. Decreases
14. Increases
15. $\beta = V_f / V_o$ = feedback signal/output signal
16. (i) voltage-series (ii) voltage-shunt
17. $Z_0/(1 + \beta A)$
18. $Z_i/(1 + \beta A)$
19. $Z_0(1+\beta A)$
20. $Z_i/(1 + \beta A)$

Answers to Multiple Choice Questions

1. b **2.** a **3.** d **4.** b **5.** c **6.** c **7.** a

8. b **9.** d **10.** a

Chapter - 7

Answers to Objective Type Questions

1. Amplifier needs external A.C. input. Oscillator circuit doesn't need external A.C. input.
2. Internally generated Noise Signal
3. Positive feedback
4. Zero or 360^0
5. $|\beta A| \geq 1$
6. Few 100 Hz to KHz range
7. Tuned oscillator circuit
8. 5 Hz to 1 MHz
9. Hartley oscillator
10. Collpitts oscillator circuit
11. Quartz and Rochelle salt
12. 1. Lithium sulphate
    2. Rochelle salt
    3. PZT (Lead Zirconate Titanate)
    4. Barrium titanate ($Batio_3$)
13. 10 : 1, 3 : 1
14. Positive, Oxides of Nickel and Manganese
15. L = 137 H
    C = 0.0235 μF
    R = 15K
    Q = 5500
16. $f_0 = \frac{1}{2\pi RC\sqrt{6}}$; $\beta = \frac{1}{29}$
17. $f_o = \frac{1}{2\pi\sqrt{R_1C_1R_2C_2}}$
18. $\frac{1}{2\pi\sqrt{LCeq}}$

    where $Ceq = \frac{C_1C_2}{C_1 + C_2}$
19. $\frac{1}{2\pi\sqrt{LeqC}}$

    where $Leq = L_1 + L_2 + 2M$
20. $1/_{R_TC_T}\ l_n\ [1/(1-n)]$

Answers to Multiple Choice Questions

1.	c	**2.**	a	**3.**	d	**4.**	c	**5.**	a	**6.**	b	7.	b
8.	a	**9.**	b	**10.**	b								

MCQs from Gate Examination, Yearwise from 1996 to 2012

GATE 2012

1. The diodes and capacitors in the circuit shown are ideal. The voltage v(t) across the diode D1 is

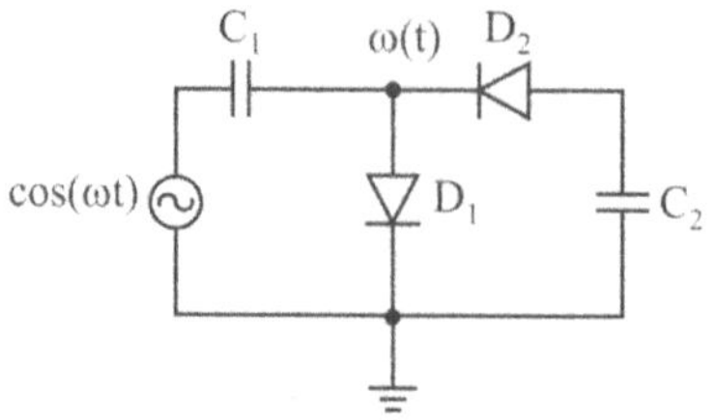

(a) cos(wt) – 1 (b) sin(wt)

(c) 1 – cos(wt) (d) 1 – sin(wt)

Sol:

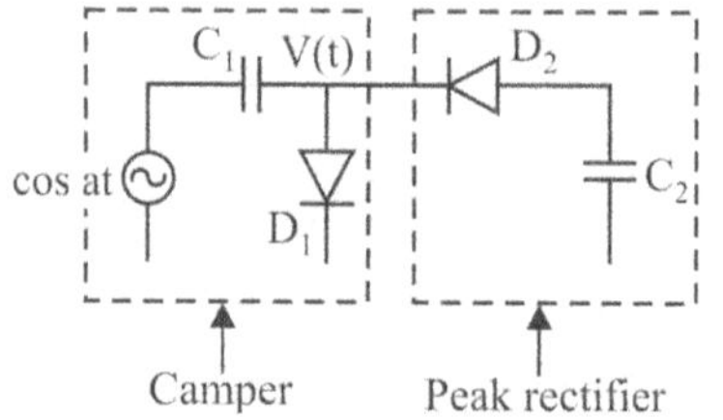

When excited by cos(wt), the clamping section clamp the positive peak to 0 volts and negative peak to – 2 volts. So whole cos(wt) is lower by –1 volts.

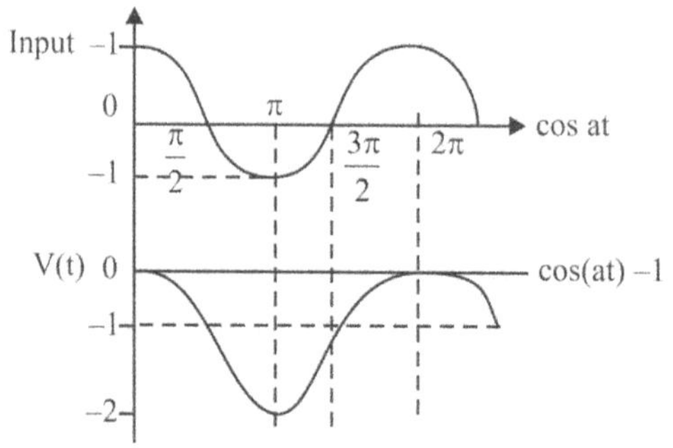

Hence (C) is correct option.

2. The i-v characteristics of the diode in the circuit given below are

$$i = \left(\frac{v - 0.7}{500}\right) A \quad \text{for} \quad v \geq 0.7V$$

$$= 0\ A, \qquad \text{for} \quad v < 0.7V$$

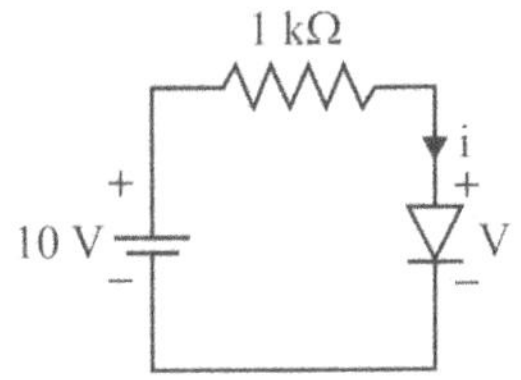

The current in the circuit is

(a) 10mA (b) 9.3mA (c) 6.67mA (d) 6.2mA

Sol:

$$i = \frac{V - 0.7}{500}$$

$$\frac{di}{dv} = \frac{1}{500}$$

$$r_d = 500\Omega$$

Since diode will be forward biased voltage across diode will be 0.7V

$$I = \frac{10 - 0.7}{500 + 1000} = \frac{9.3}{1500} = 6.2\text{mA}$$

Hence (D) is correct option.

2011 ONE MARKS

3. Drift current in the semiconductors depends upon

(A) only the electric field

(B) only the carrier concentration gradient

(C) both the electric field and the carrier concentration

(D) both the electric field and the carrier concentration gradient

Sol: Drift current $I_d = qn\mu_n E$

It depends upon Electric field E and carrier concentration n

Hence (C) is correct option.

4. A Zener diode, when used in voltage stabilization circuits, is biased in

(A) reverse bias region below the breakdown voltage

(B) reverse breakdown region

(C) forward bias region

(D) forward bias constant current mode

Sol: Zener diode operates in reverse breakdown region

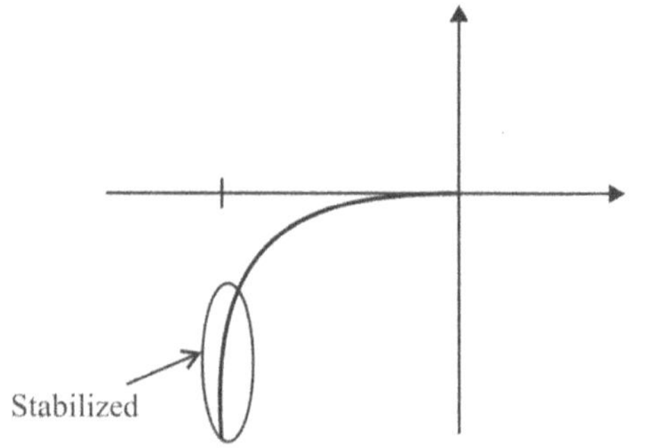

Hence (B) is correct option

5. A silicon PN junction is forward biased with a constant current at room temperature. When the temperature is increased by 10 °C, the forward bias voltage across the PN junction

(A) increases by 60 mV

(B) decreases by 60 mV

(C) increases by 25 mV

(D) decreases by 25 mV

Sol: For every 1 °C increase in temperature, forward bias voltage across diode decreases by 2.5 mV. Thus for 10 °C increase, there is 25 mV decreases.

Hence (D) is correct option.

COMMON DATA QUESTIONS: 2 MARKS EACH

The channel resistance of an N-channel JFET shown in the figure below is 600 Ω when the full channel thickness (t_{ch}) of10 μ/m is available for conduction. The built-in voltage of the gate P^+N junction (V_{bi}) is −1V. When the gate to source voltage (V_{GS}) is 0 V, the channel is depleted by 1 μm on each side due to the built in voltage and hence the thickness available for conduction is only 8 μm

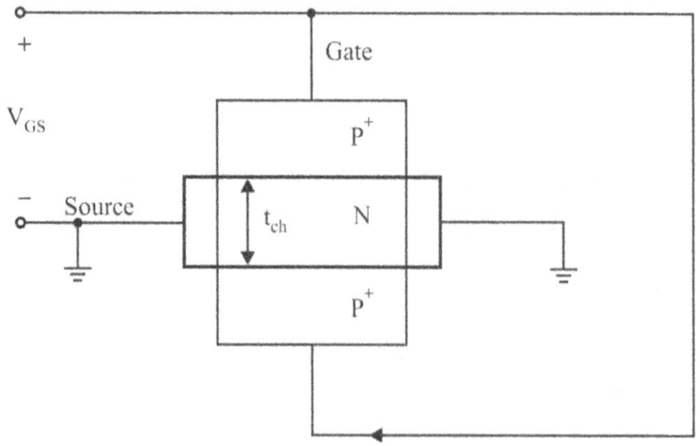

6. The channel resistance when $V_{GS} = -3$ V is

(A) 360 Ω (B) 917 Ω (C) 1000 Ω (D) 3000 Ω

Sol: Full channel resistance is

$$r\frac{\rho \times L}{W \times a} = 600\ \Omega \qquad .....(1)$$

If V_{GS} is applied, Channel resistance is

$$r' = \frac{\rho \times L}{W \times b} \qquad \text{where} \quad b = a\left(1 - \sqrt{\frac{V_{GS}}{V_p}}\right)$$

Pinch off voltage,

$$|V_p| = \frac{qN_D}{2\in} a^2 \qquad .....(2)$$

If depletion on each side is d = 1 mm at $V_{GS} = 0$.

$$V_j = \frac{qN_D}{2\in} d^2$$

or $$1 = \frac{qN_D}{2\in}\left(1 \times 10^{-6}\right)^2 \Rightarrow \frac{qN_D}{2\in} = 10^{12}$$

Now from equation (2), we have

$$|V_p| = 10^{12} \times \left(5 \times 10^{-6}\right)^2$$

or $$V_p = -25\ \text{V}$$

At $V_{GS} = -3$ V;

$$b = 5\left(1 - \sqrt{\frac{-3}{-25}}\right)\mu\text{m} = 3.26\ \mu\text{m}$$

$$r' = \frac{\rho L}{W \times b} = \frac{\rho L}{Wa} \times \frac{a}{b} = 600 \times \frac{5}{3.26} = 917\ \Omega$$

Hence (B) is correct option.

7. The channel resistance when $V_{GS} = 0$ V is

(A) 480 Ω (B) 600 Ω (C) 750 Ω (D) 1000 Ω

Sol: At $V_{GS} = 0$ V, $b = 4\ \mu$m since $2b = 8\ \mu$m

Thus $$r' = \frac{\rho L}{Wa} \times \frac{a}{b} = 600 \times \frac{5}{4} = 750\ \Omega$$

Hence (C) is correct option.

2010 ONE MARK

8. At room temperature, a possible value for the mobility of electrons in the inversion layer of a silicon n-channel MOSFET is

(A) 450 $cm^2/V\text{–}s$ (B) 1350 $cm^2/V\text{–}s$

(C) 1800 $cm^2/V\text{–}s$ (D) 3600 $cm^2/V\text{–}s$

Sol: At room temperature mobility of electrons for Si sample is given as $\mu_n = 1350\ cm^2/Vs$. For an n-channel MOSFET to create an inversion layer of electrons, a large positive gate voltage is to be applied. Therefore, induced electric field increases and mobility decreases. So, Mobility $\mu_n < 1350\ cm^2/Vs$ for n-channel MOSFET

Hence (A) is correct option.

9. In the silicon BJT circuit shown below, assume that the emitter area of transistor Q1 is half that of transistor Q2.

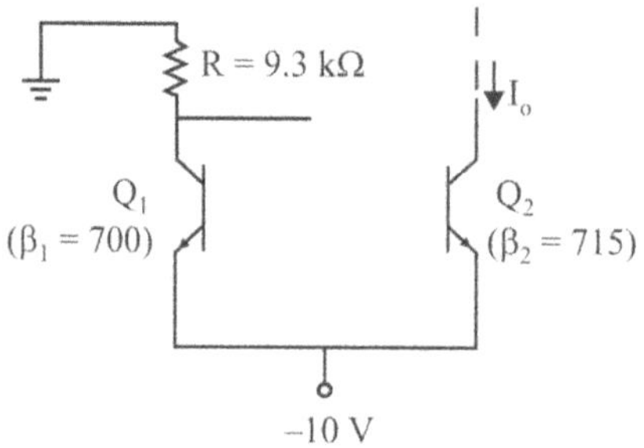

The value of current I_o is approximately

(a) 0.5 mA (b) 2 mA (c) 9.3 mA (d) 15 mA

Sol: Let both transistors are in active region therefore voltage at Q_1 (base)

$$(V_{Base})Q_1 = 0.7 - 10 = -9.3$$

$$I_R = \frac{-9.3}{9.3} = I_C$$

Since emitter area of $Q_1 = \frac{1}{2}$(emitter area of Q_2)

$$A_{Q_1} = \frac{A_{Q_2}}{2}$$

$\therefore$ (β) effective $= 2 \times (\beta_2) = 2 \times 715 = 1430$

Since effective β of Q_2 is double of Q_1, so collector current will also be double of Q_1

$\therefore$ $I_{c_2} = 2 \times I_{c_1} = 2$ mA

Hence (B) is correct option.

10. Thin gate oxide in a CMOS process is preferably grown using

(A) Wet oxidation (B) Dry oxidation

(C) Epitaxial oxidation (D) Ion implantation

Sol: Dry oxidation is used to achieve high quality oxide growth.

Hence (B) is correct option.

2010 TWO MARKS

11. In a uniformly doped BJT, assume that N_E, N_B and N_C are the emitter, base and collector doping in atoms/cm^3, respectively. If the emitter injection efficiency of the BJT is close to unity, which one of the following condition is TRUE?

(A) $N_E = N_B = N_C$ (B) $N_E >> N_B$ and $N_B > N_C$

(C) $N_E = N_B$ and $N_B < N_C$ (D) $N_E < N_B < N_C$

Sol: Emitter injection efficiency is given as

$$\gamma = \frac{1}{1+\frac{N_B}{N_C}}$$

To achieve $\gamma = 1$, $N_E >> N_B$

Hence (B) is correct option

12. Compared to a p-n junction with $N_A = N_D = 10^{14}/\text{cm}^3$, Which one of the following statements is TRUE for a p – n junction with $N_A = N_D = 10^{20}/\text{cm}^3$

(A) Reverse breakdown voltage is lower and depletion capacitance is lower

(B) Reverse breakdown voltage is higher and depletion capacitance is lower

(C) Reverse breakdown voltage is lower and depletion capacitance is higher

(D) Reverse breakdown voltage is higher and depletion capacitance is higher

Sol: Reverse bias breakdown or Zener effect occurs in highly doped PN junction through tunneling mechanism. In a highly doped PN junction, the conduction and valence bands on opposite sides of the junction are sufficiently close during reverse bias that electron may tunnel directly from the valence band on the p-side into the conduction band on n-side.

Breakdown voltage $$V_B \propto \frac{1}{N_A N_D}$$

So, breakdown voltage decreases as concentration increases

Depletion capacitance $$C = \left\{\frac{e\varepsilon_a N_A N_D}{2(V_{bi}+V_R)(N_A+N_D)}\right\}^{\frac{1}{2}}$$

Thus $C \propto N_A N_B$

Depletion capacitance increases as concentration increases

Hence (C) is correct option.

Statements for Linked Answer Question

The silicon sample with unit cross-sectional area shown below is in thermal equilibrium. The following information is given: $T = 300$ K electronic charge 1.6×10^{-19} C, thermal voltage = 26 mV and electron mobility = 1350 $cm^2/V\text{–}s$

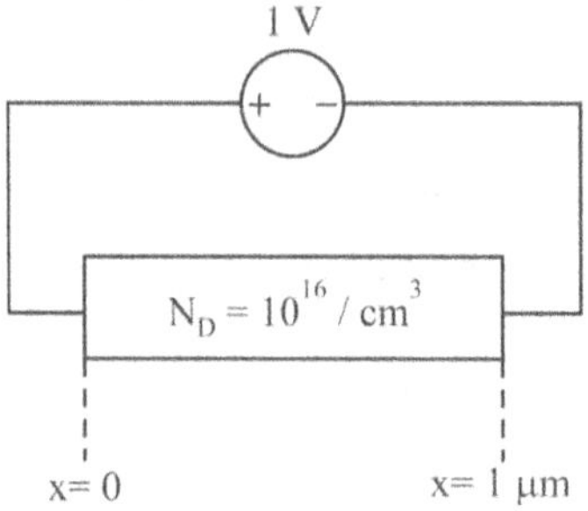

13. The magnitude of the electric field at $x = 0.5$ μm is

(A) 1 kV/cm (B) 5 kV/cm (C) 10 kV/cm (D) 26 kV/cm

Sol: Sample is in thermal equilibrium so, electric field

E =1/1 μ m= 10 kV/cm

Hence (C) is correct option.

14. The magnitude of the electron of the electron drifts current density at $x = 0.5$ μm is

(A) 2.16×10^4 A/cm^2 (B) 1.08×10^4 A/m^2

(C) 4.32×10^3 A/cm^2 (D) 6.48×10^2 A/cm^2

Sol: Electron drift current density

$$J_d = N_D \mu_n eE$$
$$= 10^{16} \times 1350 \times 1.6 \times 10^{-19} \times 10 \times 10^{13}$$
$$= 2.16 \times 10^4 \text{ A/cm}^2$$

Hence (A) is correct option.

2009 ONE MARK

15. In an n-type silicon crystal at room temperature, which of the following can have a concentration of 4×10^{19} cm^{-3}?

(A) Silicon atoms (B) Holes

(C) Dopant atoms (D) Valence electrons

Sol: Only dopant atoms can have concentration of 4×10^{19} cm^{-3} in n – type silicon at room temperature.

Hence option (C) is correct.

16. The ratio of the mobility to the diffusion coefficient in a semiconductor has the units

(A) V^{-1} (B) cm.V^1 (C) V.cm^{-1} (D) V.s

Sol: Unit of mobility μ_n is $= \dfrac{cm^2}{V.sec}$

Unit of diffusion current D_n is $= \dfrac{cm^2}{sec}$

Thus unit of $\dfrac{\mu_n}{D_n}$ is $= \dfrac{cm^2}{V.sec} \bigg/ \dfrac{cm^2}{sec} = 1/V = V^{-1}$

Hence option (A) is correct.

17. In the circuit below is ideal. The voltage V is given by

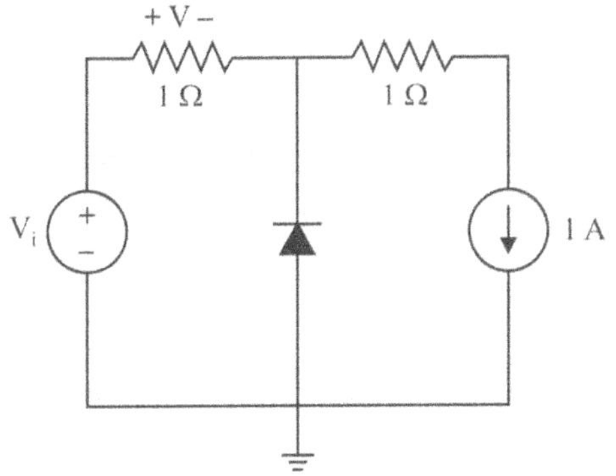

(a) Min (V_i, 1) (b) max (V_i, 1)

(c) min (–V_i, 1) (d) max (–V_i, 1)

Sol: Voltage at 1Ω is in negative direction of V_i and it is maximum value of (–V_i, 1) because diode is in reverse biased.

Hence option (D) is correct.

2009 TWO MARKS

18. Consider the following two statements about the internal conditions in a n –channel MOSFET operating in the active region.

S_1: The inversion charge decreases from source to drain

S_2: The channel potential increases from source to drain.

Which of the following is correct?

(A) Only S_2 is true

(B) Both S_1 and S_2 are false

(C) Both S_1 and S_2 are true, but S_2 is not a reason for S_1

(D) Both S_1 and S_2 are true, and S_2 is a reason for S_1

Sol: Both S_1 and S_2 are true and S_2 is a reason for S_1.

Hence option (D) is correct.

Common Data for Question 19 and 20

Consider a silicon p – n junction at room temperature having the following parameters:

Doping on the n-side = 1×10^{17} cm^{-3}

Depletion width on the n-side = 0.1 μm

Depletion width on the p – side = 1.0 μm

Intrinsic carrier concentration $n_i = 1.4 \times 10^{10}$ cm^{-3}

Thermal voltage = 26 mV

Permittivity of free space = 8.85×10^{14} $F.cm^{-1}$

Dielectric constant of silicon = 12

19. The built-in potential of the junction

(A) is 0.70 V

(B) is 0.76 V

(C) is 0.82 V

(D) Cannot be estimated from the data given

Sol: We know that

$$N_A W_P = N_D W_N$$

or
$$N_A = \frac{N_D W_N}{W_P} = \frac{1\times10^{1} \times 0.1\times10^{-6}}{1\times10^{-6}} = 1\times10^{16}$$

The built-in potential is

$$V_{bi} = V_T \ln\left(\frac{N_A N_D}{n_i^2}\right) = 0.76$$

Hence option (B) is correct.

20. The peak electric field in the device is

(A) 0.15 MV. cm^{-1}, directed from p-region to n-region

(B) 0.15 MV. cm^{-1}, directed from n-region to p-region

(C) 1.80 MV. cm^{-1}, directed from p-region to n-region

(D) 1.80 MV. cm^{-1}, directed from n-region to p-region

Sol: The peak electric field in device is directed from p to n and is

$$E = -\frac{eN_D x_n}{\varepsilon_S} \text{ from p to n}$$

$$E = \frac{eN_D x_n}{\varepsilon_S} \text{ from n to p}$$

$$= \frac{1.6\times10^{-19} \times 1\times10^{17} \times 1\times10^{-5}}{8.85\times10^{-14} \times 12}$$

$$= 0.15 \text{ MV/cm}$$

Hence option (B) is correct

2008 ONE MARK

21. Which of the following is NOT associated with a p – n junction?

(A) Junction Capacitance (B) Charge Storage Capacitance

(C) Depletion Capacitance (D) Channel Length Modulations

Sol: Channel length modulation is not associated with a p – n junction. It is being associated with MOSFET in which effective channel length decreases, producing the phenomenon called channel length modulation.

Hence option (D) is correct.

22. Which of the following is true?

(A) A silicon wafer heavily doped with boron is a p+ substrate

(B) A silicon wafer lightly doped with boron is a p+ substrate

(C) A silicon wafer heavily doped with arsenic is a p+ substrate

(D) A silicon wafer lightly doped with arsenic is a p+ substrate

Sol: It is being associated with MOSFET in which effective channel length decreases, producing the phenomenon called channel length modulation.

Hence option (A) is correct.

23. A silicon wafer has 100 nm of oxide on it and is furnace at a temperature above 1000 °C for further oxidation in dry oxygen. The oxidation rate

(A) is independent of current oxide thickness and temperature

(B) is independent of current oxide thickness but depends on temperature

(C) slows down as the oxide grows

(D) is zero as the existing oxide prevents further oxidation

Sol: Oxidation rate is zero because the existing oxide prevent the further oxidation.

Hence option (D) is correct.

24. The drain current of MOSFET in saturation is given by $I_D = K(V_{GS} - V_T)^2$ Where K is a constant. The magnitude of the transconductance g_m is

(A) $\frac{K(V_{GS} - V_T)^2}{V_{DS}}$ (B) $2K(V_{GS} - V_T)$

(C) $\frac{I_d}{V_{GS} - V_{DS}}$ (D) $\frac{K(V_{GS} - V_T)^2}{V_{GS}}$

Sol: $$g_m = \frac{\partial I_P}{\partial V_{GS}} = \frac{\partial}{\partial V_{GS}} K(V_{GS} - V_T)^2 = 2K(V_{GS} - V_T)$$

Hence option (B) is correct.

25. In the following limiter circuit, an input voltage $V_i = 10 \sin 100\pi t$ applied. Assume that the diode drop is 0.7V when it is forward biased. The Zener breakdown voltage is 6.8V.

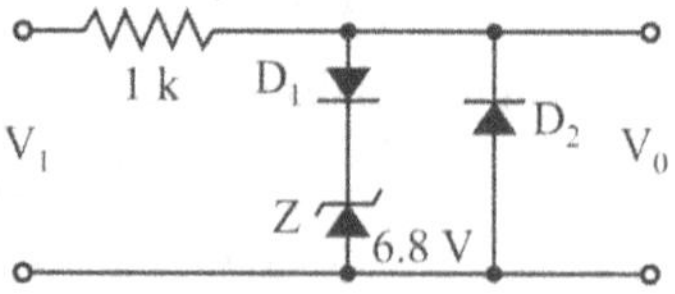

The maximum and minimum values of the output voltage respectively are

(a) 6.1V, – 0.7V (b) 0.7V, – 7.5V

(c) 7.5V, – 0.7V (d) 7.5V, – 7.5V

Sol: During positive cycle, D_1 is ON and D_2 is OFF,

$$V_o = V_Z + V_{D_1} = 6.8 + 0.7 = 7.5V$$

During negative cycle, D_1 is OFF and D_2 is ON

$$V_o = -V_{D_2} = -0.7V$$

Hence option (C) is correct.

26. The measured transconductance g_m of an NMOS transistor operating in the linear region is plotted against the gate voltage V_G at a constant drain voltage V_D. Which of the following figures represents the expected dependence of gm on V_G?

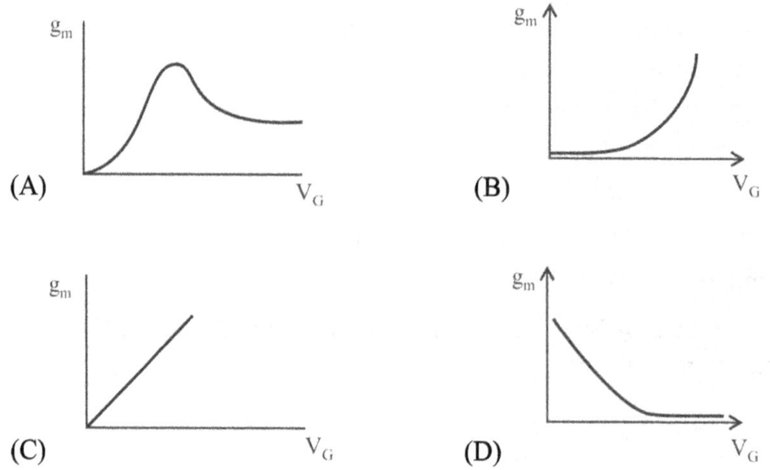

Sol: As V_D = constant

Thus $g_m \propto (V_{GS} - V_T)$ Which is straight line.

Hence option (C) is correct.

2008 TWO MARKS

27. Silicon is doped with boron to a concentration of 4×10^{17} atoms cm^3. Assume the intrinsic carrier concentration of silicon to be $1.5 \times 10^{10} / cm^3$ and the value of kT/q to be 25 mV at 300 K. Compared to undopped silicon, the fermi level of doped silicon

(A) goes down by 0.31 eV (B) goes up by 0.13 eV

(C) goes down by 0.427 eV (D) goes up by 0.427 eV

Sol: $E_2 - E_1 = kT \ln \frac{N_A}{n_i} = 0.427 \text{ eV}$

Hence fermi level goes down by 0.427 eV as silicon is doped with boron.

Hence option (C) is correct.

28. The cross section of a JFET is shown in the following figure. Let Vc be −2 V and let V_P be the initial pinch-off voltage. If the width W is doubled (with other geometrical parameters and doping levels remaining the same), then the ratio between the mutual transconductances of the initial and the modified JFET is

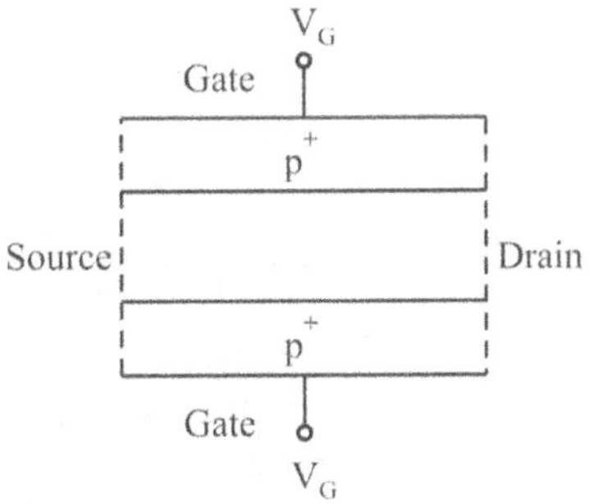

(A) 4

(B) $\frac{1}{2}\left(\frac{1-\sqrt{2/V_P}}{1-\sqrt{1/2V_P}}\right)$

(C) $\left(\frac{1-\sqrt{2/V_P}}{1-\sqrt{1/2V_P}}\right)$

(D) $\frac{1-\left(2-\sqrt{V_P}\right)}{1-\left[1\left(2\sqrt{V_P}\right)\right]}$

Sol: Pinch of voltage $V_P = eW^2N_D/\varepsilon_s$

Let $V_P = V_{P_1}$

Now $\frac{V_{P_1}}{V_{P_2}} = \frac{W_1^2}{W_2^2} = \frac{W^2}{(2W)^2}$ or $4V_{P_1} = V_{P_2}$

Initial transconductance

$$g_m = K_n\left[1-\sqrt{\frac{V_{bi}-V_{GS}}{V_P}}\right]$$

For first condition $g_{m_1} = K_n\left[1-\sqrt{\frac{0-(-2)}{V_{P_1}}}\right] = K_n\left[1-\sqrt{\frac{2}{V_{P_1}}}\right]$

For second condition $g_{m_2} = K_n\left[1-\sqrt{\frac{0-(-2)}{V_{P_2}}}\right] = K_n\left[1-\sqrt{\frac{2}{4V_{P_1}}}\right]$

Dividing g_{m_1} and g_{m_2} we get $V_P = V_{P_1}$

Hence option (C) is correct

29. Consider the following assertions.

S_1 : For Zener effect to occur, a very abrupt junction is required.

S_2 : For quantum tunneling to occur, a very narrow energy barrier is required.

Which of the following is correct?

(A) Only S_2 is true

(B) S_1 and S_2 are both true but S_2 is not a reason for S_1

(C) S_1 and S_2 are both true but S_2 is not a reason for S_1

(D) Both S_1 and S_2 are false

Sol: Option (A) is correct.

2007 ONE MARK

30. The electron and hole concentrations in an intrinsic semiconductor are n_i per cm^3 at 300 K. Now, if acceptor impurities are introduced with a concentration of N_A per cm^3 (where $N_A >> n_i$, the electron concentration per cm^3 at 300 K will be)

(A) n_i (B) $n_i + N_A$

(C) $N_A - n_i$ (D) $\frac{n_i^2}{N_A}$

Sol: As per mass action law

$$np = n_i^2$$

If acceptor impurities are introduced

$$p = N_A$$

Thus $nN_A = n_i^2$

or $n = \frac{n_i^2}{N_A}$

Hence option (D) is correct.

31. In a p^+ n junction diode under reverse biased the magnitude of electric field is maximum at

(A) the edge of the depletion region on the p-side

(B) the edge of the depletion region on the n-side

(C) the p^+ n junction

(D) the centre of the depletion region on the n-side

Sol: The electric field has the maximum value at the junction of p^+ n.

Hence option (C) is correct.

32. The correct full wave rectifier circuit is

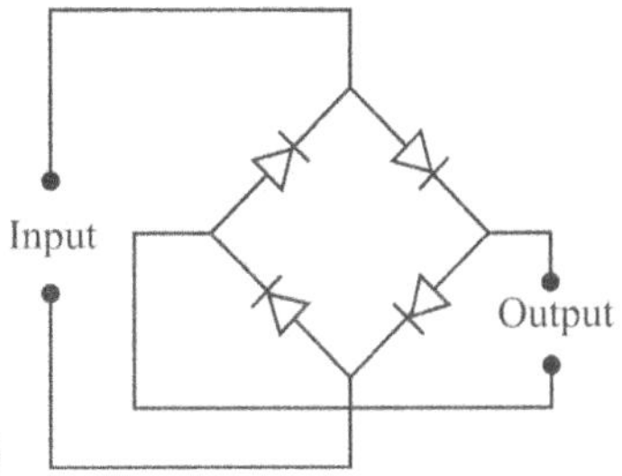

(A)

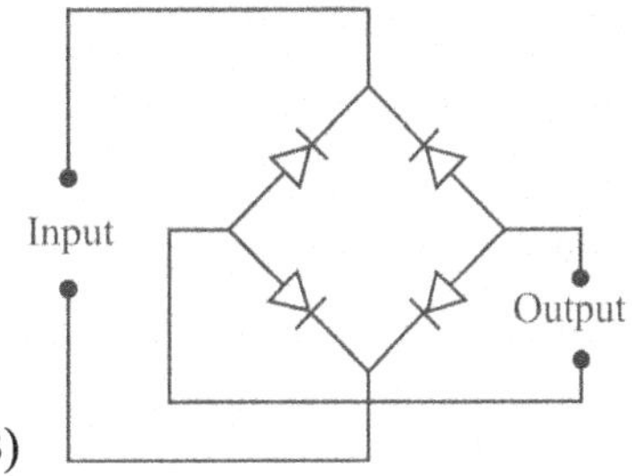

(B)

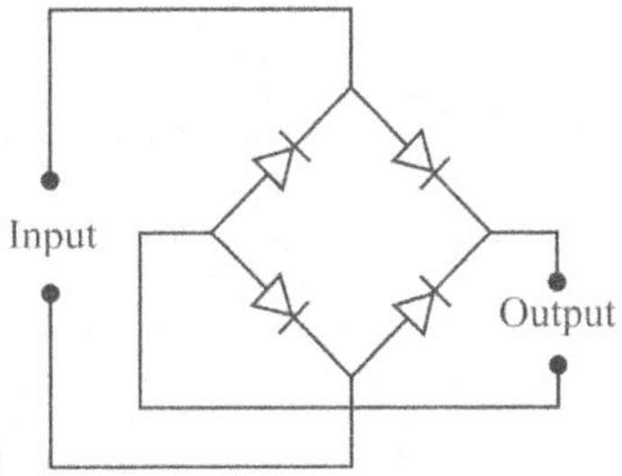

(C)

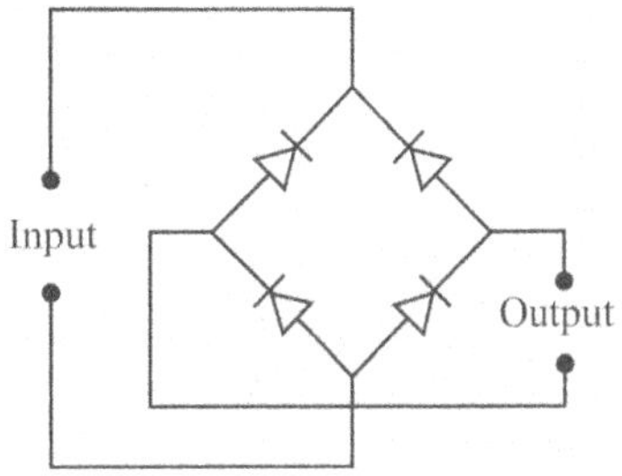

(D)

Sol: During positive cycle of input diodes D_2 and D_4 are forward biased and a complete path exists from input to output. During negative cycle of input diodes D_1 and D_3 are forward biased and a complete path exists from input to output. In both cases current passes through R_L is in the same direction.

Hence option (C) is correct.

2007 TWO MARKS

33. Group I lists four types of p − n junction diodes. Match each device in Group I with one of the option in Group II to indicate the bias condition of the device in its normal mode of operation.

Group - I		Group-II	
(P)	Zener Diode	(1)	Forward bias
(Q)	Solar cell	(2)	Reverse bias
(R)	LASER diode		
(S)	Avalanche Photodiode		

(A) P - 1, Q - 2, R - 1, S - 2 (B) P - 2, Q - 1, R - 1, S - 2

(C) P - 2, Q - 2, R - 1, S - 2 (D) P - 2, Q - 1, R - 2, S - 2

Sol: Zener diode and Avalanche diode works in the reverse bias and laser diode works in forward bias. Solar cell diode works in forward bias but photo current is in reverse direction. Thus

Zener diode : Reverse Bias

Solar Cell : Forward Bias

Laser Diode : Forward Bias

Avalanche Photo diode : Reverse Bias

Hence option (B) is correct.

34. Group I lists four different semiconductor devices. Match each device in Group I with its characteristic property in Group II

	Group-I		Group-II
(P)	BJT	(1)	Population inversion
(Q)	MOS capacitor	(2)	Pinch-off voltage
(R)	LASER diode	(3)	Early effect
(S)	JFET	(4)	Flat-band voltage

(A) P - 3, Q - 1, R - 4, S - 2 (B) P - 1, Q - 4, R - 3, S - 2

(C) P - 3, Q - 4, R - 1, S - 2 (D) P - 3, Q - 2, R - 1, S - 4

Sol: In BJT as the B-C reverse bias voltage increases, the B-C space charge region width increases which x_B (i.e., neutral base width) > A change in neutral base width will change the collector current. A reduction in base width will causes the gradient in minority carrier concentration to increase, which in turn causes an increase in the diffusion current. This effect is known as base modulation and Early effect.

In JFET the gate to source voltage that must be applied to achieve pinch off voltage is described as pinch off voltage and is also called as turn voltage or threshold voltage.

In LASER population inversion occurs on the condition when concentration of electrons in one energy state is greater than that in lower energy state, i.e., a non equilibrium condition.

In MOS capacitor, flat band voltage is the gate voltage that must be applied to create flat band condition in which there is no space charge region in semiconductor under oxide. Therefore

BJT : Early effect

MOS capacitor : Flat-band voltage

LASER diode : Population inversion

JFET : Pinch-off voltage

Hence option (C) is correct.

35. For the Zener diode shown in figure, the Zener voltage at knee is 7V, the knee current is negligible and the Zener dynamic resistance is 10Ω. If the voltage (v_i) range is from 10 to 16V, the output voltage (v_o) range from

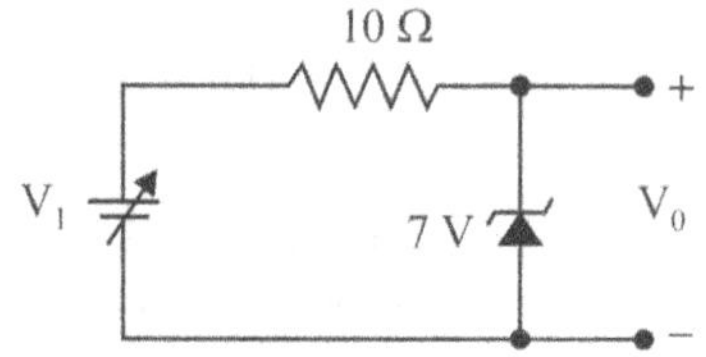

(A) 7.00 to 7.29 (B) 7.14 to 7.29 (C) 7.14 to 7.43 (D) 7.29 to 7.43

Sol: Given, Zener voltage = 7V

Zener dynamic resistance = 10 Ω

When $V_i = 10V$, $i = \frac{10-7}{210} = \frac{3}{210}$ amp

$$V_{01} = 7 + 10i = 7 + 10 \times \frac{3}{210} = 7.14 \text{ volts}$$

When $V_i = 16V$, $i = \frac{16-7}{210} = \frac{9}{210}$ amp

$$V_{02} = 7 + 10i = 7 + 10 \times \frac{9}{210} = 7.43 \text{ volts}$$

Hence option (C) is correct.

36. For the BJT circuit shown, assume that the β of the transistor is very large and $V_{BE} = 0.7V$. the mode of operation the BJT is:

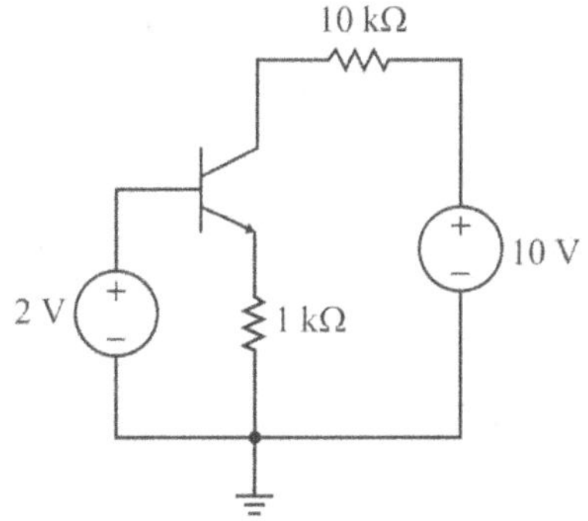

(A) Cut-off (B) Saturation
(C) Normal active (D) Reverse active

Sol: In saturation region $I_C = I_E$,

$$I_C = \frac{V_{CC}}{R_1 + R_2} = \frac{10}{10+1} = \frac{10}{11}$$

Apply KVL in base circuit

$$V_{BB} = V_{BE} + I_C R_2 2 = V_{BE} + \frac{10}{11}$$

$$V_{BE} = 1.017$$

$$V_{BE} > 0.7V$$

So, operate in saturation region.

Hence option (B) is correct.

37. A p^+ n junction has a built-in potential of 0.8 V. The depletion layer width for a reverse bias of 1.2 V is 2 μm. For a reverse bias of 7.2 V, the depletion layer width will be

(A) 4 μm (B) 4.9 μm (C) 8 μm (D) 12 μm

Sol: $$W = K\sqrt{V + V_R}$$

Now $$2\mu = K\sqrt{0.8 + 1.2}$$

From above two equation we get

$$\frac{W}{2\mu} = 2$$

or $$W_2 = 4\ \mu m$$

Hence option (A) is correct.

38. The DC current gain (β) of a BJT is 50. Assuming that the emitter injection efficiency is 0.995, the base transport factor is

(A) 0.980 (B) 0.985 (C) 0.990 (D) 0.995

Sol: $$\alpha = \frac{\beta}{\beta + 1} = \frac{50}{50 + 1} = \frac{50}{51}$$

Current Gain = Base Transport Factor X Emitter injection Efficiency

$$\alpha = \beta_1 \times \beta_2$$

or $$\beta_1 = \frac{\alpha}{\beta_2} = 0.985$$

Hence option (B) is correct.

Common Data Question

The figure shows the high-frequency capacitance – voltage characteristics of Metal/SiO_2/silicon (MOS) capacitor having an area of 1×10^{-4} cm^2. Assume that the permittivities ($\varepsilon_0\varepsilon_r$) of silicon and SiO_2 are 1×10^{-12} F/cm and 3.5×10^{-13} F/cm respectively.

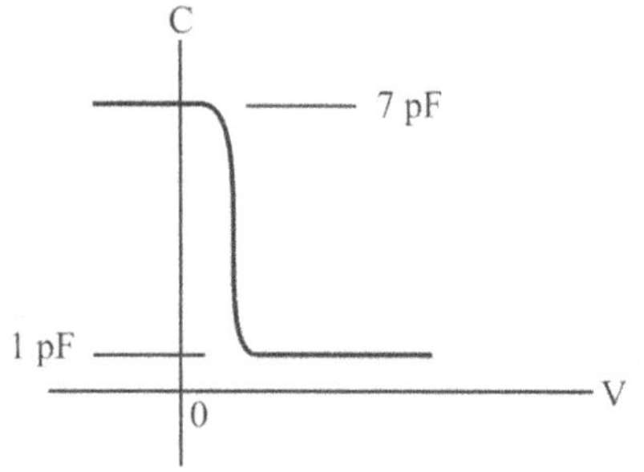

39. The gate oxide thickness in the MOS capacitor is

(A) 50 nm (B) 143 nm (C) 350 nm (D) 1 μm

Sol: At low voltage when there is no depletion region and capacitance is decided by SiO_2 thickness only,

$$C = \frac{\varepsilon_0 \varepsilon_{r_1} A}{D}$$

or $$D = \frac{\varepsilon_0 \varepsilon_{r_1} A}{C} = \frac{3.5 \times 10^{-13} \times 10^{-4}}{7 \times 10^{-12}} = 50$$

Hence option (A) is correct.

40. The maximum depletion layer width in silicon is

(A) 0.143 μm (B) 0.857 μm (C) 1 μm (D) 1.143 μm

Sol: The construction of given capacitor is shown in fig below

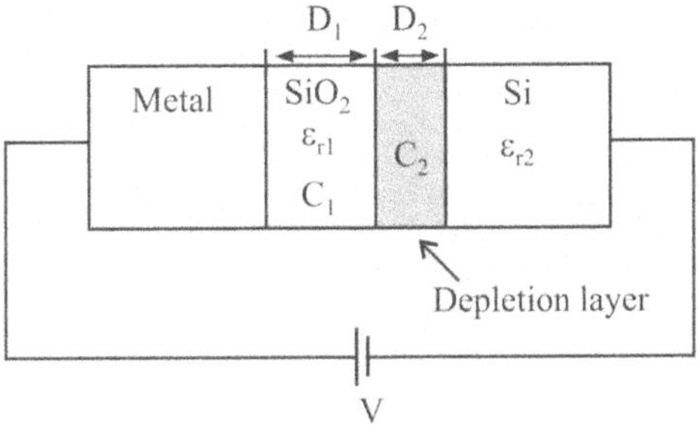

$$C_T = 1 \text{ pF}$$

$$\frac{1}{C_T} = \frac{1}{C_1} + \frac{1}{C_2}$$

Substituting values of C_T and C_1 we get

$$C_T = \frac{7}{6} \text{pF}$$

Now $$D_2 = \frac{\varepsilon_0 \varepsilon_{r2} A}{C_2} = \frac{1 \times 10^{-12} \times 10^{-4}}{\frac{7}{6} \times 10^{-12}} = 0.857 \ \mu\text{m}$$

Hence option (B) is correct.

41. Consider the following statements about the C – V characteristics plot:

S_1 : The MOS capacitor has as n-type substrate

S_2 : If positive charges are introduced in the oxide, the C – V plot will shift to the left.

Then which of the following is true?

(A) Both S_1 and S_2 are true (B) S_1 is true and S_2 is false

(C) S_1 is false and S_2 is true (D) Both S_1 and S_2 are false

Sol: Depletion region will not be formed if the MOS capacitor has n type substrate but from C-V characteristics, C reduces if V is increased. Thus depletion region must be formed. Hence S_1

is false If positive charges is introduced in the oxide layer, then to equalize the effect the applied voltage V must be reduced. Thus the C – V plot moves to the left. Hence S_2 is true.

Hence option (C) is correct.

2006 ONE MARK

42. The values of voltage (V_D) across a tunnel-diode corresponding to peak and valley currents are V_p, V_D respectively. The range of tunnel diode voltage for V which the slope of its I - V_D characteristics is negative would be

(A) $V_D < 0$ (B) $0 \le V_D < V_p$

(C) $V_p \le V_D < V_v$ (D) $V_D \ge V_v$

Sol: For the case of negative slope it is the negative resistance region

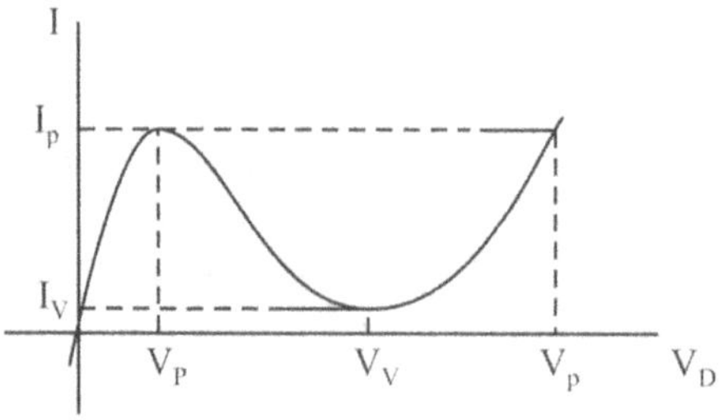

Hence option (C) is correct.

43. The concentration of minority carriers in an extrinsic semiconductor under equilibrium is

(A) Directly proportional to doping concentration

(B) Inversely proportional to the doping concentration

(C) Directly proportional to the intrinsic concentration

(D) Inversely proportional to the intrinsic concentration

Sol: For n-type 'p' is minority carrier concentration

np = Constant Since n_i^2 is constant

$$p \propto \frac{1}{n}$$

Thus 'p' is inversely proportional to n.

Hence option (A) is correct.

44. Under low level injection assumption, the injected minority carrier current for an extrinsic semiconductor is essentially the

(A) Diffusion current (B) Drift current

(C) Recombination current (D) Induced current

Sol: Diffusion current, since the drift current is negligible for minority carrier.

Hence option (A) is correct.

45. The phenomenon known as "Early Effect" in a bipolar transistor refers to a reduction of the effective base-width caused by

(A) Electron - hole recombination at the base

(B) The reverse biasing of the base - collector junction

(C) The forward biasing of emitter-base junction

(D) The early removal of stored base charge during saturation-to-cut off switching

Sol: In BJT as the B-C reverse bias voltage increases, the B-C space charge region width increases which x_B (i.e., neutral base width) > A change in neutral base width will change the collector current. A reduction in base width will causes the gradient in minority carrier concentration to increases, which in turn causes an increases in the diffusion current. This effect is known as base modulation as 'early effect'.

Hence option (B) is correct.

2006 TWO MARKS

46. In the circuit shown below, the switch was connected to position 1 at $t < 0$ and at $t = 0$, it is changed to position 2. Assume that the diode has zero voltage drop and a storage time t_s. For $0 < t \le t_s$, υ_R is given by (all in Volts)

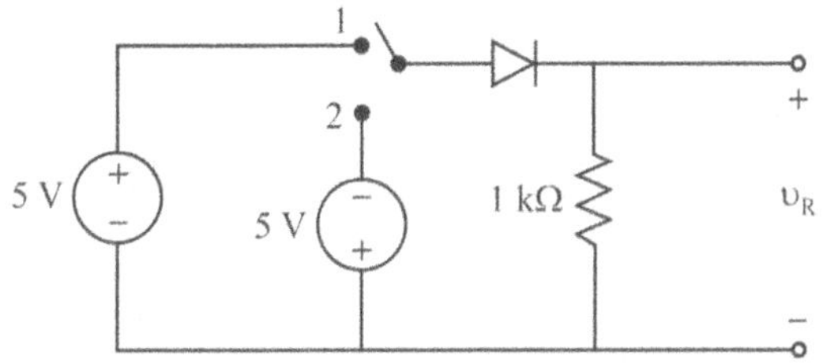

(A) $\upsilon_R = -5$ (B) $\upsilon_R = +5$ (C) $0 \le \upsilon_R < 5$ (D) $-5 \le \upsilon_R < 0$

Sol: For $t < 0$ diode forward biased and $V_R = 5$. At $t = 0$ diode abruptly changes to reverse biased and current across resistor must be 0. But in storage time $0 < t < t_s$ diode retain its resistance of forward biased. Thus for $0 < t < t_s$ it will be ON and

$$\upsilon_R = -5 \text{ V}$$

Hence option (A) is correct

47. The majority carriers in an n-type semiconductor have an average drift velocity v in a direction perpendicular to a uniform magnetic field B. The electric field E induced due to Hall effect acts in the direction

(A) $v \times B$ (B) $B \times v$ (C) along v (D) opposite to v

Sol: According to Hall effect the direction of electric field is same as that of direction of force exerted.

$$E = -v \times B$$

or $$E = B \times v$$

Hence option (B) is correct.

48. Find the correct match between Group 1 and Group 2

Group 1	Group 2
E - Varactor diode	1. Voltage reference
F - PIN diode	2. High frequency switch
G - Zener diode	3. Tuned circuits
H - Schottky diode	4. Current controlled attenuator

(A) E - 4, F - 2, G - 1, H - 3 (B) E - 3, F - 4, G - 1, H - 3

(C) E - 2, F - 4, G - 1, H - 2 (D) E - 1, F - 3, G - 2, H - 4

Sol: The varacter diode is used in tuned circuit as it can provide frequency stability.

PIN diode is used as a current controlled attenuator.

Zener diode is used in regulated voltage supply or fixed voltage reference.

Schottkey diode has metal-semiconductor function. So it has fast switching action. So it is used as high frequency switch

Varactor diode: Tuned circuits

PIN Diode: Current controlled attenuator

Zener diode: Voltage reference

Schottky diode: High frequency switch

Hence option (B) is correct

49. A heavily doped n- type semiconductor has the following data:

Hole-electron ratio : 0.4

Doping concentration : 4.2×10^8 atoms/m^3

Intrinsic concentration : 1.5×10^4 atoms/m^3

The ratio of conductance of the n –type semiconductor to that of the intrinsic semiconductor of same material and at same temperature is given by

(A) 0.00005 (B) 2000 (C) 10000 (D) 20000

Sol: We have $$\frac{\mu_n}{\mu_p} = 0.4$$

Conductance of n type semiconductor

$$\sigma_n = nq\mu_n$$

Conductance of intrinsic semiconductor

$$\sigma_i = n_i q(\mu_n + \mu_p)$$

Ratio is $$\frac{\sigma_n}{\sigma_i} = \frac{n}{n_i\left(1+\frac{\mu_r}{\mu_n}\right)} = 2\times10^4 = 20000$$

Hence option (D) is correct.

2005 ONE MARK

50. The bandgap of Silicon at room temperature is

(A) 1.3 eV (B) 0.7 eV (C) 1.1 eV (D) 1.4 eV

Sol: For silicon at 0 °K,

$$E_{g0} = 1.21 \text{ eV}$$

At any temperature

$$E_{gT} = E_{g0} - 3.6 \times 10^{-4}\,T$$

At $T = 300$ K,

$$E_{g300} = 1.21 - 3.6 \times 10^{-4} \times 300 = 1.1 \text{ eV}$$

This is standard value, that must be remembered.

Hence option (C) is correct.

51. A Silicon PN junction at a temperature of 20 °C has a reverse saturation current of 10 pico-Amperes (pA). The reverse saturation current at 40 °C for the same bias is approximately

(A) 30 pA (B) 40 pA (C) 50 pA (D) 60 pA

Sol: The reverse saturation current doubles for every 10°C rise in temperature as follows:

$$I_0(T) = I_{01} \times 2^{(T - T_1)/10}$$

Thus at 40 °C, $I_0 = 40$ pA

Hence option (B) is correct.

52. The primary reason for the widespread use of Silicon in semiconductor device technology is

(A) abundance of Silicon on the surface of the Earth.

(B) larger bandgap of Silicon in comparison to Germanium.

(C) favorable properties of Silicon - dioxide (SiO_2)

(D) lower melting point

Sol: Silicon is abundant on the surface of earth in the form of SiO_2.

Hence option (A) is correct.

2005 TWO MARKS

53. A Silicon sample A is doped with 10^{18} atoms/cm^3 of boron. Another sample B of identical dimension is doped with 10^{18} atoms/cm^3 phosphorus. The ratio of electron to hole mobility is 3. The ratio of conductivity of the sample A to B is

(A) 3 (B) 1/3 (C) 2/3 (D) 3/2

Sol:

$$\sigma_n = nq\mu_n$$

$$\sigma_p = pq\mu_p \qquad (n = p)$$

$$\frac{\sigma_p}{\sigma_n} = \frac{\mu_p}{\mu_n} = \frac{1}{3}$$

Hence option (B) is correct.

54. A Silicon PN junction diode under reverse bias has depletion region of width 10 μm. The relative permittivity of Silicon, $\varepsilon_r = 11.7$ and the permittivity of free space $\varepsilon_0 = 8.85 \times 10^{12}$ F/m. The depletion Capacitance of the diode per square meter is

(A) 100 μF (B) 10 μF (C) 1 μF (D) 20 μF

Sol:
$$C = \frac{\varepsilon_0 \varepsilon_r A}{d}$$

or
$$\frac{C}{A} = \frac{\varepsilon_0 \varepsilon_r}{d} = 10.35\ \mu F$$

Hence option (B) is correct.

55. A MOS capacitor made using p type substrate is in the accumulation mode. The dominant charge in the channel is due to the presence of

(A) holes (B) electrons

(C) positively charged ions (D) negatively charged ions

Sol: In accumulation mode for NMOS having p–substrate, when positive voltage is applied at the gate, this will induce negative charge near p–type surface beneath the gate. When V_{GS} is made sufficiently large, an inversion of electrons is formed and this in effect forms the n–channel.

Hence option (B) is correct.

56. For an n-channel MOSFET and its transfer curve shown in the figure,

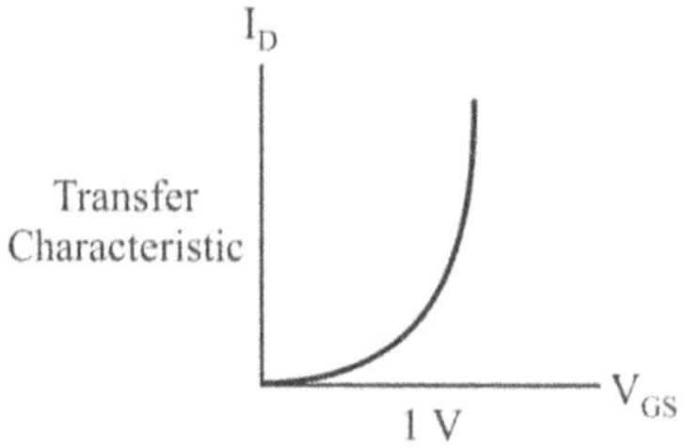

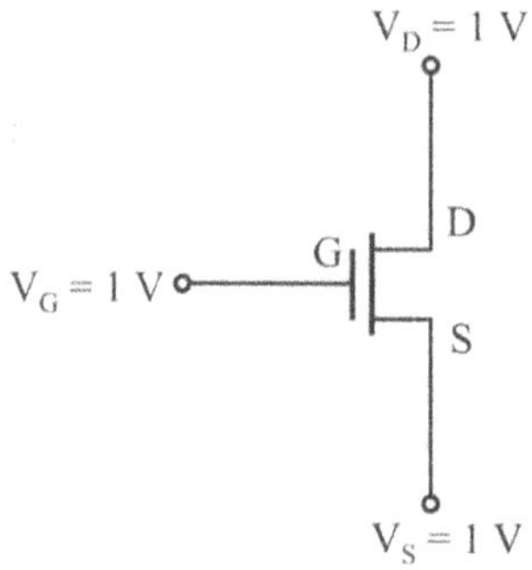

(A) 1 V and the device is in active region

(B) −1 V and the device is in saturation region

(C) 1 V and the device is in saturation region

(D) −1 V and the device is an active region

Sol: From the graph it can be easily seen that $V_{th} = 1$ V

Now $V_{GS} = 3 - 1 = 2$ V

and $V_{DS} = 5 - 1 = 4$ V

Since $V_{DS} > V_{GS}$ & $V_{DS} > V_{GS} - V_{th}$

Thus MOSFET is in saturation region.

Hence option (C) is correct.

2004 ONE MARK

57. The impurity commonly used for realizing the base region of a silicon n − p − n transistor is

(A) Gallium (B) Indium (C) Boron (D) Phosphorus

Sol: Trivalent impurities are used for making p type semiconductor. Boron is trivalent.

Hence option (C) is correct.

58. If for a silicon n-p-n transistor, the base-to-emitter voltage (V_{BE}) is 0.7 V and the collector-to-base voltage (V_{CB}) is 0.2 V, then the transistor is operating in the

(A) normal active mode (B) saturation mode

(C) inverse active mode (D) cutoff mode

Sol: Here emitter base junction is forward biased and base collector junction is reversed biased. Thus transistor is operating in normal active region.

Hence option (A) is correct.

59. Consider the following statements S_1 and S_2.

S_1 : The β of a bipolar transistor reduces if the base width is increased.

S_2 : The β of a bipolar transistor increases if the doping concentration in the base is increased.

Which remarks of the following is correct?

(A) S_1 is FALSE and S_2 is TRUE (B) Both S_1 and S_2 are TRUE

(C) Both S_1 and S_2 are FALSE (D) S_1 is TRUE and S_2 is FALSE

Sol: We have $$\beta = \frac{\alpha}{1-\alpha}$$

Thus $$\alpha \uparrow \rightarrow \beta \uparrow$$

$$\alpha \downarrow \rightarrow \beta \downarrow$$

If the base width increases, recombination of carrier in base region increases and α decreases & hence β decreases. If doping in base region increases, recombination of carrier in base increases and α decreases thereby decreasing β. Thus S_1 is true and S_2 is false.

Hence option (D) is correct.

60. Given figure is the voltage transfer characteristic of

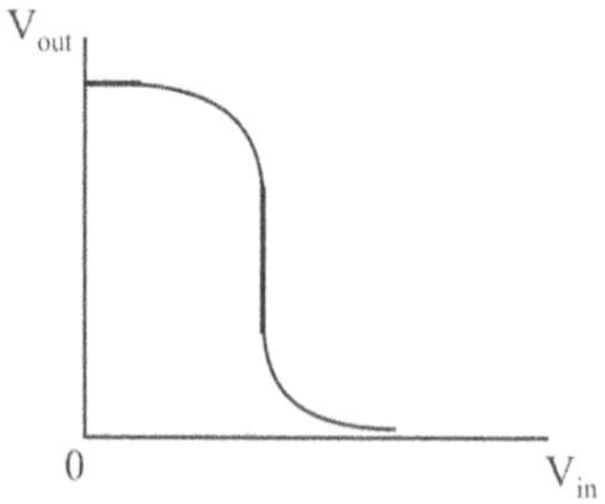

(A) an NOMS inverter with enhancement mode transistor as load
(B) an NMOS inverter with depletion mode transistor as load
(C) a CMOS inverter
(D) a BJT inverter

Sol: Hence option (C) is correct.

61. Assuming $V_{CEsat} = 0.2$ V and $\beta = 50$, the minimum base current (I_B) required to drive the transistor in the figure to saturation is

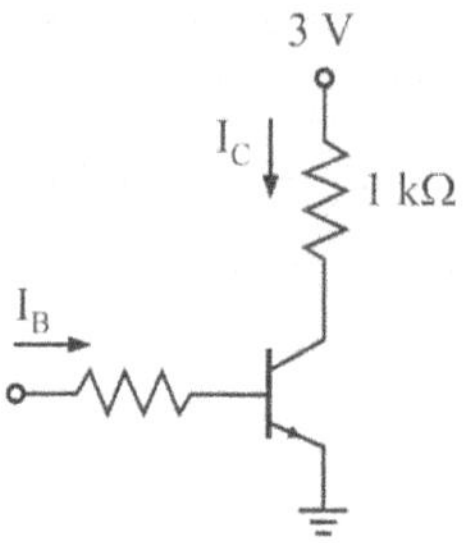

(A) 56 μA (B) 140 mA (C) 60 mA (D) 3 mA

Sol: Applying KVL we get

$$V_{CC} - I_C R_C - V_{CE} = 0$$

or $$I_C = \frac{V_{CC} - V_{CE}}{R_C} = \frac{3 - 0.2}{1\text{ k}} = 2.8\text{ mA}$$

Now $$I_B = \frac{I_C}{\beta} = \frac{2.8\text{ m}}{50} = 56\ \mu\text{A}$$

Hence option (A) is correct

2004 TWO MARKS

62. In an abrupt p – n junction, the doping concentrations on the p –side and n-side are $N_A = 9 \times 10^{16}/\text{cm}^3$ respectively. The p – n junction is reverse biased and the total depletion width is 3 μm. The depletion width on the p –side is

(A) 2.7 μm (B) 0.3 μm (C) 2.25 μm (D) 0.75 μm

Sol: We know that

$$W_pN_A = W_nN_D$$

or $$W_p = \frac{W_n \times N_D}{N_A} = \frac{3\mu \times 10^{16}}{9 \times 10^{16}} = 0.3\ \mu m$$

Hence option (B) is correct.

63. The resistivity of a uniformly doped n–type silicon sample is 0.5Ω – cm. If the electron mobility (μ_n) is 1250 cm²/V-sec and the charge of an electron is 1.6×10^{-19} Coulomb, the donor impurity concentration (N_D) in the sample is

(A) 2×10^{16} /cm³ (B) 1×10^{16}/cm³

(C) 2.5×10^{15} /cm³ (D) 5×10^{15} /cm³

Sol: Conductivity $\sigma = nq\mu_n$

or resistivity $$\rho = \frac{1}{\sigma} = \frac{1}{nq\mu_n}$$

Thus $$n = \frac{1}{q\rho\mu_n} = \frac{1}{1.6 \times 10^{-19} \times 0.5 \times 1250} = 10^{16}/cm^3$$

For n type semiconductor $n = N_D$

Hence option (B) is correct.

64. Consider an abrupt p – n junction. Let V_{bi} be the built-in potential of this junction and V_R be the applied reverse bias. If the junction capacitance (C_j) is 1 pF for $V_{bi} + V_R = 1$ V, then for $V_{bi} + V_R = 4$ V, C_j will be

(A) 4 pF (B) 2 pF (C) 0.25 pF (D) 0.5 pF

Sol: We know that

$$C_j = \left|\frac{e \in_S N_A N_D}{2(V_{bi} + V_R)(N_A + N_D)}\right|$$

Thus $$C_j \propto \sqrt{\frac{1}{(V_{bi} + V_R)}}$$

Now $$\frac{C_{j2}}{C_{j1}} = \sqrt{\frac{(V_{bi} + V_R)_1}{(V_{bi} + V_R)_2}} = \sqrt{\frac{1}{4}} = \frac{1}{2}$$

or $$C_{j2} = \frac{C_{j2}}{2} = \frac{1}{2} = 0.5\ pF$$

Hence option (D) is correct.

65. Consider the following statements S_1 and S_2.

S_1: The threshold voltage (V_T) of MOS capacitor decreases with increase in gate oxide thickness.

S_2: The threshold voltage (V_T) of a MOS capacitor decreases with increase in substrate doping concentration.

Which Marks of the following is correct?

(A) S_1 is FALSE and S_2 is TRUE

(B) Both S_1 and S_2 are TRUE

(C) Both S_1 and S_2 are FALSE

(D) S_1 is TRUE and S_2 is FALSE

Sol: Increase in gate oxide thickness makes it difficult to induce charges in channel. Thus V_T increases if we increases gate oxide thickness. Hence S_1 is false.

Increase in substrate doping concentration require more gate voltage because initially induced charges will get combined in substrate. Thus V_T increases if we increase substrate doping concentration. Hence S_2 is false.

Hence option (C) is correct.

66. The drain of an n-channel MOSFET is shorted to the gate, so that $V_{GS} = V_{DS}$. The threshold voltage (V_T) of the MOSFET is 1 V. If the drain current (I_D) is 1 mA for $V_{GS} = 2$ V, then for $V_{GS} = 3$ V, I_D is

(A) 2 mA (B) 3 mA (C) 9 mA (D) 4 mA

Sol: We know that

$$I_D = K(V_{GS} - V_T)^2$$

Thus

$$\frac{I_{DS}}{I_{DI}} = \frac{(V_{GS_2} - V_T)^2}{(V_{GS_1} - V_T)^2}$$

Substituting the values we have

$$\frac{I_{DS}}{I_{DI}} = 4$$

or

$$I_{D_2} = 4I_{D_1} = 4 \text{ mA}$$

Hence option (D) is correct.

67. The longest wavelength that can be absorbed by silicon, which has the bandgap of 1.12 eV, is 1.1 μm. If the longest wavelength that can be absorbed by another material is 0.87 μm, then bandgap of this material is

(A) 1.416 A/cm^2 (B) 0.886 eV

(C) 0.854 eV (D) 0.706 eV

Sol:

$$E_g \propto \frac{1}{\lambda}$$

Thus $$\frac{E_{g2}}{E_{g1}} = \frac{\lambda_1}{\lambda_2} = \frac{1.1}{0.87}$$

or $$E_{g2} = \frac{1.1}{0.87} \times 1.12 = 1.416 \text{ eV}$$

Hence option (A) is correct.

68. The neutral base width of a bipolar transistor, biased in the active region, is 0.5 μm. The maximum electron concentration and the diffusion constant in the base are $10^{14}/cm^3$ and $D_n = 25\ cm^2/sec$ respectively. Assuming negligible recombination in the base, the collector current density is (the electron charge is 1.6×10^{-19} Coulomb)

(A) 800 A/cm^2 (B) 8 A/cm^2 (C) 200 A/cm^2 (D) 2 A/cm^2

Sol: Concentration gradient

$$\frac{d_n}{d_x} = \frac{10^{14}}{0.5 \times 10^{-4}} = 2 \times 10^{18}$$

$$q = 1.6 \times 10^{-19}\ C$$

$$D_n = 25$$

$$J_C = qD_n \frac{dn}{dx} = 8 A/cm^2$$

Hence option (B) is correct.

2003 ONE MARK

69. n-type silicon is obtained by doping silicon with

(A) Germanium (B) Aluminium

(C) Boron (D) Phosphorus

Sol: Pentavalent make n –type semiconductor and phosphorous is pentavalent.

Hence option (D) is correct.

70. The Bandgap of silicon at 300 °K is

(A) 1.36 eV (B) 1.10 eV (C) 0.80 eV (D) 0.67 eV

Sol: For silicon at 0 °K $E_{g0} = 1.21$ eV

At any temperature

$$E_{gT} = E_{g0} - 3.6 \times 10^{-4}\ T$$

At $T = 300$ °K,

$$E_{g300} = 1.21 - 3.6 \times 10^{-4} \times 300 = 1.1 \text{ eV}$$

This is standard value, that must be remembered.

Hence option (B) is correct.

71. The intrinsic carrier concentration of silicon sample at 300 °K is $1.5 \times 10^{16}/m^3$. If after doping, the number of majority carriers is $5 \times 10^{20}/m^3$, the minority carrier density is

(A) $4.50 \times 10^{11} /m^3$ (B) $3.333 \times 10^4 /m^3$

(C) $5.00 \times 10^{20} /m^3$ (D) $3.00 \times 10^{-5} /m^3$

Sol: By Mass action law,

$$np = n_i^2$$

$$p = \frac{n_i^2}{n} = \frac{1.5\times10^{16}\times1.5\times10^{16}}{5\times10^{20}} = 4.5\times10^{11}$$

Hence option (A) is correct.

72. Choose proper substitutes for X and Y to make the following statement correct Tunnel diode and Avalanche photo diode are operated in X bias ad Y bias respectively

(A) X: reverse, Y: reverse (B) X: reverse, Y: forward

(C) X: forward, Y: reverse (D) X: forward, Y: forward

Sol: Tunnel diode shows the negative characteristics in forward bias. It is used in forward bias. Avalanche photo diode is used in reverse bias.

Hence option (C) is correct.

73. For an n – channel enhancement type MOSFET, if the source is connected at a higher potential than that of the bulk (i.e. $V_{SB} > 0$), the threshold voltage V_T of the MOSFET will

(A) remain unchanged (B) decrease

(C) change polarity (D) increase

Sol: Hence option (D) is correct.

2003 TWO MARKS

74. An n –type silicon bar 0.1 cm long and 100 μm^2 cross-sectional area has a majority carrier concentration of $5 \times 10^{20} /m^2$ and the carrier mobility is 0.13 m^2/V-s at 300 °K. If the charge of an electron is 1.5×10^{-19} coulomb, then the resistance of the bar is

(A) 10^6 Ohm (B) 10^4 Ohm (C) 10^{-1} Ohm (D) 10^{-4} Ohm

Sol: We that $R = \frac{\rho l}{A}, \rho = \frac{1}{\sigma}$ and $\sigma = nq\mu_n$

From above relation we have

$$R = \frac{1}{nq\mu_n A} = \frac{0.1\times10^{-2}}{5\times10^{20}\times1.6\times10^{-19}\times0.13\times100\times10^{-12}} = 10^6\Omega$$

Hence option (A) is correct.

75. The electron concentration in a sample of uniformly doped n-type silicon at 300 °K varies linearly from $10^{17}/cm^3$ at x = 0 to $6 \times 10^{16}/cm^3$ at x = 2 μm. Assume a situation that electrons are supplied to keep this concentration gradient constant with time. If electronic charge is

1.6×10^{-19} coulomb and the diffusion constant $D_n = 35 cm^2/s$, the current density in the silicon, if no electric field is present, is

(A) zero (B) $-112 A/cm^2$

(C) $+1120 A/cm^2$ (D) $-1120 A/cm^2$

Sol:
$$\frac{d_n}{d_x} = \frac{6\times10^{14} - 10^{17}}{2\times10^{-4} - 0} = -2\times10^{20}$$

Since no electric field is present, E = 0 and we get

So,
$$J_n = qD_n\frac{dn}{dx} = 1.6\times35\times\left(-2\times10^{20}\right) = -1120\,A/cm^2$$

Hence option (D) is correct.

76. Match items in Group 1 with items in Group 2, most suitably.

Group 1	Group 2
P. LED	1. Heavy doping
Q. Avalanche photo diode	2. Coherent radiation
R. Tunnel diode	3. Spontaneous emission
S. LASER	4. Current gain

(A) P - 1, Q - 2, R - 4, S - 3 (B) P - 2, Q - 3, R - 1, S - 4

(C) P - 3 Q - 4, R - 1, S - 2 (D) P - 2, Q - 1, R - 4, S - 3

Sol: LED works on the principal of spontaneous emission.

In the avalanche photo diode due to the avalanche effect there is large current gain.

Tunnel diode has very large doping.

LASER diode are used for coherent radiation.

Hence option (C) is correct.

77. At 300 °K, for a diode current of 1 mA, a certain germanium diode requires a forward bias of 0.1435 V, whereas a certain silicon diode requires a forward bias of 0.718 V. Under the conditions stated above, the closest approximation of the ratio of reverse saturation current in germanium diode to that in silicon diode is

(A) 1 (B) 5 (C) 4×10^3 (D) 8×10^3

Sol: Hence option (C) is correct.

78. A particular green LED emits light of wavelength 5490 Å. The energy bandgap of the semiconductor material used there is (Plank's constant = 6.626×10^{-34} J- s)

(A) 2.26 eV (B) 1.98 eV (C) 1.17 eV (D) 0.74 eV

Sol:
$$E_g = \frac{hc}{\lambda} = \frac{6.626\times10^{-34}\times3\times10^{8}}{54900\times10^{-10}} = 3.62\ J$$

In eV
$$E_g(eV) = \frac{E_g(J)}{e} = \frac{3.62\times10^{-19}}{1.6\times10^{-19}} = 2.26\ eV$$

Alternatively

$$E_g = \frac{1.24}{\lambda(\mu m)} eV = \frac{1.24}{5490 \times 10^{-4} \mu m} = 2.26 \text{ eV}$$

Hence option (A) is correct.

79. When the gate-to-source voltage (V_{GS}) of a MOSFET with threshold voltage of 400 mV, working in saturation is 900 mV, the drain current is observed to be 1 mA. Neglecting the channel width modulation effect and assuming that the MOSFET is operating at saturation, the drain current for an applied V_{GS} of 1400 mV is

(A) 0.5 mA (B) 2.0 mA (C) 3.5 mA (D) 4.0 mA

Sol: We know that

$$I_D = K(V_{GS} - V_T)^2$$

Thus

$$\frac{I_{D_2}}{I_{D_1}} = \frac{(V_{GS_2} - V_T)^2}{(V_{GS_1} - V_T)^2}$$

Substituting the values we have

$$\frac{I_{D_2}}{I_{D_1}} = \frac{(1.4 - 0.4)^2}{(0.9 - 0.4)^2} = 4$$

or

$$I_{D_2} = 4I_{D_1} = 4 \text{ mA}$$

Hence option (D) is correct.

80. If P is Passivation, Q is n – well implant, R is metallization and S is source/drain diffusion, then the order in which they are carried out in a standard n – well CMOS fabrication process, is

(A) P – Q – R – S (B) Q – S – R – P

(C) R – P – S – Q (D) S – R – Q – P

Sol: In n – well CMOS fabrication following are the steps:

(A) n – well implant (B) Source drain diffusion

(C) Metalization (D) Passivation

Hence option (B) is correct.

81. The action of JFET in its equivalent circuit can best be represented as a

(A) Current controlled current source (B) Current controlled voltage source

(C) Voltage controlled voltage source (D) Voltage controlled current source

Sol: For a JFET in active region we have

$$I_{DS} = I_{DSS}\left(1 - \frac{V_{GS}}{V_D}\right)^2$$

From above equation it is clear that the action of a JFET is voltage controlled current source.

Hence option (D) is correct.

2002 ONE MARK

82. In the figure, silicon diode is carrying a constant current of 1 mA. When the temperature of the diode is 20 °C, V_D is found to be 700 mV. If the temperature rises to 40 °C, V_D becomes approximately equal to

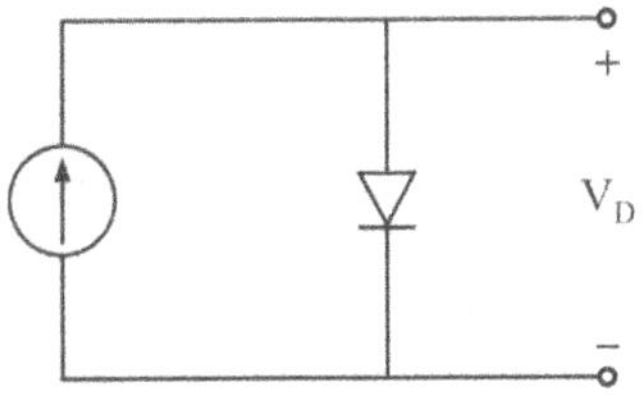

(A) 740 mV (B) 660 mV (C) 680 mV (D) 700 mV

Sol: At constant current the rate of change of voltage with respect to temperature is $dT/dV = -2.5$ mV per degree centigrade

Here $\Delta T = T_2 - T_1$

$= 40 - 20 = 20$ °C

Thus $\Delta V_D = -2.5 \times 20 = 50$ mV

Therefore, $V_D = 700 - 50 = 650$ mV

Hence option (B) is correct.

83. If the transistor in the figure is in saturation, then

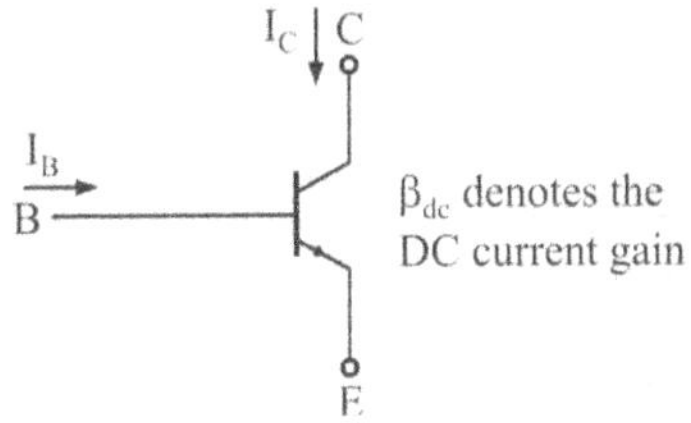

(A) I_C is always equal to $\beta_{dc} I_B$

(B) I_C is always equal to $-\beta_{de} I_B$

(C) I_C is greater than or equal to $\beta_{dc} I_B$

(D) I_C is less than or equal to $\beta_{dc} I_B$

Sol: Condition for saturation is $I_C < \beta I_B$

Hence option (D) is correct.

2001 ONE MARK

84. MOSFET can be used as a

(A) current controlled capacitor

(B) voltage controlled capacitor

(C) current controlled inductor

(D) voltage controlled inductor

Sol: The metal area of the gate in conjunction with the insulating dielectric oxide layer and semiconductor channel, form a parallel plate capacitor.

It is voltage controlled capacitor because in active region the current voltage relationship is given by

$$I_{DS} = K\,(V_{GS} - V_T)^2$$

Hence option (B) is correct

85. The effective channel length of MOSFET in saturation decreases with increase in

(A) gate voltage (B) drain voltage

(C) source voltage (D) body voltage

Sol: In MOSFET the body (substrate) is connected to power supply in such a way to maintain the body (substrate) to channel junction in cutoff condition. The resulting reverse bias voltage between source and body will have an effect on device function. The reverse bias will widen the depletion region resulting the reduction in channel length.

Hence option (D) is correct.

1999 ONE MARK

86. The early effect in a bipolar junction transistor is caused by

(A) fast turn-on (B) fast turn-off

(C) large collector-base reverse bias (D) large emitter-base forward bias

Sol: At a given value of v_{BE}, increasing the reverse-bias voltage on the collector-base junction and thus increases the width of the depletion region of this junction. This in turn results in a decrease in the effective base width W. Since I_S is inversely proportional to W, I_S increases and that I_C increases proportionally. This is early effect.

Hence option (C) is correct.

1999 TWO MARKS

87. An n-channel JEFT has $I_{DSS} = 2\text{mA}$ and $Vp = -4$ V. Its transconductance g_m (in milliohm) for an applied gate-to-source voltage V_{GS} of -2 V is

(A) 0.25 (B) 0.5 (C) 0.75 (D) 1.0

Sol: For an n-channel JEFT trans-conductance is

$$g_m = \frac{-2I_{DSS}}{V_P}\left(1 - \frac{V_{GS}}{V_p}\right)$$

$$= \frac{-2 \times 2 \times 10^{-3}}{-4}\left[1 - \frac{(-2)}{(-4)}\right]$$

$$= 0.5 \text{ mho}$$

Hence option (B) is correct.

88. An npn transistor (with C = 0.3 pF) has a unity-gain cutoff frequency f_T of 400MHz at a dc bias current I_C = 1mA. The value of its C_μ(in pF) is approximately (V_T = 26 mV).

(A) 15 (B) 30 (C) 50 (D) 96

Sol: We have $$g_m = \frac{I_C}{V_T} = \frac{1}{26}$$

Now

$$f_T = \frac{g_m}{2\pi(C_\pi + C_\mu)}$$

or $$400 = \frac{1/26}{2\pi(0.3\times10^{-12}\times C_\mu)}$$

or $$(0.3\times10^{-12}\times C_\mu) = \frac{1}{2\pi\times26\times400} = 15.3\times10^{-12}$$

or $$C_\mu = 15.3\times10^{-12} - 0.3\times10^{-12} = 15\times10^{-12} = 15\text{ pF}$$

Hence option (A) is correct.

1998 ONE MARK

89. The electron and hole concentrations in an intrinsic semiconductor are n_i and p_i respectively. When doped with a p-type material, these change to n and p, respectively, Then

(A) $n + p = n_i + p_i$ (B) $n + n_i = p + p_i$

(C) $np_i = n_i p$ (D) $np = n_i p_i$

Sol: For any semiconductor (Intrinsic or extrinsic) the product n p remains constant at a given temperature so here

$$np = n_i p_i$$

Hence option (D) is correct.

90. The f_T of a BJT is related to its g_m, C_π and C_μ as follows

(A) $f_T = \frac{C_\pi + C_\mu}{g_m}$ (B) $f_T = \frac{2\pi(C_\pi + C_\mu)}{g_m}$

(C) $f_T = \frac{g_m}{C_\pi + C_\mu}$ (D) $f_T = \frac{g_m}{2\pi(C_\pi + C_\mu)}$

Sol: $$f_T = \frac{g_m}{2\pi(C_\pi + C_\mu)}$$

Hence option (D) is correct.

91. The static characteristic of an adequately forward biased p-n junction is a straight line, if the plot is of

(A) log I vs log V (B) log I vs V

(C) I vs log V (D) I vs V

Sol: For a Forward Bias p-n junction, current equation

$$I = I_0\left(e^{V/kT} - 1\right)$$

So $$kT\log\left(\frac{I}{I_0} + 1\right) = V$$

So if we plot log I vs V we get a straight line.

Hence option (B) is correct.

92. A long specimen of p-type semiconductor material

(A) is positively charged

(B) is electrically neutral

(C) has an electric field directed along its length

(D) acts as a dipole

Sol: A specimen of p – type or n – type is always electrically neutral.

Hence option (B) is correct.

93. Two identical FETs, each characterized by the parameters g_m and r_d are connected in parallel. The composite FET is then characterized by the parameters

(A) $\frac{g_m}{2}$ and $2r_d$ (B) $\frac{g_m}{2}$ and $\frac{r_d}{2}$

(C) $2g_m$ and $\frac{r_d}{2}$ (D) $2g_m$ and $2r_d$

Sol: Hence option (C) is correct.

94. The units of q /kT are

(A) V (B) V^{-1} (C) J (D) J/K

Sol: The unit of q is e and unit of kT is eV. Thus unit of $\frac{e}{KT}$ is $\frac{e}{eV} = V^{-1}$.

Hence option (B) is correct.

1997 ONE MARK

95. For a MOS capacitor fabricated on a p-type semiconductor, strong inversion occurs when

(A) surface potential is equal to Fermi potential

(B) surface potential is zero

(C) surface potential is negative and equal to Fermi potential in magnitude

(D) surface potential is positive and equal to twice the Fermi potential

Sol: Hence option (D) is correct.

96. The intrinsic carrier density at 300 K is $1.5 \times 10^{10}/cm^3$, in silicon. For n-type silicon doped to 2.25×10^{15} atoms/cm^3, the equilibrium electron and hole densities are

(A) $n = 1.5 \times 10^{15}/cm^3, p = 1.5 \times 10^{10}/cm^3$

(B) $n = 1.5 \times 10^{10}/cm^3, p = 2.25 \times 10^{15}/cm^3$

(C) $n = 2.25 \times 10^{15}/cm^3, p = 1.0 \times 10^{15}/cm^3$

(D) $n = 1.5 \times 10^{10}/cm^3, p = 1.5 \times 10^{10}/cm^3$

Sol: We have $$n_i = 1.5\times10^{10}/cm^3$$

$$N_d = 2.25\times10^{15} \text{ atom/cm}^3$$

For n type doping we have electron concentration

$$n = N_d = 2.25\times10^{15} \text{ atom/cm}^3$$

For a given temperature

$$np = n_i^2$$

Hole concentration $p_n = \dfrac{n_i^2}{n} = 1.0\times10^5/cm^3$

Hence option (C) is correct.

1996 ONE MARK

97. The p-type substrate in a conventional *pn*-junction isolated integrated circuit should be connected to

(A) nowhere, i.e. left floating

(B) a DC ground potential

(C) the most positive potential available in the circuit

(D) the most negative potential available in the circuit

Sol: In pn-junction isolated circuit we should have high impedance, so that pn junction should be kept in reverse bias. (So connect p to negative potential in the circuit)

Hence option (D) is correct.

1996 TWO MARKS

98. If a transistor is operating with both of its junctions forward biased, but with the collector base forward biased greater than the emitter base forward bias, then it is Operating in the

(A) forward active mode

(B) reverse saturation mode

(C) reverse active mode

(D) forward saturation mode

Sol:

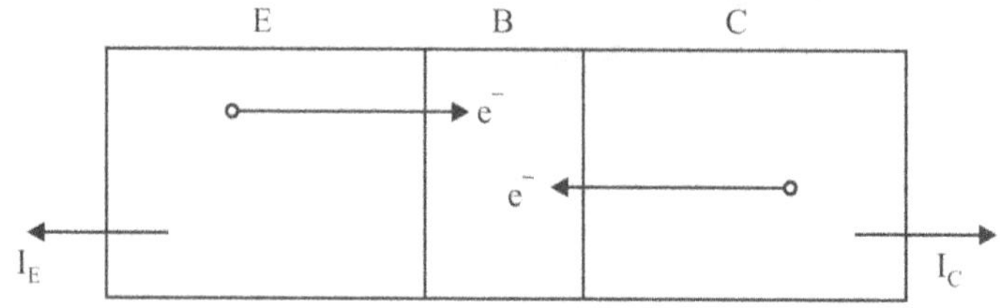

If both junctions are forward biased and collector base junction is more forward biased then I_C will be flowing out wards (opposite direction to normal mode) the collector and it will be in reverse saturation mode.

Hence option (B) is correct.

99. The common-emitter short-circuit current gain β of a transistor

(A) is a monotonically increasing function of the collector current I_C

(B) is a monotonically decreasing function of I_C

(C) increase with I_C , for low I_C , reaches a maximum and then decreases with further increase in I_C

(D) is not a function of I_C

Sol: For normal active mode we have

$$\frac{I_C}{I_B} = \beta$$

For small values of I_C, if we increases I_C, β also increases until we reach (I_C) saturation. Further increases in I_C (since transistor is in saturation mode know) will increases I_B and β decreases.

Hence option (C) is correct.

100. A n-channel silicon (Eg = 1.1 eV) MOSFET was fabricated using n^+ poly-silicon gate and the threshold voltage was found to be 1V. Now, if the gate is changed to p+ poly-silicon, other things remaining the same, the new threshold voltage should be

(A) − 0.1V (B) 0 V (C) 1.0 V (D) 2.1V

Sol: For n-channel MOSFET thresholds voltage is given by

$$V_{TN} = V_{GS} - V_{DS}\,(sat)$$

for p-channel [p+ polysilicon used in gate]

$$V_{TP} = V_{SD}\,(sat) - V_{GS}$$

So $$V_{TP} = -V_{DS}\,(sat) + V_{GS}$$

∴ Threshold voltage will be the same.

Hence option (C) is correct.

101. In a bipolar transistor at room temperature, if the emitter current is doubled the voltage across its base-emitter junction

(A) Doubles (B) halves

(C) Increases by about 20 mV (D) decreases by about 20 mV

Sol: Emitter current is given

$$I_E = I_0\left(e^{V_{BE}/kT} - 1\right)$$

or $$I_E = I_0\, e^{V_{BE}/kT} \qquad e^{V_{BE}/kT} >> 1$$

or $$V_{BE} = kT\ln\left(\frac{I_E}{I_0}\right)$$

Now $$(V_{BE})_1 = kT\ln\left(\frac{I_{E1}}{I_0}\right)$$

$$(V_{BE})_2 = kT\ln\left(\frac{I_{E2}}{I_0}\right)$$

or $$(V_{BE})_2 - (V_{BE})_1 = kT\left[\ln\left(\frac{I_{E_2}}{I_{E_1}}\right)\right] = kT\ln\left(\frac{2I_{E_1}}{I_{E_1}}\right)$$

Now if emitter current is doubled i.e., $I_{E_2} = 2I_{E_1}$

$$(V_{BE})_2 = (V_{BE})_1 + (25\times 0.60)\,\text{m volt}$$

$$= (V_{BE})_1 = 15\,\text{m volt}$$

Thus if emitter current is doubled the base emitter junction voltage is increased by 15 mV.

Hence option (C) is correct.

102. An npn transistor has a beta cut-off frequency f_β of 1MHz and common emitter short circuit low-frequency current gain β_o of 200, its unity gain frequency f_T and the alpha cut-off frequency f_α respectively are

(A) 200MHz, 201MHz (B) 200MHz, 199MHz

(C) 199MHz, 200MHz (D) 201MHz, 200MHz

Sol: Unity gain frequency is given by

$$f_T = f_B \times \beta = 10^6 \times 200 = 200\text{MHz}$$

α-cutoff frequency is given by

$$f_T = \frac{f_\beta}{1-\alpha} = \frac{f_\beta}{1-\dfrac{\beta}{\beta+1}} = f_\beta(\beta+1) = 10^6 \times (200+1) = 201\text{ MHz}$$

Hence option (A) is correct.

103. A silicon n MOSFET has a threshold voltage of 1V and oxide thickness of Ao. [ε_r (SiO_2) = 3.9, ε_0 = 8.854 × 10^{-14} F/cm, q = 1.6 × 10^{-19} C]

The region under the gate is ion implanted for threshold voltage tailoring. The dose and type of the implant (assumed to be a sheet charge at the interface) required to shift the threshold voltage to −1V are

(A) 1.08 × 10^{12}/cm^2, p-type (B) 1.08 × 10^{12}/cm^2, n-type

(C) 5.4 × 10^{11}/cm^2, p-type (D) 5.4 × 10^{11}/cm^2, n-type

Sol: Option (A) is correct.

Index

A

B

C

D

E

F

G

H

I

J

L

M

www.ingramcontent.com/pod-product-compliance
Lightning Source LLC
Chambersburg PA
CBHW081458190326
41458CB00015B/5277
* 9 7 8 9 3 8 5 4 3 3 6 5 8 *